D1759319

The Nature and
Origin of Granite

Frontispiece Stoping in the Pativilca Pluton illustrated in the granite cliff of Llanachupan, Pativilca Valley, Ancash, Peru. Blocks of basaltic andesite engulfed in a coarse-grained, biotite granite. Roof of pluton seen far left in distance. Height of cliff approximately 250–300m. Drawing by K. Lancaster. An alternative explanation involves the disruption of synplutonic mafic dykes.

The Nature and Origin of Granite

Second edition
Wallace Spencer Pitcher
Emeritus Professor of Geology
University of Liverpool

CHAPMAN & HALL
London · Weinheim · New York · Tokyo · Melbourne · Madras

Published by Chapman & Hall, 2–6 Boundary Row, London SE1 8HN

Chapman & Hall, 2–6 Boundary Row, London SE1 8HN, UK

Chapman & Hall GmbH, Pappelallee 3, 69469 Weinheim, Germany

Chapman & Hall USA, 115 Fifth Avenue, New York, NY 10003, USA

Chapman & Hall Japan, ITP-Japan, Kyowa Building, 3F, 2-2-1 Hirakawacho, Chiyoda-ku, Tokyo 102, Japan

Chapman & Hall Australia, 102 Dodds Street, South Melbourne, Victoria 3205, Australia

Chapman & Hall India, R. Seshadri, 32 Second Main Road, CIT East, Madras 600 035, India

First edition 1993
Reprinted 1995
Second edition 1997

© 1997 Chapman & Hall

Typeset on 10/12 Palatino by Best-Set Typesetters Ltd., Hong Kong
Printed in Great Britain by T.J. International, Padstow, Cornwall

ISBN 0 412 75860 1

A catalogue record for this book is available from the British Library

Library of Congress Catalog Card Number: 97-65818

∞ Printed on permanent acid-free text paper, manufactured in accordance with ANSI/NISO Z39.48-1992 and ANSI/NISO Z39.48-1984 (Permanence of Paper).

'Granite is not a rock which was simple in its origin but might be produced in more ways than one'. Joseph Beete Jukes in 1863, as Director of the Irish Geological Survey, attempting to arbitrate in a vigorous debate on the origin of granite in the mid-19th century.

Contents

Preface

The origin of granite has for long fascinated geologists though serious debate on the topic may be said to date from a famous meeting of the Geological Society of France in 1847. My own introduction to the subject began exactly one hundred years later when, in an interview with Professor H.H. Read, I entered his study as an amateur fossil collector and left it as a committed granite petrologist – after just ten minutes! I can hardly aspire to convert my reader in so dramatic a way, yet this book is an attempt, however inadequate, to pass on the enthusiasm that I inherited, and which has been reinforced by innumerable discussions on the outcrop with granitologists of many nationalities and of many shades of opinion.

Since the 1960s, interest in granites has been greatly stimulated by the thesis that granites image their source rocks in the inaccessible deep crust, and that their diversity is the result of varying global tectonic context. So great a body of new data and new ideas has accumulated that my attempt to review the whole field of granite studies must carry with it a possible charge of arrogance, especially as I have adopted the teaching device of presenting the material from a personal point of view with its thinly disguised prejudices. This seemed the only way to escape from a textbook format to one of self-contained essays, each covering a specific facet or problem, yet contributing to a connected story highlighting the central themes of tectonic control, source rock imaging and the multifactorial nature of all those processes that produce 'granites and granites'. Despite this personal approach, discussion is based on actual examples, many of which I have examined in the field and laboratory under the guidance of the particular researchers. Furthermore I have tried to avoid unnecessary jargon and to provide a path through the ever-accumulating thicket of literature which threatens to overwhelm the very science it aims to report. I can only hope that this overall, holistic approach will intrigue the specialist, stimulate the student and interest the general reader.

I am grateful to the many friends who have helped to curb some of the worst of my conceptual errors and prejudices, with particular thanks due to Michael Atherton, Charles Bacon, Paul Bateman, Geoff Brown, Michael

Brown, Michael Cheadle, John Clemens, John Cobbing, Ilmari Haapala, Bernard Leake, Scott Rogers, Edryd Stephens and Ron Vernon. The publishers encouraged me to start writing, the librarians of the Geological Society, London, researched and copied papers galore, Chris Amstutz translated parts of Goethe, Hilary Davies assisted with the primary editing and indexing, and my wife, Stella, continued, as always, to provide her sure support.

This second edition has given me the opportunity to reconsider and develop certain of the central themes of this book in the light of much new research, especially as presented at recent keynote lectures and international conferences. Rapid and continuing advances in our knowledge lead me to hope for a general solution to the granite problem, but it will not be fully revealed without a proper appreciation of the complexity of the real processes involved and a continuous 'back to the rocks' testing of those dreamland models. And I continue to fear that we might be overwhelmed by the torrent of commissioned research papers!

Wallace S. Pitcher
Emeritus Professor of Geology
University of Liverpool

The historical perspective: an ever changing emphasis

1

In these hurried days, geologists will take no harm from a quiet contemplation of the history of even this small part of their science.
  Herbert Harold Read (1957) The Granite Controversy, p. xiii.

THE BEGINNING

Where do we begin this story of the stony, granular rock known as granite? Romantically, perhaps, in seeking the derivation of the name in the ancient gaelic of Wales and Cornwall, as *gwenith faen*, a wheatstone for the grinding of flour; or in the late medieval *granito* of the Italian Cesalpinas. More logically, of course, in the 18th century when the intellectual demands of the Age of Enlightenment encouraged the search for secular explanations for natural things. An early consensus, the Wernerian view that granite was a chemical precipitate from a primordial, universal ocean, was only slowly replaced by the contrary claim that it was produced by the consolidation of matter made fluid by heat. Indeed, this radical new theory was at first regarded as nothing less than a blasphemy by Johann Wolfgang von Goethe, Germany's Minister of Mines, who had himself written a 'Neptunian' essay on granite in 1784 and, reacting as well a poet might, later complained in his *Xeniae Tamed*

  Scarce noble Werner turns about
  Poseidon's realm falls prey to loot.

Every geologist knows of the leading role of James Hutton of Edinburgh in this late 18th century debate between the Neptunists and Plutonists, but it is salutary to learn from his own pen that 'all the granite I had ever seen when I wrote my *Theory of the Earth* was some at Peterhead and Aberdeen' (*Theory of the Earth*, 1795, p. 241), and, further, 'at the time, however, I was not particularly decided in my opinion concerning granite, whether it was to be considered as a body which had been originally stratified . . . and afterwards consolidated by the fusion of these materials, or whether it were not rather a body transferred from subterraneous regions, and made to break and invade the strata' (*Transactions of the Royal Society of Edinburgh*, 1790, 3(1), p.77). It is hardly possible to state more succinctly the theme of the debate which was to dominate petrological thought for the next century and a half.

Only after the initial communication of his general theory in 1785 did Hutton find it necessary to go into the field specifically in search of proof, finding it in full measure and with great joy in his Scottish glens, in the form of granite veining in Glen Tilt (Figure 1.1), and in the contact relationship of the Cairnsmore of Fleet Granite (1790, p.79). Hutton's walk along this latter contact, and that of his friend James Hall, must surely rank as the first purposeful examination of a granite mass. It was this same James Hall who later successfully experimented with the melting of granite, joining his contemporary Thomas Beddoes in 1791 in affirming that the chemical affinity of basalt and granite was good evidence for the igneous origin of the latter!

We might speculate whether James Hutton would ever have published his epic work if he had not been goaded into doing so by a fierce public attack from Richard Kirwan of Dublin, who sarcastically dubbed Hutton's thesis as that 'plutonic theory', dismissing its 'conception of almost limit-

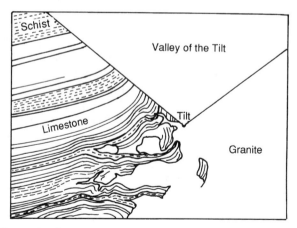

Figure 1.1 Copy of coloured engraving by McCulloch showing his interpretation (in 1813) of the contact relationships earlier described by Hutton in a section across Glen Tilt, Perthshire, Scotland (Plate 20, Figure 4, *Transactions of the Geological Society*, 1816, p.3). In a prescient account of the way in which the granite intrudes limestone, he writes, 'We have seen that all the irregularities of the beds take place wherever the granite comes into contact with them.' (*loc. cit.*, p. 278).

less time (as) an abyss from which human reaction recoils'. It is certain that, immediately on reading of this riposte, Hutton rose from his sick bed and wrote about granite within the broad context of a new concept of Earth history. His plutonism was, in fact, a philosophical triumph, with prediction coming before discovery.

THE INFLUENCE OF LYELL

Possibly due to the general rejection of theory and speculation that Derek Flinn tells me was the fashion for several decades after 1800, it took another 50 years for these new ideas to be generally accepted. It is really to Charles Lyell, no matter how second hand his observations on granite, that we owe the next general consensus of the 1830s – that is, that all granite was not primordial, that granite was formed at many successive periods, and that it was of igneous origin formed at great depth in the Earth from subterranean lava, there to crystallize with great slowness. Furthermore, it was Lyell's fourfold classification of rocks into aqueous, volcanic, plutonic and metamorphic, with its distinction between the volcanic and plutonic classes, which heralded the next great debate – that is, whether the origins of basalt and granite were separate, or not.

However, Lyell himself was never dogmatic on this issue but was inclined to 'consider whether many granites and other rocks of this (plutonic) class may not sometimes represent merely the extreme of a similar slow metamorphism' (*Students Elements*, 1871, p. 570). Nevertheless, he approvingly quoted from the seminal work of Poulett Scrope, who published his classic works on volcanoes and magmas in 1825 and 1862. It was Scrope who provided the first scientific basis for the study of a 'magma of granules or imperfect crystals enveloped in a liquid', and who showed a remarkable prescience in describing magmas as crystal suspensions with an essential vapour content and a viscosity controlled by composition. Scrope also suspected that their differences were due 'to some chemical process in an internal reservoir'.

The influence of Lyell's masterly exposition was immediate at this remarkable time when geology stirred the public imagination to a degree undreamed of today, as shown by the fact that no less than 4500 ticket applications were made for Lyell's lectures at Boston in 1841! How fortunate it was that Charles Darwin took the first edition of Lyell's *Principles of Geology* on his memorable voyage on the *Beagle* (1831–6), for it was he who, during his excursions in the Volcanic Islands, realized that the settling of crystals under gravity could be important in producing the observable differences in the composition of the volcanic rocks. Enthusiastically he writes: 'I clambered over the mountains of Ascension with a bounding step and made the volcanic rocks resound under my geological hammer.'

Of course, new ideas about igneous rocks, and in particular granite, were not confined to the English speaking community. Thus Baltazar Keilhau, a pioneer of Norwegian petrology, was studying the Drammen Granite near Oslo, and it was on its outcrop that he and Lyell had a lively discussion in 1835, signalling the beginnings of yet another great debate, that of the room problem. Keilhau had already introduced the concepts of metamorphism and metasomatism, and now put forward a thesis of transmutation of the country rocks by a process of 'granitification'. This Lyell countered by invoking forceful intrusion, a difference of opinion only partly resolved 20 years later by Theodor Kjerulf, who envisaged that the space was created by the swallowing up and assimilation of the country rocks.

In Germany at the same time chemical geology was being established by Gustav Bischof of Bonn, and Robert von Bunsen was writing his treatise on the chemistry of the volcanic rocks of Iceland. Bunsen convincingly showed the coexistence of acid and basic lavas, considering that they represented separate rock melts whose mixing might give rise to rocks of transitional composition. He concluded that it was impossible for high potassium granite to derive from the same primary magma as basalt, so it followed that granite and basalt were of different origins. This minority view was repeatedly restated in the years to come, with granite deriving from the crust, and basalt from depths later identified with Reginald Daly's world-wide basaltic shell.

THE FRENCH TRADITION

In French geological circles a very different tradition was emerging whereby Lyell's plutonism and metamorphism were seen to be convergent. In 1824 we find Ami Boué, a founder of the Geological Society of France, describing what we would now call granitization and advocating an eventual 'liquefaction', while not much later Joseph Fournet was distinguishing those metamorphic rocks formed by simple recrystallization from other rocks in which partial melting had occurred. The essence of the eventual French consensus is contained in the proceedings of a meeting of the Geological Society of France in 1847 devoted largely to the subject of granite. The volcanic and plutonic classes were considered to be separate, and within the latter class granites were thought to arise by 'granitification' of the country rocks by a process of 'permeation' or 'imbibition' aided by 'agents mineralisateurs'. This hypothesis soon became so fashionable that in 1858, and later in 1869, Achille Delesse could confidently conclude that granite was the result of the metamorphic growth of granite minerals in the country rocks, and that 'les roches plutoniques . . . representent le maximum d'intensité ou le terme extrême du metamorphism général' (Read, 1957, p. 102). He also considered that such granites might become mobile and, driven by orogenic forces, intrude, losing all evidence of their origin.

The school of granitization that blossomed in France in the last quarter of the 19th century, as represented by the works of Michel-Lévy, Lacroix, Barrois, Duparc and others, was clearly built on this early established tradition in which injection, imbibition, transformation and assimilation had become part of the French geological vocabulary. One has only to read the classical work of Charles Barrois on the granites of Brittany, with its demonstration of the depth control of the form and nature of granite masses, to appreciate the sophistication of French thinking on granite problems at the turn of the century. It reached a high point in the bold and enthusiastic writings of Pierre Termier, as summarized in the latter's contribution to a second conference milestone in the granite saga, the meeting of the 1910 International Geological Congress in Stockholm.

Termier imaginatively involved 'colonnes filtrantes' in his permeation and transformations. However, of more importance is the fact that he could now take for granted the interrelationship between the granite-making process and orogeny, and opine that concordant granite masses, 'les granites d'anatexie', were generated as an end-product of orogenic metamorphism, whereas the discordant masses, 'les granites en massifs circonscrites', evolved by the bulk mobilization of the anatexites.

Later, French geologists were among the first to recognize that the nature of granites depends on their environment, and also that granites occur outside orogenic belts and within rigid cratons when they have certain distinctive characteristics. In general, the French have followed one of two major, parallel strands in the development of petrological thought, and one contrasted with that of the contemporary German schools.

SEDERHOLM TO READ: A PROPOSITION SUSTAINED

With granite forming so great a part of the bedrock of Finland and Scandinavia it is not surprising that the native geologists should also agonize over this granite problem. The master was Johannes Sederholm, the head of the Finnish Geological Survey, who, in striving to apply concepts of geological time to this ancient basement, needed first to understand the complexities of the granite gneisses. It was he who provided the now familiar nomenclature in his classic *Om granite och gneiss* (1907), and opened debate with his Swedish colleague Per Johan Holmquist on the origin of the potassium feldspar-bearing granitic *lits* of *lit-par-lit* gneisses – that is, whether they emanate from adjacent granite masses or from the country rocks. Such keywords as palingenesis, anatexis, migmatite and ptygmatite entered the language, and from then on were persistently abused until Kenneth Mehnert's careful redefinitions of 1968.

Of Sederholm, Eskola wrote, 'One may ask how he, a pupil of Harry Rosenbusch, the extreme magmatist, became the announcer of granitisation' (Sederholm, 1967, p. 587). The answer lay in his 'emotional enthusiasm' on examining the migmatites exposed in the skerries off the southern coast of Finland during an enforced summer holiday in 1906!

Building upon Sederholm's pioneering work, his fellow Finns sought to establish a genetic series within a context of orogenic time, categorized as synkinematic, late-kinematic and post-kinematic, a general thesis propounded in a famous essay on magma-tectonic grouping by Pentti Eskola in 1932 (Table 1.1). The latter argued that the synkinematic granites were the most likely to be of palingenetic origin and, as a consequence, were not naturally associated with basic rocks. On the other hand, the later groups in this tectonic sequence were the most likely to be of conventionally magmatic origin and would therefore be accompanied by basic rocks. Eskola emphasized the importance of partial melting in the origin of the former group, and also opined that this process would have become increasingly important with the advance of geological time.

We shall see that Eskola's propositions are very similar to today's consensus views, and even so committed a magmatist as Norman Bowen could applaud this essay as 'an eminently sane discussion of the various processes' (1948, p. 87). I would remark at this point that many of the Fennoscandians were equally familiar with geochemistry and petrogenetic theory as with field studies; Eskola, for example, studied with, and was greatly influenced by, the great Victor Goldschmidt, and Sederholm himself had been a student of Rosenbusch.

Eskola's contemporary, the Swiss tectonician Cesare Eugen Wegmann, on the basis of extensive studies in Finland and Greenland, extended the then current thesis to embrace all the elements of orogenesis – that is,

Table 1.1 Various nomenclatures expressing the time and place of intrusion in relation to the stages of orogenesis. The first three on this list refer directly to orogenic time, the others more to style of intrusion

Reference	Terms used		
Sederholm (1891) ⎫ Eskola (1932) ⎭	Synkinematic	Late-kinematic	Post-kinematic
Scheumann (1924)	Syntectonic		Apotectonic
Stephansson (1975)	Catatectonic	Mesotectonic	Epitectonic
Wegmann (1935)	Infracrustal	Transition level	Supercrustal
Jung & Roques (1938)	Anatectic	Subautochthonous	Intrusive
Read (1939)	Autochthonous	Parautochthonous	Intrusive
Buddington (1959)	Catazonal	Mesozonal	Epizonal

tectonic shortening and squeezing, high grade metamorphism and granitization, and the ultimate mobilization with diapiric intrusion. Moreover, Wegmann envisaged tectonic transfer of heat and materials on a massive scale from the reactive infrastructure of the deep crust, into the 'passive' superstructure (Figure 1.2). According to Wegmann, depth within the orogen was all important in determining the granite–crust relationship, with the corollary that at great depth there would be none of the sharp distinction between magmatic and non-magmatic rocks that is so obvious at shallow levels in the crust.

Within this orogenic model he ascribed a major role to regional metasomatism. Along with Krank, Wegmann not only envisaged the long range migration of elements through the crust, a view long held by the French, but sought to show that this occurs in the form of waves or 'fronts' of the different elements, commonly represented by a wave of alkalis constituting a feldspathizing front, displacing one of calcium, iron and magnesium, a mafic or basic front. Perhaps such extreme views were purposely advocated to be heard by the orthodoxy of the time. Certainly Wegmann's colleague, Eskola, suspected that an article *Zur Deutung der Migmatite*, published in 1935, was to an extent making 'fun of the world's

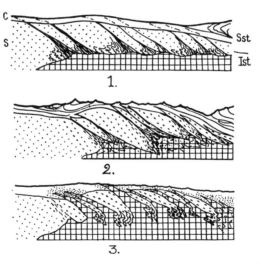

Figure 1.2 Wegmann's concept of the activation of deep crustal material during compression. (1) The hot subcrust is activated and begins to flow upwards into zones of schistosity and higher still into thrust zones. (2) The activated material is injected between old crustal blocks, relative to which it is a mobile magma. (3) Moving upwards, this 'magma' fills the arches and thrusts in the superstructure and its cover. (C) Cover and (S) basement, with (Sst) superstructure and (Ist) infrastructure. Copy of Figure 9.1, *Geologische Rundschau*, 1935, **26**, 339.

wise geologists', and this was likely to have included his fellow Swiss Paul Niggli, whose stern logic rebutted these transformist ideas. Hans Cloos once said that it was tragic that the two foremost geologists of Switzerland did not understand each other's 'language'.

Although substantial parts of these concepts relating tectonics, metamorphism and granite formation can be accommodated in present day models, the concept of long range diffusion and chemical fronts did not survive long in the face of strong criticism from Bowen, and can now be recognized as one of those blind alleys into which scientific thought is prone to drift. Even in its heyday Herbert Read, in one of his masterly syntheses of *The Granite Controversy* written between 1939 and 1955, cannily remarked, 'if it is demonstrated that there is a zone of Fe, Mg enrichment around a granite mass, its explanation may be an entertaining matter' (Read, 1957, p. 185). He quipped that it might also be considered an affront by many.

Read clearly felt on safer ground in his defence of the central unifying concept itself and, expanding on that long held tenet of the French that there is a granite series, he synthesized his own definitive version – that is to say, not only were there granites and granites but all were genetically connected in time and place (Figure 1.3). He, in his turn, was carried away with the sheer elegance of the concept and overstressed, I believe, the genetic detachment of volcanism and plutonism with its separation of acid and basic rocks, a position which, as his student, I joined others in referring to as Read's 'dark chasm'.

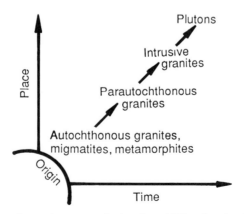

Figure 1.3 The granite series according to Read (*The Granite Controversy*, 1957, p. 335).

A UNIVERSAL DEBATE

In various measure and at various times this swirling debate on the origin of granite surfaced in every geological community, often stimulated by the iconoclasm of a single individual. In Britain, for example, the authoritative position was settled in 1866 by the declaration by Andrew Ramsay, Director of the Geological Survey, that granite was a metamorphic rock, a decision only reversed in 1901 when James Geikie finally persuaded a new Director to recant. In Ireland in the 1860s the debate was characteristically more vigorous, the origin of the Main Donegal Granite, for example, providing a sometimes acrimonious discussion centred, as usual, on metamorphism versus magmatism. It must be said that now that we can better appreciate the overall complexity and multifactorial character of the processes involved in the emplacement of granite in Donegal, it is easy to see how each of the disputants could find evidence for their particular point of view, and equally easy to agree with the Irish Director, Beete Jukes, who rationalized, in 1863, 'that granite is not a rock which was simple in its origin but might be produced in more ways than one'. And, still in Ireland, we can marvel at the lone contribution of Samuel Haughton, who devised and used an entirely novel and comprehensive chemical system for the classification and comparison of granites, published in the *Quarterly Journal of the Geological Society*, as early as 1862!

ROSENBUSCH TO BOWEN: THE ALTERNATIVE

A second strand in this growth of ideas on the origin of granite stemmed from the first examinations of thin sections of rocks. Particularly well known is the story of Lyell's estimable 'Mr Sorby'; how this remarkable Yorkshireman applied a thin section technique to igneous rocks, specifically to the study of fluid inclusions in granites, deducing thereby that water was an essential component of granite magma. Equally well known is the consequence of his 1862 meeting with Ferdinand Zirkel, later Professor at Leipzig, and the latter's immediate application of such microscopical studies. These quickly bore fruit in his seminal *Lehrbuch de Petrographie* (1866), which must have so captivated Heinrich Rosenbusch of Heidelberg that he devoted his life to a field of research which was to change the course of the study of igneous rocks. After the publication of Rosenbusch's influential *Die microskopische Physiographie der massigen Gesteine* (1877) and leaving aside the claims and counterclaims of Rosenbusch and his French contemporaries Ferdinand Fouqué and August Michel-Lévy as to priorities in publication, we can say that petrography was truly born.

The order and logic of the new discipline, especially as expressed in the publications of the micropetrology school of Heidelberg, focused opinion towards a magmatic origin of all igneous rocks. By the time of Rosen-busch's fullest statement of his position, in his *Elemente der Gesteinlehre* (1901), Bunsen's two magma hypothesis had been all but swept aside by a thesis of a single primary magma with an ordered crystallization history. This was embodied in the Kerntheorie of Rosenbusch, which involved a set course of differentiation by an as yet unspecified process of splitting apart of a primary magma, this leading to an enrichment in alkalis and silica, and with an order of intrusion corresponding to this evolution.

In the last quarter of the 19th century the initial phase of microscopical description gave way to an increasing awareness of the significance of the natural interrelationships between igneous rocks, with the notion that there was a general connection between magma composition and geologi-cal causality in space and time. Thus in 1886 William Judd introduced the concept of petrographic provinces, while the consanguinity of rock suites was demonstrated a little later by Joseph Iddings, then by Alfred Harker and Waldemar Brögger. Each of these petrologists provided beautifully detailed field studies to support their proposals, though the recognition of Brögger's seminal work on the Oslo rocks may have been muted by its publication in Norwegian journals.

Outside France only a few petrologists resisted the appeal of this el-egant model involving an ordered evolution through crystallization, nota-ble among them being the Russian Franz Yulevich Loewinson-Lessing, who, in his masterly expositions of 1897 and 1911 (Loewinson-Lessing, 1936), upheld the contrary view that the mixing of separate 'acid' and 'basic' magmas (his terminology) led to a transitional class of 'neutral' rocks. We should remember that as long ago as 1857 Durocher had pro-posed that hybrids might be produced by the commingling of these two different magmas, and at the turn of the century Alfred Harker, in a masterly study of the Tertiary centre on Skye, detailed the processes involved in the mixing of basic and acid magmas and the formation of hybrids. However, he clearly regarded such hybridization as merely a 'disturbing factor' in the overall process of magmatic differentiation (Harker, 1909, p.356), and despite powerful voices in its support the mixing of magmas was never to move to the centre stage of petrological theory.

Increasingly, physical chemical concepts were called upon to explain these consanguineous relationships. For example, the English chemist Frederick Guthrie was the first to suggest the importance of eutectics in the crystallization of granite, an application which enabled Jethro Teal, in 1888, to solve the mystery of graphic granite. However, the real founder of modern physicochemical petrology was the Norwegian Carl Vogt, who, between 1884 and 1888, from his work on slags, concluded that the most abundant igneous rocks corresponded in composition to eutectic mix-tures. It was Vogt who first suggested that granitic magma was of the nature of a eutectic residue, a rest magma from the crystallization of the basic magma. He also expressed what always ought to have been the proper balance between the magmatist and transformist points of view when he wrote 'Perhaps the future decision in these matters should above all discredit the view that all granites have originated in the same way, and on the contrary prove that more processes exist whereby sedimentary rocks on the one hand, and really eruptive rocks on the other, have led to the same petrographical result that we call granite' (translation by Den Tex, 1990, p.217).

Despite repeated restatements of this rational point of view the magmatists and transformists remained in confrontation for another 70 years. The French relied, in particular, on field evidence for replacement such as that provided by 'les dents de cheval', the growth of big feldspars in both granite and country rock, and Lacroix was plainly astonished at Rosenbusch's stubborn refusal to accept what he regarded as the obvious explanation. The protagonists, as in all major debates of this nature, even divided into subfactions, the transformists, for example, splitting between the 'wets', who, like the Helsinki group, favoured pore fluids as the transporting agents, and the 'drys', led by the Frenchmen René Perrin and Marcel Roubalt, who, in the late 1930s, advocated solid diffusion. There was much bandying of literary aspersions with personal references (by Wegmann) to the 'pontiffs' and (by Bowen) to the 'soaks', as part of a sometimes barren debate. Eskola's shrewd reflections on these inter-changes are epitomized in the jingle attributed to Gustaf Fröding:

What's settled truth in Leipzig and Jena
is just a stupid joke in Heidelberg

Returning to the central theme, the end of the 19th century had seen the study of igneous rocks sufficiently advanced to engender the publication of the basic texts: in Germany by Rosenbusch himself (1888), three years later in England by Jethro Teal, in France by August Michel-Lévy (1889), and in Russia by Loewinson-Lessing (1899). Particularly influential in the United States was a paper (1892) and a textbook (1909) by Joseph Iddings, who had been one of the first outside Europe to examine igneous rocks in thin section, and who had himself studied in Heidelberg. Indeed, it could be said that he initiated the 'grand tour' to Europe, and when we learn that most of the American masters, including Daly, Cross, Pirsson, Washington and Bowen, made the same pilgrimage to either Heidelberg or Leipzig, we can understand how deep rooted became the thesis, in the United States, of the single differentiating magma.

The appointment of Norman Levi Bowen, in 1910, to the newly estab-lished geophysical laboratory of the Carnegie Institution in Washington not only ushered in the modern era of the experimental study of igneous rocks, but soon provided a proof of the widely accepted model that such

rocks were derived 'from a common stock through some systematic process of differentiation' (1915, p. 90). As to its explanation, the realization, by
Bowen in particular, that chemical diffusion – the Soret effect – was too
slow to separate magmas into fractions resembling the common igneous
rocks, and with both Greig and Bowen showing that immiscibility was not
a feature of melts with common rock compositions, the search continued
for an alternative mechanism.

We should remember that the several possiblilities – diffusion, immiscibility, crystal settling, filter pressing and, to a lesser extent, magma
mixing – were familiar concepts by the turn of the century and entered
into discussions by Schweig, Harker and Daly, among others. For example, as early as 1897 Becke was discussing the possible connection
between fractional crystallization and convection currents, and filter
pressing was favoured by many workers, being particularly refined by
Herbert Thomas and Edward Bailey of the Geological Survey of Great
Britain who, in 1924, envisaged the rest magma being squeezed out of the
crystal mush by the weight of the crystal overburden. As a further illustration of the wit of our geological ancestors, every present day petrologist
should read the final chapters of Reginald Daly's memoir on the *Geology of
the North American Cordillera* (1912), not only to guard against any unwit-
ting plagiarism, but to appreciate what may seem to us the remarkably
advanced state of the art of that time.

Nevertheless, it has to be said that the various current hypotheses were
largely unsupported by real physicochemical data, and these were to be
provided in full measure by the exactly scientific experiments of Bowen. It
was the ordered elegance of these results which convinced him and many
of his contemporaries that not only was fractional crystallization of a
primary magma the dominant process, but the crystals settled out under
gravity, leaving a residual liquid gradually evolving to the composition of
a granite. In his classic publication *The Evolution of the Igneous Rocks* (1928)
Bowen did enter the important caveat that 'many granite magmas may
have their immediate origin in the remelting, say by deep burial, of a
granite . . . derived in more remote times from basic material' (p. 319), but
the central thesis remained inviolate.

We need to remember that these experiments were carried out on dry
melts, a limitation of which Bowen was well aware, and was later amply
to rectify in his further studies with Frank Tuttle, leading to the publication, in 1958, of another of petrology's milestone texts, *Origin of Granite in
the Light of Experimental Studies in the System $NaAlSi_3O_8$–$KAlSi_3O_8$–SiO_2–
H_2O*, in which they pay tribute to the work, 20 years before, of their
colleague Robert Goranson on the *Silicate–Water Systems*.

In the early 1950s, powerful support was given to the Rosenbusch–
Bowen model by the pioneering work of Stephen Nockolds and his co-
workers on the trace element distribution between the constituent

minerals of granitic rocks, showing that these distributions follow a pattern entirely consonant with the progressive separation of crystals from a liquid. Nockolds not only addressed the problem of the apparent lack of intermediate rocks but also probed the modifying role of assimilation.

Thus, in the first half of this century this elegant model assumed a dominance which was difficult to counter, especially in the absence of experimental work of equal calibre on those other possible processes already mentioned – that is, partial remelting, mixing, immiscibility and diffusion. As Walton writes, 'the question to which American geologists especially were, in general, somewhat insensitive was whether this magmatic differentiation model outlined an actual sequence of events which really contributed in nature to produce granite' (1965, p. 635). There were, however, always some acute observers, quite apart from the transformists themselves, who were aware of the incompleteness of the model. Clarence Norman Fenner in 1926, Frank Grout in 1926, Arthur Holmes in 1931 and 1936, Reginald Daly in 1933 and Tom Barth in 1952 were among those who commented on the obvious inconsistency between the volumes of granite and granodiorite, relative to gabbro, as predicted and in actuality. An explanation was sought in the pure melting of the sialic crust deep in the orogen, perhaps brought about by depressing the 'sial' into the hot 'sima' or, according to Holmes and Daly, by the rise of hot basaltic magma into the crust, this latter possibility also explaining the contrasted association of basic and acid rocks. And Harold Read continued to ask why granitic magmas are only voluminous within the thickened crustal welts of the orogens.

DALY TO CLOOS: CONSIDERATIONS OF SPACE

Turning from the genesis of granitic magmas to their mode of emplacement it is clear that the room problem has preoccupied geologists from the occasion of that famous debate between Lyell and Keilhau. We have seen that *in situ* transformation – granitization – satisfied one school of thought, but with all its nuances, from mass assimilation by an active granitic magma to grain by grain, metasomatic replacement by means of the granitic juices, ichors or mineralisateurs, which were thought to soak through the crust, carrying new materials, enhancing chemical reactions, and feldspathizing the country rocks. Such a solution to the room problem led, almost inevitably, to the taking up of extreme positions, as, for example, Helge Backland (1938) arguing for the replacement of red sandstone by Rapakivi granite, and Doris Reynolds (1944) envisaging the granitization of metagreywacke with the expulsion of a basic front in the form of appinitic diorites and gabbros, each apparently heedless of the enormous energies required by such processes at high levels in the crust.

However, even from the earliest chapters of this history there were perceptive field geologists searching for a more structural approach to this central problem. As examples the Scot Charles Lyell, writing in his *Principles* (1865, p.723), considered that the Arran Granite had shouldered aside its wall rocks, whereas the Irishman William Sollas, describing the geology around Dublin in 1895, interpreted the Leinster Granite as a domal sheet emplaced during folding in the rising arch of its cover.

It was the North Americans Joseph Barrell and Reginald Daly whose detailed field studies seemed to demand a purely mechanical explanation for the emplacement of certain granites. They particularly favoured the foundering of crustal blocks, to which mechanism Daly gave the name of 'magmatic stoping', though he correctly recognized that this central idea had already impressed itself on 'some few of his contemporaries'. How human it was of Daly to admit so readily, concerning the New Hampshire stocks that, to quote from his monumental *Igneous Rocks and the Depths of the Earth*, 'for nine years the writer was baffled in the attempt to explain their mode of intrusion' (1933, p.267).

Even so, it still required the development of granite tectonics, particularly by the German Hans Cloos in the 1920s and 1930s, to address the structural issues in measured and detailed case histories. It was Cloos who, in his *Das Batholithenproblem* (1923), accepted Iddings' contention that batholiths follow crustal discontinuities, and who relegated stoping to the incidental in favour of other mechanisms involving crustal accommodation. He strongly believed that it was the structural interaction between intruding magma and the deforming envelope of country rock that held the real key to emplacement problems.

With hindsight it seems extraordinary how Hans Cloos, one of the most brotherly of men, had to face the criticisms, even displeasure, of the father of the Viennese school of petrofabric studies, Bruno Sander, who considered that Cloos' work lacked originality! But this is a commonplace reaction of authority. And we might wonder whether granite tectonics might have languished in its original high German text if Hans's brother, Ernst, had not emigrated to the United States and, further, had not that most perceptive of Americans, Frank Grout, commissioned Hans Cloos' fellow student Robert Balk, another émigré, to summarize the state of the art in yet another milestone memoir, the *Structural Behaviour of Igneous Rocks* (1937).

We should remember that Grout had himself begun to model the rise of magma blobs, no doubt spurred on by Wegmann's earlier (1930) comparison of certain deep-seated granite masses with salt domes or diapirs.

It is also of great interest to appreciate that by this time the granite problem was sufficiently in vogue to warrant the establishment of a high powered batholith committee of the US National Research Council, chaired by this same Frank Grout. It was the corporate meditations of this august body which led to an important international debate, in 1947 in Ottawa, on the 'Origin of Granite'. From the record of the discussions (Gilluly, 1948) it is clear that no consensus had yet been reached between the long established factions, though I suspect that many in that largely American audience would have agreed with Frank Buddington's estimation that over 85% of northwest Adirondack granitic rocks were the product of consolidation and differentiation of magma intruded as sheets or phacoliths, and that less than 15% were the product of migmatization and granitization of metasediments and amphibolite, a conclusion later extended to the whole of the granitic rocks of North America. They might also have favoured Buddington's acceptance and restatement of Wegmann's position on the influence of crustal depth.

I am tempted to tell a personal story in illustration of the depth of this disagreement and its effect on personalities. At the International Geological Congress in London in 1948 I travelled up in a lift alone but for Read and Bowen; the human silence was tangible, neither authority spoke, nor was I introduced!

THE SHIFT FROM THE GREAT DEBATE

So far in this history there has been little reference to the physical nature of granitic magma, and indeed other than the original prescient comments of Scrope and innumerable references to an almost mystical broth, mash, bouille or Brei, any real understanding of magma rheology had to await the seminal experimental work of Donald Lacy in England in 1960, and of Herbert Shaw in the United States in 1965. Equally our knowledge of the fluid dynamics of magmas and their rise due to buoyant forces, although discussed in general terms by Frank Grout in 1945, had to await the important experiments devised by Hans Ramberg. Since 1963 the latter has attempted to model the interaction between magma and country rock in a series of beautifully executed experiments simulating the effect of gravity by the use of a giant centrifuge. The results, coupled with field studies, have allowed him to construct a structural series in parallel with the granite series: diapirs rise away from deep seated melts and force entry into the upper brittle crust. All in all, such an approach has changed the course of ideas on granite emplacement.

The thermal effect of granitic intrusions, the presence of an 'aura granitica', was recognized early, particularly by the French and Fennoscandians. Indeed, for some, granite represented an almost magical source of heat, a veritable driving agent of granitization. For example, George Barrow, who was the first to establish the zonal nature of regional metamorphism in Scotland (1893), considered it 'reasonable to attribute both the minerals and the crystallisation to the thermometamorphism of the intrusion', so causing his colleagues to quip

This intrusion the schists all around it does roast
And when it is absent it alters them most.

No doubt that, in the theoretical unification of plutonic processes, it was logical to see a coherent relationship between metamorphic processes and the granite series. The synkinematic granites belonged with regional metamorphism, but as their anatexitic products moved upwards through the crust their thermal effect gradually became more subdued, more local and superposed, finally to be presented as simple contact aureoles. Moreover, and most remarkably, almost the whole theoretical basis of the mineral chemistry of metamorphic rocks rested, until relatively recently, on some few, though notable, studies of such aureoles: the essentially isochemical nature of the thermal metamorphic process itself on Rosenbusch's seminal study at Barr-Andlau in the Vosges (1877), the counter view of metasomatism and the first statement of the doctrine of chemical equilibrium on Goldschmidt's studies in the Oslo area (1911) and the concept of metamorphic facies on Eskola's work at Orijavi (1914, 1915).

Finally, concerning rock texture, we have seen how the Heidelberg school of micropetrology revelled in the description of rocks, though largely from the magmatist point of view. The tradition of such studies continues to this day though with special emphasis on work in ore microscopy, but it is to earlier research by O.H. Erdmannsdörffer of Heidelberg, from 1912 onwards, and also by F.K. Drescher Kaden, from 1927 onwards, that we owe the early development in the understanding of rock textures. Not surprisingly, such studies could only make limited progress in the absence of experiments and the formulation of theories of crystal growth: the seminal work of Fouqué and Michel-Lévy, in 1882, on the cooling and crystallization of basalt had for long remained but a glorious memory. However, we should never forget that Brögger's famous conclusion – that the sequence of crystallization reflects the sequence of differentiation in a cooling magma – was experimentally based. Neither should we forget that Erdmannsdörffer always emphasized the separate, though overlapping, roles of metablastic and endoblastic growth, so that petrologists ought not to have been surprised to find that granitic rocks can often show a metamorphic texture without being of primary metamorphic origin: a source of much acrimony in the past.

The more recent history of the development of granite studies is the theme of this present collection of essays. I shall start from a base represented by an important group of international texts dealing directly with the granitic rocks and published around the middle of this century: Arthur Holmes' *Natural History of Granite* (1945); the Geological Society of America's *Memoir* No. 28, *Origin of Granite* (Gilluly, 1948) with its two key papers, one by N.L. Bowen, the other by H.H. Read; E. Raguin's *Géologie du Granite* (1946); H.H. Read's *The Granite Controversy* (1957); O.F. Tuttle and N.L. Bowen's *Origin of Granite in the Light of Experimental Studies* (1958); K.E. Mehnert's *Migmatites and the Origin of Granitic Rocks* (1968); and Marmo's *Granite Petrology and the Granite Problem* (1971). We shall see how the entrenched and apparently irreconcilable positions, as recorded in the proceedings of the 1947 Ottawa conference, were soon forced into a tolerable coexistence by a veritable deluge of new information.

There is now a much greater willingness to accept multifactorial explanations though little enthusiasm remains for wholesale granitization – a thesis largely superseded by bulk melting at depth. Furthermore, the advent of the plate tectonic theory has encouraged a more holistic approach. In particular very detailed, field based petrographic studies have awakened a fresh interest in rock suites and the realization that magma mixing is an important process in their production. Structural research is providing elegant solutions of the room problem which emphasize the interconnection between granite emplacement and the local and global tectonic regimes. Cleverly designed melting experiments, carried out in the presence of water and using a variety of synthetic and natural materials, have constrained fairly closely what is and what is not possible from the physical chemical point of view. Not least, rheological theory is providing insights into the nature of granitic magma itself, and ingenious simulations of the crystallization history of magma in deep chambers has radically changed views on magma movement and crystal segregation.

Perhaps of greatest promise is the new light on the problem of the ultimate origin of granitic rocks which has come with the realization that, geochemically, the various types of granite can be regarded as images of their source. However, with some notable exceptions, the thrust of recent research has been fuelled more by a desire to understand the nature of the continental lithosphere than to understand the granitic intrusions themselves. Also there is a danger that petrological theory is in danger of driving itself by circular argument and the historical weight of Heidelberg and Washington, to the conclusion that most granite is born, albeit by a staged process, in the uppermost mantle. In what follows I propose to sift through a great variety and volume of evidence in order to judge whether this is true or false.

REFERENCES AND FURTHER READING

NOTE: The references listed below are key references for the history of petrology. Some other important references are given in the Bibliography at the end of the book, while many of the earlier references can be found in those listed below.

Atherton, M.P. (1996) Granite magmatism. In: Le Bas, M.J. (ed.), *Milestones in Geology, Geological Society of London, Memoir No. 16*, pp. 221–235.

Bailey, E.B. (1958) Some aspects of igneous geology, 1908–1958. *Transactions of the Geological Society of Glasgow*, **23**, 29–52.

Eskola, P. (1955) About the granite problem and some masters of the study of granite. *Suomen Geologinen Seura*, **28**, 117–130. Geologisk Sällskapet, Finland.

Gilluly, J.L. (Chairman) (1948) *Origin of Granite. Geological Society of America Memoir No. 28* (conference of the Geological Society of America held in Ottawa, Canada, 30 December 1947).

Holtedahl, O. (1963) *Studies on the Igneous Rock Complex of the Oslo Region*: 19. Skrifter Norske Videnskaps-Acad. Oslo, Mat. Naturv. Kl **12**, 1–24. (Refers to Lyell's visit to Norway in 1837, with remarks on the 'Granite Problem'.)

Loewinson-Lessing, F.Y. (1936) *A Historical Survey of Petrology*. (Translation by Tomkeieff, S.I.) Oliver and Boyd, Edinburgh, 112 pp.

Marsh, B.D. (1996) Solidification fronts and magmatic evolution. *Mineralogical Magazine*, **60**, 5–40 (pp. 20–23).

Read, H.H. (1957) *The Granite Controversy*. Thomas Murby and Co., London, 430 pp.

The categories of granitic rocks: the search for a genetic typology

2

We can classify rocks, for petrological purposes, exactly, definitely, and strictly only by creating arbitrary divisions, cutting them up by sharp planes and putting them into man-devised pigeon-holes. Such a classification is a pis-aller, a makeshift, a classification of convenience; in may or may not correspond to the evolution of igneous rocks as it really is.

Henry S. Washington (1992) Bulletin of Geological Society
of America, **33**, p.801.

THE GRANITIC ROCKS

In his role as a disputant in the *Granite Controversy*, Herbert Read saw a need to include a wide variety of quartz-bearing igneous rocks within the ambit of his discussions. He pretended to excuse himself, writing, 'the attentive and pugnacious listener may interject that I am using the term granite in a scandalously loose fashion', but he had no doubt that he was dealing with a natural family of rocks.

The main reason for this family relationship between the various combinations of quartz, two feldspars and mica or hornblende, is because they represent the various stages in the convergence to a ternary minimum in what Norman Bowen regarded as petrogeny's residua system. This remains true for a range of genetic processes involving fractional melting, fractional crystallization, metasomatism or, in a more complex way, the mixing of two magmas. Nevertheless, the coherence of this family of rocks and particularly whether any boundary line might be drawn between them and the gabbros, are questions rooted deep in petrological theory. For the moment I will concern myself solely with this boundary and the possibility that there is some truth in Read's claim that there is a natural hiatus between the two fundamentally different rock groups – that is to say, between the series quartz diorite–tonalite–granite on the one hand and the series peridotite–anorthosite–noritic gabbro–diorite on the other.

Granites and rhyolites, gabbros and basalts are certainly associated in space and time, but the relative volumes are very different in the various contrasted geological environments. Different, too, is the importance of the 'Daly gap', representing the relative lack of intermediate rock types. Thus although the intrusive mafic rocks predominate in the island arcs and remain important in the marginal arcs – with the compositional gap hardly apparent – they are greatly reduced in proportion within the collisional orogenic belts, where the gap itself widens. In the anorogenic centred complexes the volumetric relationships are various, but the Daly gap is now very evident. Even more important, in all environments, is the time difference, for in innumerable igneous complexes the mafic rocks herald the main plutonic event, and the gabbros are universally the precursors of the granites. Naturally this hiatus is represented geochemically, albeit rather subtly expressed in arc environments, and in compositional diagrams it takes the form of a step, an inflexion, or a change in population density. It is at this point of change that water enters as an essential component, marked by the exchange of hornblende for pyroxene.

In granitic terranes I have never found any difficulty in identifying this natural gap between what is clearly gabbro or meladiorite and what is clearly quartz diorite, tonalite or other granitic rock, and this despite its frequent disguise by the mixing and mingling of the two components and the concomitant mineral changes. I allow myself to be quietly amused with the thought that, as long ago as 1891, D'Aubisson could perhaps do the same, for it was he who coined the term 'diorite' from the Greek word meaning to distinguish!

IS CATEGORIZATION POSSIBLE?

It could well be argued that any attempt to categorize the granite family on a natural basis is doomed to failure given the virtually infinite number of different types which might be generated in response to a variety of generative processes and possible source rock compositions. However, experience shows that it is possible to recognize both a degree of order and a natural division into several distinct groups of granitic rocks.

The problems arise in the comparative process itself, and even more so in agreeing to a universal nomenclature. Mathematical geologists, such as Whitten and co-workers in 1986, have demonstrated that present nomenclatures are based, as a mere convenience, on a partitioning procedure that evolved arbitrary pattern classes, and only fortuitously do these have a petrogenetic significance. In effect, the variables have been specified by the definitions themselves, and different sets of variables would be likely to yield different groupings: in short, groupings may only be of significance if the variables used were prescribed by a genetic model. Furthermore, there is even dispute on the nature and form of the representation of the data, especially in the use of Harker diagrams in discriminating between rival petrogenetic hypotheses (Whitten, 1996).

Probably as a result of such criticisms the present emphasis is on the use of cluster analysis to determine objectively the real groupings, but it would obviously be better to classify granites using variables chosen on the basis of petrological understanding. In fact, in any one region, a good case could be made for adopting just those discriminants best suited for highlighting some special feature such as the mineral resource potential. An example of this specific approach is provided by Ramsay, Stoeser and Drysdall in their description of granite magmatism in the Arabian Shield.

We do need some universally acceptable framework even though, manifestly, classifications may have different purposes. However, although a proper order is obviously required for description and comparison, the resulting arrangements are wholly static, often artificial, and lead nowhere along the path of understanding. This was acceptable in the past but now we need a genetic, process based, dynamic classification to run in parallel with the descriptive form. Such a natural classification already exists for both the sedimentary and metamorphic rocks, but still remains to be established for the granites.

At this point it is of interest to recall that some of the early classifications did have the logic of being process based. Thus Becker, in 1901, and Vogt, in 1908, constructed classifications on a concept of eutectic crystallization, while Shand, in his 1927 edition of *Eruptive Rocks*, introduced his silica saturation concept, seeking to use both critical phase boundaries and cooling history. In 1958, Tuttle and Bowen classified granites as hypersolvus or subsolvus on the basis of the original temperature control of feldspar unmixing. Moreover, since the time of Zirkel, granularity has been regarded as a discriminant, allowing the fine-grained silicic volcanic rocks to be separated from the coarse-grained granitic rocks on the supposition of different cooling histories; indeed, this justified, in part, Zirkel and Rosenbusch's original classifications. However, such a process based approach clearly depends on the reality of the underpinning thesis.

CATEGORIZATION BY MODE

Appreciating the complications and realizing that igneous petrologists were not yet ready to follow this ideal path, Streckeisen in 1976 used the historical approach in seeking an international consensus for an acceptable convenient classification based on easily determined modal data (Figure 2.1a). We ought not to forget, however, that for the granitic rocks the guidelines for this proposal, based essentially on the feldspar ratio, were established by Lindgren at the turn of the century when he and his

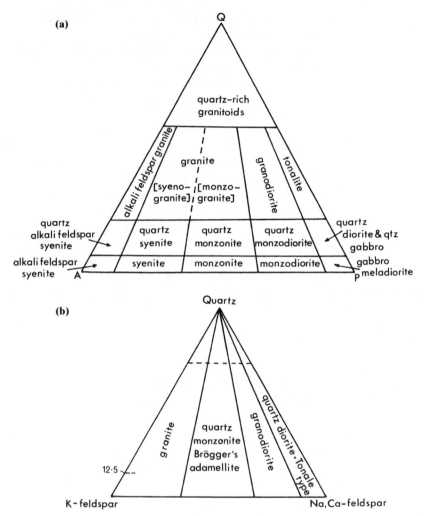

Figure 2.1 (a) IUGS–Streckeisen modal classification and nomenclature of the plutonic rocks. (b) Representation of the scheme that Lindgren, Becker and Turner would have used in the 1890s during their mapping of the granitic rocks of the Sierra Nevada, California. Early in the 20th century a $\frac{1}{8}$ division came into fashion to separate the granites from the syenites and diorites. Data from Lindgren (1900).

co-workers needed to define exactly 'granodiorite' for the good reason that this was the most abundant granitic rock of the Sierra Nevada (Figure 2.1b).

It is easy to be critical, especially as the mafic minerals are not directly represented and species boundaries too often pass through natural populations. Also, as La Roche rather harshly claimed in 1986, the Streckeisen grid is 'more geometrically attractive than geologically sound'. Furthermore, there is but a poor discrimination of such common granitic rocks as granodiorite, tonalite and trondhjemite (leucotonalite). Despite these reservations, we have to be grateful that Streckeisen restored order to igneous rock nomenclature in accepting, as did Washington and afterwards Bowen, that 'a loosely quantitative classification based on mineralogy and texture is entirely adequate and is as natural a system as one could expect' (Bowen, 1928, p.322). After all, greater exactitude can be obtained by adopting the norm-based refinement of Streckeisen and Le Maitre, or the AbAnOr normative plot earlier introduced by O'Connor, but in each system the disadvantages almost outweigh the advantages.

I consider it unprofitable to review here all the ingenious ways by which rock compositions can be displayed and classified, though a sufficient guide is provided in the selected references and in a review, in 1984, by Bowden and his colleagues. Nevertheless, it would be churlish not to comment on a system originally devised by La Roche in 1986 and modified by Debon and Le Fort in 1988. In seeking to involve all the major elements and to avoid dependence on crystallization models, both the mode and the norm have been abandoned in favour of cationic plots. The latter discriminate well between the different petrogenetic series and, moreover, the formal nomenclature can be accommodated within them, the separate compositional fields being determined empirically using 'large data files'. The resultant curvilinear grid does have a more natural appearance, but this appeal to the 'files' does not convince me that the boundary lines are any less subjective, or that the diagrams themselves are greatly superior to the simpler systems already in use.

Of one thing I am sure, that it is pointless to return to the outcrop for guidance on the original definitions.

CATEGORIZATION BY ROCK SERIES

Notwithstanding the criticisms of mathematical geologists, it seems that no matter how the compositions of a field association of granitic rocks are represented, fairly simple plots show that there are systematic groupings, and that these are often ordered into discrete series or suites, each reflecting the order of intrusion. Almost invariably there is a general trend towards a granitic composition, be it tonalite, granodiorite or granite. The more geologically controlled the sampling, the more specifically defined is the series and its trend.

In general terms this was understood long ago. Thus Iddings, in 1892, defined 'series' as having a sequential order with 'a gradual transition in chemical composition'. In doing so he recognized a consanguinity which carried the implication that the members of the series evolved one from the other. Certainly to Brögger, writing in 1894, and to Harker, in his seminal treatise *The Natural History of Igneous Rocks* of 1909, the use of series signified the underlying idea of a differentiation process such that each member of the series was derived from its compositional antecedent. However, now that we know that rock series may arise by crystal fractionation, mixing or fractional melting, it would be more objective to accept 'series', 'suite' or 'association' as lacking a specific genetic connotation. It is not always easy to be so thoroughly reasoned, but in my use of 'series' I do not necessarily imply that the generative process is evolutionary in the unidirectional sense of Brögger and Harker. As for the terminology itself I consider 'series' and 'suite' to be synonyms.

As an example of the present altogether higher degree of sophistication concerning the recognition of suites, we can note that Lameyre and his colleagues recognized nine distinct rock series simply by the use of the Streckeisen–IUGS system (1982, 1991; Figure 2.2), whereas Debon and Le Fort (1988), on the basis of their own 'Q-P' system, defined seven such series – that is, tholeiitic, calc-alkaline–trondhjemitic, calc-alkaline–granodioritic, monzonitic (dark mineral-rich and dark mineral-poor) and alkaline (quartz-rich and quartz-poor).

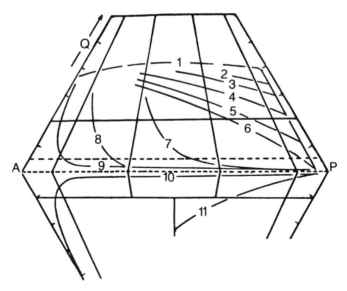

Figure 2.2 Main trends of some plutonic type series based on APQ modal compositions. (1) Tholeiitic series; (2) calc-alkaline–trondhjemitic series; (3–6) various calc-alkaline–granodiorite series; (7) monzonitic series; (8–11) various alkaline series. Data and reference in Lameyre and Bowden (1982) and Lameyre and Bonin (1991).

Each of the rock series discriminated on the basis of major elements or mineral proportions also shows many other individual characteristics, ranging from differences in the trace element content and ratios, and in mineral associations, mineral species and texture, to such outcrop features as the nature of the enclaves, associated co-magmatic dykes, gabbros, lavas and, not least, the metal mineralization. The ability to objectively integrate these many factors may be improved by new, computer-based techniques such as the Expert System for the Characterisation of Rock Types, otherwise dubbed ESCORT, as developed by Pearce (1987; see also Nicholls and Russell, 1990).

Specifically, in relation to particular granite terranes, rock series often show this identity and individuality to a remarkable degree, with chemical and mineralogical features diagnostic of each particular series. Such a signature is not only maintained within a single complex, but can be duplicated regionally in many separate zoned plutons. In this way it defines the supersuites of Chappell and White (Chappell, White and Hine, 1988, p. 507), or the superunits of Cobbing and myself (1972), which I later reason are due to the variations having been established elsewhere, perhaps in some deep-seated melt cell.

When well-defined suites are deemed to have arisen by fractional crystallization it ought to be possible to describe their lineage with reference to the experimentally derived system An–Ab–Or–Qtz–H_2O, on the basis that many granites contain 80% or more of normative albite, anorthite, orthoclase and quartz. However, the involvement of the mafic components can hardly be trivial in the case of the granodiorites and tonalites, being important in the control of f_{O_2}; also the precipitation of hornblende will inevitably govern melt composition in the early stages of evolution. We will find this reservation strengthened in later discussions concerning both chemical and textural evolution, so that it seems that such modelling can, at best, only be regarded as semiquantitative (p. 74 and Figure 5.3).

CATEGORIZATION BY SOURCE ROCK

Of all the possible discriminants the chemical composition is the most objective. Most popular is the use of the abundances and ratios of the rare earth elements (REEs), coupled with the isotopic proportions of certain key elements. A notable example of the latter is the ratio $^{143}Nd/^{144}Nd$, now often expressed in relation to a chondritic reservoir standard, in the form of ε_{Nd} units. Yet another highly significant isotopic ratio is that of $^{87}Sr/^{86}Sr$ recalculated to the time of crystallization, this initial value being represented by the abbreviation Sr_i. Undoubtedly these methods have provided potent clues as to the sources of the granitic rocks, particularly when based on the suspected inertness, in the generative processes, of elements such as zirconium, yttrium and niobium.

Table 2.1 Comparison of the main petrogenetic classifications of granitic rocks according to Barbarin (1990). MA or AM refer to magmatic associations. La Roche abbreviations are as follows: (A) alumino-; (K) potassic; (L) leucogranite; (G) granodiorite; (SA) subalkaline; (CA) calc-alkaline; (TH) tholeiitic; (A-PA) alkaline-peralkaline. Other abbreviations are: (IA) island arc; (CA) continental arc; (CC) continental collision; (PO) post-orogenic; (RR) rift related; (CE) continental epeirogenic; (CEU) continental epeirogenic uplift; (OP) oceanic plagiogranite. S and I indicate derivation from sedimentary and igneous sources, respectively. In the proposal by Chaoqun: (MM) metamorphism–metasomatism; (CR) crust remelting; (MS) mixed source; and (MD) magma source differentiation. In Barbarin's suggested synthetic classification the acronyms signify: (C_ST) crustal shearing and thrusting; (C_CA, C_Cl) crustal collision autochthonous or intrusion; (H_LO) hybrid late orogenic; (H_CA) hybrid continental arc; (T_IA and T_OR) tholeiitic island arc and tholeiitic ocean ridge; (A) alkaline. Adapted from and referenced in Barbarin (1990) with the permission of the author and John Wiley and Sons

Classification basis		Origin					
		Crustal		Mixed		Mantle	
		Peraluminous rocks		Metaluminous rocks		Peralkaline rocks	
First chemical nomenclatures	Shand (1943)	Roches Calco-alc. hyperalumineuses		Roches calco-alcalines			
	Lacroix (1933)					Roches alcalines	
Petrography	Capdevila and Floor (1970) Capdevila et al. (1973)	Granites mesocrustaux		Granites mixtes	Granites basicrustaux		
	Orsini (1976, 1979)		AM sub-alc. alumineux	AM sub-alc. hypoalum	AM calco-alc.		
	Yang Chaoqun (1982)	MM-type		CR-type	MS-type	MD-type	
	Tischendorf and Pälchen (1985)	S_I	S_s	$S_?$ / I_{KK}	I_{MT}	I_{OK}	I_{MA}
Enclaves	Didier and Lameyre (1969) Didier et al. (1982)	C-type (Crustal) (Leucogranites)		M-type (Mixed or mantle) (monzogranites and granodiorites)			
Mineralogy (QAP system)	Lameyre (1980) Lameyre and Bowden (1982)	(Leucogranites) (Crustal fusion)		Calc-alkaline series (High K, medium K or low K)		Tholeiitic series	(Per) alkaline series
Mafic minerals	Rossi and Chevremont (1987)	AM aluminopotassique (s.s. ou composites)		AM monzonitique	AM calco-alcaline	AM tholeiitique	AM (PSer) alcaline
Biotite composition	Nachit et al. (1985)	Lignées alumino-potassiques		Lignées calcoalcalines et subalcalines		Lignées alcalines et hyperalcalines	
Zircon morphology	Pupin (1980, 1985)	Type 1	Type 2	Type 3	Type 4 & 5	Type 6	Type 7

	Ilmenite series			Magnetite series					
Opaque oxides — Ishihara (1977), Czamanske et al. (1981)	Ilmenite series			Magnetite series					
Geochemistry (major elements) — Chappell and White (1974, 1983), Collins et al. (1982), Whalen et al. (1987)			S-type		I-type		M-type		A-type
La Roche (1986), La Roche et al. (1980)	AK-L MA		AK-G MA	SA MA	CA MA		TH MA		A-PA MA
Debon and Lefort (1983, 1988)	Aluminous MA			Alumino-cafemic and cafemic MA (Subalkaline, calc-alkaline, tholeiitic and (per)alkaline)					
Maniar and Piccoli (1989)		CCG		POG	CAG	IAG	OP	RRG	CEUG
Geochemistry (trace elements) — Tauson and Kozlov (1973)	Plumasitic leucogranites		Ultra-MM granites	Palingenic granites (normal and subalkalines)		Plagiogranites			Agpaitic leucogranites
Pearce et al. (1984)			COLG – Collision Granites (syntectonic) (post-tectonic)		VAG volcanic arc granites		ORG	WPG within plate granites	
Associated mineralizations — Xu Kegin et al. (1982)	Transformation type (continental crust)			Syntexis type (transitional crust)	Mantle-derived type				
Tectonic environment — Pitcher (1983, 1987)		Hercynotype		Caledonian type	Andinotype		W.Pacific type		Nigeria type
Suggested classification by Barbarin (1990)	C_{ST}	C_{CA}	C_{CI}	H_{LO}	H_{CA}		$_{IA}T_{OR}$	A	

It is important to emphasize that the major elements often suffice by themselves to image adequately the source rocks. Shand, in 1927, came close to recognizing this when he typed granites as peraluminous, metaluminous and peralkaline on the basis of the molecular proportions of Al, Ca, Na and K, expressed in the form A/CNK >1, A/CNK ~1, A/CNK <1, respectively (Table 2.1). Such a usage has been convincingly employed by Chappell and White in their studies of the granites of the Lachlan region of southeastern Australia, summarized in 1992. There the chemical contrasts between certain granites strongly suggest that there are two main types, one a chemically evolved, relatively potassium-rich, 'S-type', derived from crustal rocks that had previously passed through the erosional–sedimentary cycle, and another, a more primitive, potassium-poor, 'I-type', derived from crustal igneous rocks that had not previously been recycled. We shall see that the source rock signatures so determined provide a powerful discriminant in distinguishing between various granite types in different tectonic environments, though never to be used in isolation from the petrological and geological criteria.

It soon becomes apparent that the granitic types so defined may just be end-member types, for there are others with transitional characteristics, either as a result of derivation from composite sources, or the result of the mixing of the remelts, or both. Indeed there are potassium-poor, calc-alkaline series of granitic rocks geochemically even more primitive than those of the I-types of Lachlan. These occur in the volcanic arcs, where, because they have exclusively mantle affinities, they may be labelled as 'M-types'. Thus it may be that the I-type of Chappell and White, pertaining to a suite of potassium feldspar-bearing granodiorites, itself represents a transition, albeit compositionally and genetically close to the natural M-type end-member.

With this concept of end-members in mind, various workers, including Barbarin in 1990, have constructed a typology on the basis of a transition, with 'S' and 'M' designating, in effect, the two extremes (Table 2.1). Unfortunately, despite the logic of this approach, the suggested labelling differs between authors, so that considerable confusion is bound to arise, especially in view of the current usage of M, S and I, and a parallel nomenclature introduced by Tischendorf and Pälchen in 1985.

Remarkably, and despite the difficulty in arriving at a satisfactory genetic nomenclature, there is wide agreement on the nature and limits of the groupings themselves, as is shown by Barbarin's well researched literature survey (Table 2.1). In addition to the purely geochemical, his compilation encompasses the great range of natural parameters previously referred to. It confirms that we cannot be wholly content with the identification of type by source alone, or indeed by any other single factor, even though by its involvement we are at last approaching the truly genetic classification. The reality is that the nature of the source must often remain conjectural, for how is it possible to take account of inhomogeneities in the mantle, let alone the crust?

CATEGORIZATION BY TECTONIC SETTING

It is obvious that primary source and geotectonic setting must be intimately interrelated in the generative processes, so that it is likely that the chemical composition will also characterize the tectonic setting. Thus, with a broad-brush approach, the granite types can be allocated on this basis. As we shall see as this narrative unfolds, when assembled together with tectonic environment each can be tentatively assigned to a magmatectonic niche which is but a first move towards an eventual genetic classification. Leaving aside the rather special plagiogranites of the constructive plate boundaries, the M-type quartz diorites and tonalites are associated with the island arcs, whereas for the most part the I-type tonalites and granodiorites and various M–I transitions are associated with the active plate margins of the continents. The S-type, peraluminous granites belong with the early stages of the evolution of continent collision zones, the classical orogens, whereas granites of distinctly mixed parentage are associated with post-orogenic uplift. All such granitic types stand in some contrast with the alkali granites of within-plate, anorogenic environments – granites that can usefully be categorized as 'A' types even though their magmas may have a predominantly mantle source. Using various discrimination procedures Maniar and Piccoli have confirmed this relationship between the major element composition of granites and their tectonic environment, but they obviously have no sympathy with the alphabetical nomenclature!

Turning to the use of those trace elements generally held to be immobile during alteration processes, such as zirconium, yttrium, niobium and certain heavy REEs, these have been plotted in all manner of ingenious ways, and in the adept hands of Floyd and Winchester (1975), Brown, Thorpe and Webb (1984), and also Pearce and his colleagues (1984), particularly by the latter's use of the ESCORT procedures, such trace elements and their ratios have become a powerful tool in the discrimination of the ambient tectonic environments, especially for the basic rocks. Perhaps in the simplest cases it is so for granites (Figure 2.3), but here we face the complexities of multiple generative processes and mixed sources, with the likelihood of superposition, with the reworking of old crust carrying the imprint of several earlier tectonomagmatic events.

In fact, every one of the constraining factors, geological, mineralogical and geochemical, has some special message for the petrologist. Few would have guessed that, as demonstrated by Pupin in 1980 and 1985, the crystal form of zircon not only reflects magmatic processes but identifies specific tectonic environments. Less remarkable, perhaps, is the demon-

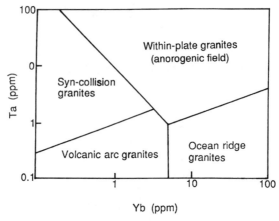

Figure 2.3 One of a range of discrimination diagrams introduced by Pearce *et al.* (1984).

stration by Ishihara in 1977 that the ilmenite–magnetite pair provides some measure of the f_{O_2} of the magma, but hardly that this might discriminate between two main granitic types. On the other hand, in view of the control of type by process and source, it is not surprising that the type of metal mineralization also proves to be a discriminating factor, as was well demonstrated by the Russians, for example Tauson in 1977, and also Zonensajn and co-workers in 1976, by the Chinese, particularly Xu Kegin and co-workers in 1984, by the English, for example Mitchell and Garson in 1981, and by the Canadian Hutchinson in 1982. I make no excuses for bringing the geotectonic aspect to the fore because it is only in this geological context that the origin of granite makes any sense to me, in just the same way as it does to Tischendorf, and did to Jean Lameyre and to our masters, Lacroix, Eskola and Read.

Undoubtedly plate tectonic theory has provided an excellent framework on which to base such a genetic classification of the granitic rocks. It is now almost conventional to define these global tectonic environments in relation to plate tectonic configurations, which I will do in all that follows. However, as we have seen, the tectonomagmatic connection was early suspected, particularly by Eskola, who associated different rock suites with their geological setting, though we can now much more clearly appreciate that to each tectonic situation, to each stage of the Wilson cycle, there is likely to be a corresponding type of granitic association. Although it is usual to make a primary distinction between orogenic and anorogenic magmatism, I later conclude that all magmatism is tectonically motivated, so that such a division is simply a matter of emphasis.

These are purely proposals for discussion because there are many misfits and reservations, but in what follows we shall see how they fare in the face of a general survey of the whole field of granite studies. They are a basis for the final genetic grid of the future, to be constructed, as we have seen to be necessary, not from composition alone but in parallel with compositional series, process, source rock and tectonic environment. As for the disputed nomenclature, I would ask the reader to regard it as a philosophical exercise to be discarded when it ceases to be useful as an aide-mémoire; certainly it should never be used as a utility for field identification. As those four pioneers Cross, Iddings, Pirsson and Washington wrote long ago, 'since it is not to be assumed that the present state of petrographical knowledge is complete, provision should be made for expansion and adjustment along lines which seem to be those in which future development will take place' (1902, p.690). I now turn to survey those developments that are already in train before attempting the impossible, choosing to wait until the conclusion of these essays before unveiling my own prejudice concerning the categorization of granites, though, of course, the impatient reader is free to turn to Table 19.1!

SELECTED REFERENCES

NOTE: For this and all further chapters key references will be listed at the end of the chapter. For details of all other references the reader should consult the bibliography at the end of the book. In the text the reference is by author alone, but a date follows whenever there is a possibility of confusion.

Barbarin, B. (1990) Granitoids: main petrogenetic classifications in relation to origin and tectonic setting. *Geological Journal*, **25**, 227–238.
Bowden, P., Batchelor, R.A., Chappell, B.W., Didier, J. and Lameyre, S. (1984) Petrological, geochemical and source criteria for the classification of granitic rocks: a discussion. *Physics of the Earth and Planetary Interiors*, **35**, 1–11.
Streckeisen, A.L. (1976) To each plutonic rock its proper name. *Earth Science Reviews*, **12**, 1–33.
Whitten, E.H.T. (1996) Molar-ratio and Harker diagrams in portraying the actual chemical variability of granitoid suites. *Journal of the Geological Society of London*, **153**, 121–125.

Granite as a chemical system: the experimental impact

3

The results of experimental petrology . . . help to distinguish between possible and impossible processes.

Peter J. Wyllie (1983) In: Migmatites, Melting and Metamorphism, p. 13.

THE POSSIBLE AND THE IMPOSSIBLE

The generation and evolution of granitic magmas are so much matters of physical chemistry that we might hope that the processes involved will eventually be fully understood. Following the lead set by Burnham, granite systems may eventually be modelled using thermodynamic data. Meanwhile, however, their investigation in the laboratory has proved to be technically difficult, the time factor frustrating and the reaction kinetics complicated. Too few laboratories are sufficiently equipped and funded for undertaking this primary research. As a result there remains considerable ambiguity in the determination of the phase relationships and the liquid compositions involved. The literature is crowded with contradictory phase diagrams for which the magic test of reversibility has not been substantiated. Indeed, for complex silicate systems with volatile components, and sealed within noble metal capsules, it may be impossible to establish reversibility, and Johannes in 1983 concluded that all melting experiments at high pressures involving granitic compositions necessarily involve metastability, which is certainly true for systems containing both sodium and calcium. He argued that such metastability is largely avoided in natural rocks by the very slowness of the several reactions which follow one another step by step. When one also learns that kinetic considerations mean that metastable as well as stable equilibria can be reversed, and that the temperature and order of precipitation can be altered by changing the cooling rate, one can surely be excused a certain scepticism.

Presented by the reality of the outcrop, Wyllie tells us that, 'Given this picture of a complex multistage process involving magmas derived from different sources with the prospect of all kinds of dynamic, non-equilibrium processes of the diffusion within melts and of chemical changes produced by circulating solutions, one may fairly ask what possible applications the crystal melt phase equilibrium experiments completed under tidy conditions in a laboratory can have for the interpretations of the great granitic cauldrons' (1983, p. 13). This was exactly why that arch-disputant Read, in one of his most provocative moods, enquired if it were ever possible to gaze into a tiny gold capsule and expect to see a granite intrusion. Of course, Wyllie answers his own rhetorical question and counters Read's challenge by asserting that the crucible most certainly can help in distinguishing between possible and impossible processes.

GRANITE SYSTEMS

This claim is amply supported by the impact of Tuttle and Bowen's seminal study (in 1958) of the haplogranite system $NaAlSi_3O_8$–$KAlSi_3O_8$–SiO_2–H_2O, which not only introduced new concepts but placed precise constraints on speculation as to the origin of granitic magmas. Of course, Bowen had always understood the need to take account of water at high pressures, but it had been left to Goranson, between 1931 and 1938 (Goranson, 1938), to develop the necessary technology, to design the first key experiments and to establish the form of the simplest feldspar–water systems.

Luth's review of 1976 showed how these early investigations engendered a plethora of similar work from which both experimentalists and petrologists made extravagant claims as to their relevance to nature. Although the use of synthetic starting materials was appropriate enough in the investigation of a specific phase relationship, and the carrying out of the experiments under water-saturated conditions was necessarily dictated by the existing technology, there was a growing realization that such experiments might bear little relationship to natural processes. Piwinskii was compelled to write: 'Because the experimental investigation was undertaken in the presence of excess water, no direct link can be established with physico-chemical processes occurring within the earth' (1973, p. 125).

Of course, much of immense value was discovered. Thus between 1957 and 1961 Winkler and his associates established that the melting of natural rocks, such as clays and greywackes, in the presence variously of water, fluorine, chlorine and ammonia, produced granitic melts at geologically realistic temperatures and pressures. Initial composition was shown to be all important in determining the melt type, and almost the entire range of granitic compositions, tonalite to aplite, was produced by mere variation in the calcite content of the charges. At the same time the 'granite system' was further elaborated by von Platen's investigation, in 1965, of the effect of the anorthite component, and it became clear that there was a need to study the more complex granodiorite system $NaAlSi_3O_8$–$CaAl_2Si_2O_8$–$KAlSi_3O_8$–SiO_2–H_2O. This task was undertaken in 1975 by Whitney, who selected synthetic compositions approximating to natural granites and also tightly controlled the water content, though not to the extent of being aware of the actual water content of the melt. Nevertheless, the form of this synthetic system is close to that later determined for natural rocks.

Some twenty years later we are now led to expect that, given sufficient data, it should eventually be possible to calculate crystal-melt phase equilibria in magmas. Lange and Carmichael have given us, in their 1990 state of the art discussion, an account of the methodology involved, and have also drawn attention to the deficiencies in the present database. This essentially thermodynamic approach is very well illustrated by the work of Burnham and his colleagues which I touch on below.

ROLE OF WATER

It is the water content that proves to be the most critical factor in all these experimental attempts to simulate nature. It was quickly realized that if granites were generated by the melting of dry metamorphic rocks deep in the crust, their melts were likely to be relatively water-deficient and vapour-absent. As Brown and Fyfe showed experimentally, the conditions of substantial melting are necessarily coupled to the release of water through dehydration reactions involving the vapour-absent breakdown of the micas. There may, of course, be a small amount of initial melting due to trace amounts of free water in the rocks.

This view of the likely water-undersaturation of natural granitic magmas was amply confirmed by the new approach initiated by Piwinskii and Wyllie in 1968 and adopted by Wyllie and his co-workers in a long series of experiments (reviewed in 1983). In essence the method involved the complete melting of a natural granite under various pressures, temperatures and concentrations of water, with the conditions adjusted so as to reproduce on cooling the original mineral assemblage in its natural order of crystallization. Despite all the possible reservations, coupled with the realization that such experiments have never yet been carried out to completion in reproducing a fully crystallized granite, the results are highly significant and, I am tempted to add, geologically realistic.

Thus, in one key example of cooling a melt obtained from biotite granite, the natural sequence of crystallization was reproduced at geologically acceptable temperatures and pressures, but with a water content as low as 2 wt.% (Figure 3.1). Furthermore, at this degree of water-undersaturation, the crystallization interval was greatly extended, being of the order of

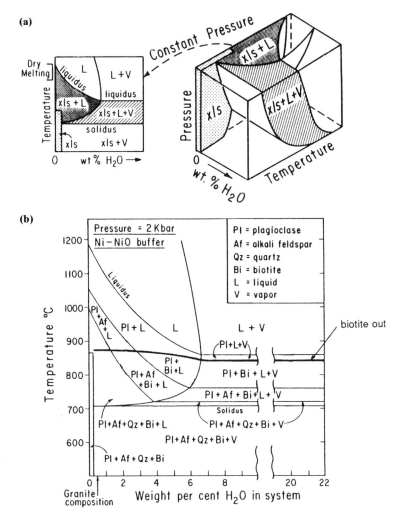

Figure 3.1 (a) Phase relationships for a single rock, represented within a P–T–X (H₂O) model, and illustrated by isobaric and isoplethal sections through the model. After Robertson and Wyllie (1971). (b) Isobaric section, corresponding to left-hand part of (a), for a biotite granite, with oxygen fugacity buffered. After Maaløe and Wyllie (1975). Perhaps the most significant features of these figures is that for bulk compositions with low water contents the assemblage crystals and liquid extends through a wide temperature interval; also that a granitic melt undersaturated with water can coexist with feldspars and quartz through several hundred degrees, and that the phase boundaries for the anhydrous silicates within the vapour-absent field are steep, with temperatures very sensitive to the water content. Reproduced from a fully referenced overview of the work of Robertson, Maaløe and Wyllie, in Wyllie (1983) with the permission of the editors of *Migmatites, Melting and Metamorphism, Shiva Geology Series.*

several hundred degrees, in sharp contrast with the very narrow temperature interval in the analogous system with excess water. From this finding alone it is easy to deduce that 'wet' granites must freeze quickly after leaving their source, implying that only 'dry' granites can rise high into the crust, both expectations which are again in accord with nature.

From their full series of experiments Maaløe and Wyllie concluded that the water content of large bodies of magma is unlikely to exceed 2 wt.% throughout a substantial part of their evolutionary history. From an overview of more recent work it seems that to duplicate the natural mineral parageneses requires about 2 wt.% for diorites and up to 4 wt.% for granites. Of course, the water content may eventually increase to the saturation point, owing to its decrease in solubility as the magma ascends into lower pressure regions and as crystallization proceeds; this is wholly consonant with the natural situation whereby water-saturation at a late stage is marked by the development of aplitic and pegmatitic variants.

In these experiments the liquid compositions were found to trend towards tonalite and diorite with increasing temperature and pressure (Figure 3.2). With the liquidus temperature for tonalite of the order of 1100°C at 2 wt.% concentration of water, it seems that the remelting of such rock compositions requires higher temperatures than are thought to be normally available in the crust. Although new experiments on the melting of appropriate crustal materials indicate that somewhat lower temperatures are sufficient (p. 44; also Johannes and Holtz, 1996, p. 298), it remains true that an extra source of heat is required for the generation of granodiorites and tonalites and this is often provided by the contemporary intrusion of basaltic magmas. It also follows that the natural melts of these rocks are most likely to be at temperatures below their liquidus and so can be expected to carry suspended crystals.

John Clemens tells me that in many experiments with water-under-saturated granitic melts pyroxene appears near the liquidi. That it is absent in natural rocks is probably due to reaction with the residual melt and indeed it is common to find magnetite-peppered cores within hornblende crystals, and also tiny aggregates of hornblende and biotite, which probably represent pseudomorphs after pyroxene.

The most important of these experiments used samples of metaluminous granites and tonalites of the cordilleran type. Nearly a decade later Clemens and Wall successfully subjected a garnet and cordierite-bearing, peraluminous granite to the same test, finding that greater water contents (3–5 wt.%) combined with a higher pressure (5 kbar) were required for reproducibility with such a composition. I think it worth noting that in some of these latter melt experiments the charges had to be seeded with both cordierite and almandine because the nucleation kinetics are so very sluggish. This may seem to detract from their value, except that in nature many S-type granites probably did contain residual seed crystals of

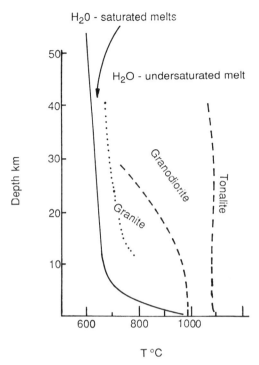

Figure 3.2 According to Wyllie (1983), and based on the experimental melting of a granite gneiss with 2 wt.% water, the estimated compositions of liquids generated from crustal gneisses are shown as a function of depth and temperature. Note that a temperature approaching 1100°C is required to reach the liquidus for tonalitic composition with 2 wt.% dissolved water (but see text). Reproduced from a fully referenced overview of the work of Robertson, Maaløe and Wyllie, in Wyllie (1983) with the permission of the editors of *Migmatites, Melting and Metamorphism, Shiva Geology Series.*

these minerals inherited from their metamorphic source rocks. Such experimental devices serve to remind us that the charge composition we select may not represent the composition of the original, primary melt.

Many other valuable deductions as to what is and is not possible can be made from these and other studies. Thus from the general phase relationships established it can be seen that the familiar rock-forming minerals remain stable at temperatures and pressures to be expected down to the base of the normal continental crust, presenting a great simplification in the modelling of the evolution of granitic magmas. It also implies that if any of the granite systems presently under discussion is to have any relevance to granite petrogenesis, then the generative processes must have operated within the crust. Furthermore, the phase relationships re-

main essentially similar over a wide range of natural granite compositions, even though the mineral phases vary in accord with the bulk composition, a latitude in the physical controls which goes far to explain why
granites of different origin are so similar modally.

This simplicity is not maintained in the thickened crust of the mobile
belts where pressures may be expected to reach 15 kbar near the crust–
mantle boundary. Thus melting experiments at these pressures by Carroll
and Wyllie indicate that garnet and clinopyroxene now become the stable
phases at the liquidus, in contrast to amphibole and plagioclase at lower
pressures. One result of this phase change is that garnet becomes the
effective agent in inhibiting iron-enrichment, as in andesites and more
evolved magmas, and there are other consequences concerning the fate of
the REEs.

Despite the complications, Wyllie and his colleagues have confirmed
Bowen's suspicion that the granite system consists of partially independent felsic and mafic components. It seems that, at temperatures near to the
appropriate solidus, quartz, orthoclase and sodic plagioclase melt to produce a eutectic-like granitic liquid that coexists with the more refractory
assemblage of calcic plagioclase and mafic minerals, and with little exchange of components through a wide temperature interval. Naney has
questioned this assumption but, nevertheless, it is exactly what we observe in rocks, for many granodiorites contain clots of early formed mafic
minerals and also plagioclases with calcic cores, together representing this
refractory assemblage, whether it be early crystallization or a resistant
residue.

ON THE CONTRIBUTION OF MINERAL CHEMISTRY

We need to turn to mineral chemistry and experimental data on individual mineral stabilities to obtain some estimate of the values of the
intensive variables at the time of crystallization of real magmas. This
encompasses an immense field of research (Clarke, 1992, table 3.2), but
I choose to illustrate its methodology and results by just one good example of the crystallization history of the composite, zoned pluton of
Ballachulish, Scotland, where a broad suite of rocks is available for study,
including diorites, monzodiorites, quartz diorites, granodiorites and
granites.

In their detailed investigations of this pluton, Weiss and Troll in 1989
used two-pyroxene geothermometry, thermobarometry based on the Fe–
Ti oxides, and also the stability of the ternary feldspars, biotite and
amphibole, to calibrate the crystallization sequence with respect to several
variables: the fractionation stage of the host magmas, the water content of
the latter, the compositions of the mineral phases, and the oxygen fugacity
(that is, the idealized partial pressure of oxygen). In doing so they as-

sumed a fluid pressure of 3 kbar, a value independently derived from the P–T–X calibration of prograde metamorphic reactions in the aureole.

Apropos of the pyroxene geothermometry, and with due reference to the reservations and complications addressed by Lindsley and Anderson, Weiss and Troll deduced that the orthopyroxene of the monzodiorites started to crystallize near to 1100–1050($\pm$60)°C at 3 kbar. There was textural evidence that it was joined by augite and pigeonite, which later suffered an extensive subsolidus re-equilibration. This temperature of the primary precipitation corresponded well with the calculation of a liquidus temperature at the appropriate magma composition and a concentration of 1 wt.% water in the melt. It also corresponded with the experimentally determined temperature of crystallization of the accompanying plagioclase, as judged from the studies of Eggler and Burnham.

The coexistence of Fe–Ti oxides provided an opportunity to apply the methods of oxygen barometry and geothermometry refined by Spencer and Lindsley using microprobe analyses to allocate Fe^{2+} and Fe^{3+} on the basis of mineral stoichiometry. Here again, there were a number of variables and complications to be taken into account, not least the modifying effects of granular exsolution, but nevertheless a range of T–f_{O_2} values were obtained representing a progression from an original oxide equilibration at 880 $\pm$ 60°C in the diorites, to changes reflecting an early influx of water derived from the country rocks, followed by the gradual concentration of water in the residual melt.

Values of T–f_{O_2} were also calculated from the compositions of the biotite using the experimental results of Wones and Eugster (1965) on the stability range of this mineral at various temperatures and oxygen fugacities. At a value of the latter near to that of an Ni–NiO buffer, the biotite of the monzodiorite would have become stable at 980–960($\pm$50)°C, whereas in the quartz diorite it would have crystallized at 910–870($\pm$40)°C, and in the granites at 900–850($\pm$40)°C.

Weiss and Troll agonized over their experience with the hornblende geobarometer involving the determination of the tetrahedral Al, a method developed by Nabelek and Lindsley, by Hammarstrom and Zen, and also by Hollister and co-workers (1985, 1986 and 1987, respectively). They experienced what I consider to be a general difficulty in that the amphibole textures suggest a late to post-magmatic crystallization, mainly at the expense of the earlier formed pyroxenes. I do not find it surprising, therefore, that the pressures recorded are relative underestimates, especially when the thermal stability of the Ballachulish amphiboles, as determined from the relationship of Ti and Al^{IV}, yield values ranging from 770–720($\pm$70)°C to 650–500°C, so covering the whole crystallization history from the late magmatic stage to subsolidus uralitization.

At this juncture it is important to insert yet another caveat concerning hornblende barometry for there are other explanations for variations of this kind as provided by Leake and Ahmed Said's study, in 1994, of the hornblendes of the Galway Batholith, Ireland. The geobarometric results revealed that hornblendes from different parts of the batholith crystallized at very different pressures, which, when translated into depth, suggests that they had subsequently been juxtaposed at the present level of erosion. Moreover some of the early part of hornblende crystallization, represented by cores of zoned crystals, took place at significantly greater depths than final crystallization at the emplacement level; also the zoning occurred during magma movement. Such findings fit very well the vision of kinematic emplacement of granitic magmas that will be advocated in this book.

Studies of the plagioclases from the Ballachulish suite of rocks provided further insights into a crystallization history involving the early inheritance of calcic xenocrysts which acted as nuclei for a first stage of near equilibrium precipitation at a time of low crystallization rates. Then, as cooling quickened and water accumulated, oscillatory growth in the suspended crystals continued along with a compositional change. Finally, a return to equilibrium crystallization followed the formation of a rigid matrix and the interstitial precipitation of alkali feldspar.

I will later return to a further consideration of the crystallization history of granites in general, but the coherent plan elucidated by Weiss and Troll for the Ballachulish pluton provides an excellent example of both the method and the range of values we may expect for the intensive variables during the evolution of a typical zoned pluton. In summary, the conclusions are that a relatively dry monzodioritic magma, with less than 1 wt.% initial magmatic water, remained water-deficient throughout its entire crystallization. In contrast, 2–3.5 wt.% water is estimated for the more fractionated quartz diorites and granites. The main crystallization interval of the orthopyroxene, clinopyroxene and plagioclase primocrysts in the diorites is of the order of 1100–950°C. Late magmatic biotite and alkali feldspar join the paragenetic sequence below 980 and 860°C, at an Ni–NiO buffer. Overall, a solidus temperature of about 900°C is inferred for this relatively 'dry' system, in which amphiboles are largely of subsolidus growth, with the crystallization intervals, stretching from magmatic to the subsolidus stages, of the order of 1050–680°C for the quartz diorites and 900–680°C for the granites – that is, at the present level of emplacement. Such values are entirely in accord with the expectation of the experimental results.

FURTHER ROLES FOR THE VOLATILES

In addition to its master role in determining the form of the granite system and the growth and nature of the mineral phases, water continues to play a vital part in the later stages of magma evolution as it is concentrated into the residual fluids with the result that, during the final stages of crystallization and exsolution, enormous energy is released. Burnham has estimated the very great vapour pressure engendered by this late stage boiling, so that it can be no surprise that fractures are now formed and hydraulically extended, and that brecciation and fluidization become so important that crystal fabrics can be disrupted and the crystal fragments entrained and transported in tuffisitic breccias.

In addition to water there are the other volatile components to take into account, namely fluorine, chlorine, boron, carbon dioxide, nitrogen and methane. Their effect is varied. According to Manning and Pichavant the depolymerizing combined effect of fluorine and boron is even greater than that of water, and may depress the solidus of the granite system to below 600°C at 1 kbar (Figure 3.3 shows the separate effect of fluorine). Furthermore, the phase relationships are sufficiently changed from those of largely water-bearing melts to produce different phase relationships, with distinctive mineral parageneses, within which quartz has an especially wide temperature interval of crystallization. Also, according to Kovalenko, sodium is preferentially enriched in the late fluid phase of such halogen-rich melts.

The effect of these additions on the physical properties of melts is dramatic. For example, according to Baker and Vaillancourt the presence of fluorine and water (1.5 wt% F, 6 wt% water) in a peralkaline melt at

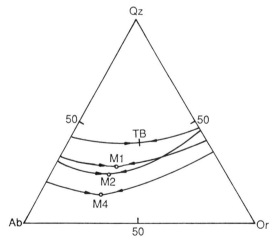

Figure 3.3 Liquidus phase relationships for the system Qz–Ab–Or–H_2O–F. (TB) Data from Tuttle and Bowen (1958) for fluorine-free system; (M) data from Manning (1981) for additions of 1, 2 and 4% F. Temperatures of minimum melt composition: TB 720°C, M1 690°C, M2 670°C, M4 630°C. P = 1 kbar.

800°C is lower by 1.5 orders of magnitude than a hydrous metaluminous melt (6 wt% water) at this same temperature. Thus it is easy to understand the reason for the unique properties of the alkali granites of the cratons, the A-types, including their exceptional fluidity, the absence of any trace of true restite, their single feldspar composition, their textural complexities, and the late-stage redistribution of elements: all can be explained as due to the evolution of a relatively anhydrous granite melt enriched in fluorine and boron.

As for the effect of carbon dioxide in silicate–water–carbon dioxide systems, the indications are that its solubility is low at crustal pressures, so that perhaps its principal effect is to simply diminish volumetrically the effect of any water present. However, according to Holloway, a carbon dioxide-bearing magma should exsolve a carbon dioxide–water fluid at a greater depth than a purely aqueous fluid in a carbon dioxide-free magma with the same water content.

The theoretical treatment of the role that these volatiles play in silicate melts has been the main thrust of studies of pegmatite systems initiated by Jahns in 1953 and subsequently carried on by Burnham and his colleagues, and later by Silver and Stolper. From such studies it seems that we are near to understanding the equilibrium properties of hydrous granitic magmas, more particularly those of the late stage pegmatites. The basic premise of the modelling, and one of particular significance, is the assumption that ideal mixing occurs in the multicomponent solutions, with the components acting as discrete and thermodynamically distinct complexes or species – water molecules, hydroxyl groups and oxgen atoms included. The non-volatile components are considered to mimic the crystalline phases that are finally precipitated, although the modelling suggests the possible existence of precursor, feldspar-like and quartz-like, polynuclear aqueous complexes. Perhaps the mechanism of water incorporation in H_2O-undersaturated melts is rather more complicated, as is indeed indicated by the experiments of Johannes and Holtz (1996, pp. 71–83) but, for the moment, I will remain with pegmatite systems.

On this basis there is a considerable simplification of what might have been expected to be a very complex system, permitting the derivation of mathematical expressions (partly empirically in 1986 by Burnham and Nekvasil, and wholly theoretically in 1985 by Silver and Stolper) representing the relationship between the activities of the melt components and their mole fractions under changing conditions of pressure, temperature and volatile content. The full explanation of such a formulation is quite challenging, but the simple fact that it can yield experimentally verifiable values for such properties as H_2O-solubility and freezing point depression attest to a certain validity. Thus it seems that hydrous haplogranite melts are so nearly thermodynamically ideal systems as not only to permit the modelling of their phase equilibria but also, even more significantly, to allow the description of the melt structure. Indeed Burnham and Nekvasil (1986, pp.252–4) were sufficiently confident in their model system to extrapolate and predict, and also to use it to investigate the shift in the haplogranite liquidus minimum in response to the addition of fluorine and boron. This enables them to conclude that the latter probably form complexes with sodium to form precursor cryolite-like and sodium tetraborate-like species, respectively. The possibility of thermodynamically modelling the evolution of a crystallizing melt, with its corollary that *intermediate* complexes may form in a crystallizing magma, is of great significance and one to bear in mind in all the discussions which follow concerning the nature of natural magmas.

The role of water in the magma is, however, only part of the story. During intrusion a 'hydrothermal cycle' operates involving the leaching of the country rock and the absorption of mineralized water into the magma, followed by its expulsion as an ore-laden fluid. By determining whether heat is conducted or convected, this hydrothermal circulation controls many granite-associated phenomena, not least, according to Parmentier and Schedl, the width and form of the metamorphic aureole. Furthermore Fenn tells us that hot water circulation may continue long after the main cooling event, being sustained by the heat generated in granites with higher than usual concentrations of radioactive elements, as in the high heat production granites. Such granites ought to be most typical of continental crustal environments, being derived from radiogenic crust, although a contrary view has been expressed by Atherton and Plant.

These are matters I must continually return to, but clearly the full understanding of this gassy, multicomponent system representing real granites lies ahead in difficult experimental terrain, bedevilled by the problem of representing the results in any easily comprehensible form. However, I believe that enough is already known for a credible petrogenetic framework to be formulated.

ON PARTIAL MELTING: A PRELIMINARY STATEMENT

Matters concerning melting must also necessarily appear in many of the discussions which follow about the origin of granitic magmas. I merely introduce the subject here in noting that there is a substantial volume of experimental work, some involving, as we have seen, both melting and crystallization studies. Most often cited are contributions by Winkler and Breitbart, Büsch, Schneider and Mehnert, and Wyllie and his associates, and I have found the 1983 essay by Johannes most enlightening.

It is apparent from these investigations that both pelitic metasediments and amphibole-bearing metaigneous rocks partially melt at geologically realistic temperatures. We learn that the low melting fractions from pelitic compositions are likely to be of granitic composition, and that the incongruent melting of micas can provide both the potassium feldspar component and the water, the latter being essential for the melting of dry crustal rocks. However, as already noted, we are still not sure of the extent to which these low temperature melts will dissolve iron and magnesium, so we cannot with certainty assume that the first melts will be purely felsic in composition.

It is the temperature range of melting that will assume particular importance in many of the discussions which follow. In Winkler's early experiments (1961) the temperatures recorded for anatexis were remarkably low, with representative model values of 700°C at 2 kbar, and 680°C at 4 kbar, for an alkali feldspar-free paragenesis. Hoschek (1976) then reported 650°C at 4 kbar for the incipient melting of a quartz + albite + biotite assemblage, and Büsch and co-workers (1974) 690–660°C at between 3 and 7 kbar, but these were all determined at water saturation. More realistic investigations of melting in the pelitic system by Vielzeuf and Holloway (1988) led to the important conclusion that it is possible that large volumes of S-type granitic liquids could be produced by fluid-absent melting at around 850°C at 10 kbar – that is, at the stage of biotite breakdown. In a similar study of the reaction biotite + plagioclase + aluminosilicate + quartz → garnet + K-feldspar + melt at 10 kbar, Le Breton and Thompson (1988) found that melting begins at between 760 and 800°C and is extensive at 850°C. Thus the melting of pelitic lithologies is likely to occur in stages, that is to say, with a first production, at about 610°C at 10 kbar, of a few per cent of felsic melt with vapour present, followed by two stages of vapour-absent melting as first muscovite and then biotite react. Melt is produced in bulk at temperatures not far above 800°C at 10 kbar. Clearly the beginning of melting temperatures and pressures of pelitic metasediments is likely to lie just within the upper energy limits of regional metamorphism, but it would seem that crustal temperatures need to be augmented before bulk melting occurs. Such augmentation is even more likely to be required for the partial melting of those most characteristic basinal sediments of plate margins, the greywackes. Thus Vielzeuf and Montel reported that experiments on the fluid-absent melting of a calcium-poor aluminous greywacke only yielded substantial melt at around 950°C at 10 kbar, that is, at a significantly higher temperature than for pelites. The reaction is of the form: biotite + plagioclase + quartz ⇌ garnet + K-feldspar + melt.

Furthermore, similar dehydration melting experiments show that amphibolites, representing the basaltic and andesitic greenstones, would form a fertile source rock for tonalitic and trondhjemitic liquids. Thus work by several researchers, for example Rushmer in 1991, and Beard and Lofgren in the same year, showed that the onset of melting occurs between 850 and 900°C at pressures of the order of 8 kbar, with substantial melt volumes being generated a few tens of degrees above the solidus temperatures (compare p. 283). Not surprisingly the lowest temperatures for the onset of melting were found when the starting materials included quartz and albitic plagioclase: the latter, of course, are the very materials produced in mafic volcanic rocks during the low-grade burial metamorphism so typical of greenstone assemblages. Perhaps the most significant finding of Beard and Lofgren is that it is dehydration melting and not water-saturated melting which produces melts that most nearly resemble those of arc-related tonalites.

Clearly such experiments can distinguish between what is possible and what is impossible, but the actuality may be very complex. Thus in their overall review in 1995 of the results of these researches, Patiño Douce and his colleagues concluded that there is no one exclusive explanation of the generation of the series tonalite–granodiorite–granite. Different magma compositions within this spectrum may reflect different melting temperatures and pressures, different source compositions (as, for example, the Fe/Mg ratio), different H_2O activities, or any combination of these. For example, by varying the pressure and water content the melting of a metagreywacke can be made to yield a liquid of tonalitic, granodioritic or granitic composition, in part a justification of the original contention of Winkler.

These melts derive from reactions producing solid residues rich in clinopyroxene and garnet ($P > 7$ kbar), a *restite* so nearly approaching that of the natural pyroxene granulites and kinzigites that it is tempting to embrace a general thesis whereby granitic melts represent the extract, and the granulitic basements the residuum. But I am not so easily tempted at this early stage of a discussion touching on the very origin of granites.

Despite the complexities, these experimental data send the clear message that substantial melting of such crustal rocks requires that temperatures be raised above those normally recorded for the highest grades of regional metamorphism. Such a conclusion is surely reinforced if we take into account the need to maintain sufficient mobility for intrusion of voluminous magma. Thus we need to seek a geological explanation for the sources of the extra heat. Even so there remains the problem of the actual extraction of such melts, particularly how they are separated from the restite, either *in situ* or in transit, and in sufficient volumes to fill the plutons and their huge batholithic assemblies.

Furthermore it is perhaps surprising that the experimental and geothermometric estimations of temperature and pressure even approximately correspond when the complications of the melting process are considered. This is because melting, even more so than crystallization, involves a variety of kinetic factors, so that time becomes the essence in any discussion of the processes involved. For example, any change of state is rate determining, and the unstable melting of intermediate plagioclase is a case in point and one that possibly accounts, in part, for the diverse results obtained for melting experiments of different durations.

Unfortunately, the kinetics of melting of rocks and the surface reactions involved are only just beginning to be addressed in this field of petrology, as, for example, by Jurewicz and Watson, though the earlier microscopical studies of Büsch, Schneider and Mehnert on the nature of incipient melting at grain boundaries provide a glimpse of its complexities. It seems that melting is not only located at particular mineral grain contacts but is preferred at certain polyphase boundaries. Jurewicz and Watson found that the juvenile melt does not automatically wet the intergrain boundaries, but remains in the form of tiny pockets at its points of origin. But is this the commonplace?

Such findings are indeed the very essence of surface physical chemistry and we can be sure that grain size, shape and orientation all play a part in influencing diffusion and reaction rates. We can confidently predict that any high concentration of defects at grain boundaries resulting from deformation will enhance reaction and melting. However, it is probably not yet possible to determine and quantify the total effect of all these factors, either the purely chemical or the kinetic, yet their appreciation, coupled with the experimental findings, does confirm this expectation that not only is crustal melting possible, it is almost inevitable, though only, I repeat, to the extent that melting will begin. We cannot therefore be suprised that migmatites are so common a feature of high grade metamorphism. On the other hand, we cannot expect that the heat of metamorphic processes will produce great volumes of granitic magmas without augmentation.

This is part of the reason for a strong prejudice in that I do not believe that metamorphic migmatites often represent the birth of bulk granitic melts. Leaving aside the problem of whether layer-parallel migmatites represent actual melting or not, and the proof of this lies not so much with the chemistry as with mass balance and textural studies, there is little direct evidence that the material of leucocomes is ever sufficiently extracted and collected for supply to plutons. Furthermore, the generation of granitic melts in bulk is rarely precisely tied to the acme of metamorphism, and is often much later. These are matters I will return to after extensive discussion of the petrogenetic processes involved. The first of these concerns the very nature of magma itself.

SELECTED REFERENCES

Burnham, C.W. and Nekvasil, H. (1986) Equilibrium properties of granite pegmatite magmas. *American Mineralogist*, **71**, 239–263.

Clemens, J.D. and Vielzeuf, D. (1987) Constraints on melting and magma production in the crust. *Earth and Planetary Science Letters*, **86**, 287–306.

Johannes, W. and Holtz, F. (1996) *Petrogenesis and Experimental Petrology of Granitic Rocks*. Springer-Verlag, Berlin.

Wyllie, P.J. (1983) Experimental studies on biotite- and muscovite-granites and some crustal magmatic sources. In: Atherton, M.P. and Gribble, C.D. (eds), *Migmatites, Melting and Metamorphism*, Shiva, Nantwich, Cheshire, pp. 12–26.

The physical nature of granitic magmas: a case of missing information

4

. . . crystalline stony types were not thoroughly fused at the time of their emission, but consisted of a granulated magma, the imperfectly formed crystals being lubricated by a more liquid base, and by intercalated steam, composing a red-hot granulated paste.

Paulett Scrope (1872) Volcanoes, *2nd edition, p. 335.*

THE NATURE OF THE GRANITE MUSH

Although there is some considerable understanding of the guiding chemical controls of the granite system, little is known with certainty about the rheology of granitic magmas: a gap in our knowledge that needs to be filled if only to underpin the advanced geochemical models of today. However, we do know that such magmas are highly polymerized, consisting of networks of discrete molecular units: also that their viscosity is strongly dependent on the proportion of bridging atoms – bridges that are broken in the presence of volatiles such as fluorine, chlorine, boron and water so as to drastically lower the viscosity. Furthermore, we continue to learn of other influences such as the effect of the degree of oxidation, as described by Scaillet and his colleagues in 1995, and the effect of the differing associations of the water molecules, as reported by Holtz and others in that same year.

It would seem that the whole concept of viscosity is difficult to handle in a kinematic system of a flowing silicate melt that undergoes phase changes during its life history. So much so that Bergantz, in timely essays on this subject, considered that it may be impossible to apply the traditional formulations of fluid mechanics to magmatic systems. Thus it behoves me to tread lightly through this complex field before turning to discuss outcrop patterns that may throw light on the real nature of real magmas.

SUSPENSIONS, MUSHES AND BINGHAM BODIES

My understanding of the nature of granitic magma is that it represents a continuum from the wholly liquid state to the stiffening suspension of crystals which heralds solidification. It may be useful later in these discussions to distinguish between those magmas generated at source as pure liquids and those originally generated at source as suspensions, as for example when partial melts are mobilized in their entirety. But whatever their nature at this early stage it has long been suspected that most granitic magmas bear crystals during ascent, conforming in varying degree with Scrope's definition. Certainly subliquidus temperatures are indicated both by the survival of restitic and early formed crystals, and by the modest scale of reaction temperatures recorded during contact metamorphism. Furthermore the presence of a significant fraction of crystals is also indicated by the finding, by Fernandez and Gasquet in 1994, that the strains recorded by the preferred orientation patterns observed in outcrop match sufficiently well with the finite strains inferred from computer simulation models as to suggest that crystals were present during much of the ascent. At these subliquidus temperatures, crystallization can be expected to proceed apace during intrusion, that is until the temperature is steadied by the exothermic nature of the process, with the important result of delaying the consolidation of the crystal mush.

The effect of loading a silicate melt with crystals is dramatic. From the work of McBirney, Murase, Marsh, Spera and Sparks, it seems likely that such suspensions become highly viscous during some part of their life history, their rheological properties changing significantly in behaviour from Newtonian to pseudoplastic. This must be especially so when the crystals are in sufficient density to begin to interact, possibly at a crystallinity of about 30%. Fernandez and Gasquet in their 1994 modelling of the physical evolution of the Tichka plutonic complex, successfully described the succession of changes in terms of an interplay of the several rheological behaviours. A key factor is the rate of shear strain, when low rates favour a Newtonian behaviour, with its linear relationship between rate and apparent viscosity, while high rates favour a pseudoplastic behaviour where the relationship is non-linear, the apparent viscosity then varying with the rate of shear according to a power law of either Bingham (having a yield strength) or a shear-thinning (lacking a yield strength) formulation.

Sparks and his colleagues alert us to the possibility that even a melt containing but a few per cent of crystals may behave in this way, a view confirmed in 1986 by Nicholls and co-workers, who found that, during the early stages of crystallization, the rates of change of the physical properties of suspensions are much in excess of those to be expected from the decrease in temperature alone. In his more general analysis of 1987,

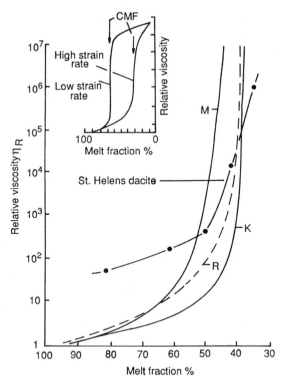

Figure 4.1 Effective viscosity as a function of the crystal–melt proportion in a granitic magma. The results are shown of the St Helens dacite experiments of McBirney and Murase (1984), together with reported calculated curves by (M) Mooney (1951), (K) Kreiger and Doogherty (1959) and (R) Roscoe (1953), all referenced in McBirney and Murase (1984). Inset is a schematic representation showing the possible effect of changing strain rate. (CMF) Critical melt fraction. After Miller *et al.* (1988) and Wickham (1987).

Wickham deduced that magmas with up to 25% volume of crystals probably behave as Newtonian fluids, though he conceded that no rigorous analytical models exist for suspensions with higher crystal content. However, Fernandez and Gasquet attempted such an analysis, settling for three main rheological states: Newtonian, at crystal volume fractions of less than 30–35%; pseudoplastic, at between 35 and 65%, and 'solid-like', beyond 65% crystallinity. In particular, the melting–deformation experiments of van der Molen and Paterson, and of Arzi, clearly indicate that between crystal concentrations of 25% and somewhere between 50 and 70%, the viscosity changes by as much as 10 orders of magnitude, implying a radical change in mechanical behaviour. In these latter experiments the most dramatic stiffening of the magma mush occurred at about 65% crystal content, at a point now generally referred to (in terms of the remaining melt) as a critical melt fraction (CMF) – that is of 35% (Figure 4.1). Presumably the CMF represents the change-over point from a suspension-controlled to a granular-controlled mechanism of deformation on the formation of a grain framework or matrix.

Although these experiments have contributed much to the present understanding of the evolutionary history of magma, I believe that the CMF concept has attained an acceptance stretching the available evidence. Just how much so was demonstrated in 1995 by Rutter and Neumann, whose particular experiments with the melting of a granite during deformation showed not only the complexity of the process but also no evidence of a critical melt fraction. Indeed Petford has argued that the concept may be meaningless because it is likely that melt is extracted via pockets into coalescing fractures, so short-circuiting the rheological barrier represented by the CMF. Furthermore we need to be aware that all experiments so far reported involved melting and not crystallization; that any rheologically critical point is likely to be compositionally dependent, with a granitic magma behaving differently from its more dioritic analogues, both in controlling the volume per cent of the critical fraction and in the rate of change across this rheological barrier. This is not only because there are a different number of mineral phases, but because these will be in different and progressively changing crystal concentrations, with unequal crystal sizes and shapes, all factors that are known to affect viscosity. Furthermore high rates of strain will significantly reduce the bonding effect of grain continuity (Figure 4.1, inset). And during these rheological changes, is the effective or apparent viscosity to be regarded as an additive quantity? Will both crystals and fluid contribute to the non-Newtonian behaviour? And how do we calculate the separate effects of changes in grain size and changes in grain mix?

There are natural features that seem to support the Bingham model for granitic magmas. Very often both phenocrysts and enclaves are seen to be strangely separate and groundmass supported, in harmony with a lack of evidence for crystals either sinking or floating – that is, except in some highly fluxed, fluorine-rich varieties. Furthermore, concentrations of crystals or small enclaves are much better explained by mechanisms of either crystal accretion or mechanical sorting than by gravity settling. Sharp or rapid transitional internal contacts between different surges find a ready explanation in that special property of pseudoplastics whereby dramatic changes in viscosity occur over small ranges of temperature. Yet another thixotropic feature is the way in which fracture and flow phenomena can coexist in granite plutons depending solely on the rate of application of stress. Finally, the very contrast between the flow phenomena in basaltic and granitic magmas could well be due to this difference in behaviour associated with the degree of polymerization.

Such reasoning suggests that when magma mushes reach a certain degree of crystal content it becomes rheologically impossible for them to erupt. From the measurements of the crystallinity of Aleutian lavas and dykes Marsh (1981) discovered that this critical point was reached, in this natural example, at crystal concentrations of around 55% for basalts and 25% for the andesites (expressed as remaining melt fractions of 45% and 75%, respectively). A related study by McBirney and Murase provides similar data. The implication is that, at these higher crystallinities, the magma was retained in the crust as a pluton. Is this then, Marsh asks, the simple reason for the classic dichotomy of voluminous basalts and voluminous granitic plutons? I would add that it is also a plausible explanation for basaltic magmas preferentially entering fissures as dykes, as well as for other natural features such as the low crystal content of many quartz porphyry dykes, and for the fact that so many extrusive rhyolites are not represented by lavas but by ash flows.

It is not easy to assess the importance of an increasing concentration of water as a countering factor to decreasing temperature and increasing crystallinity. No doubt it will have a depolymerizing effect, and indeed its presence is known to lower the viscosity of obsidian melt. Johannes and Holtz calculated the effect of decompression on a hydrous granitic magma on ascent (1996, pp. 120–123). Under the adiabatic conditions we might expect where bulk volumes are in movement, and taking into account the opposing effects of pressure-release melting and the temperature decrease due to this melting, it seems that adiabatic ascent produces relatively little variation in temperature and melt proportion until pressures fall below 3 kbar, and, moreover, little change in viscosity! Of course, a wholly different set of conditions must apply during the decidedly non-adiabatic ascent of magma in dykes. These are complex but nevertheless important matters and I urge the reader to follow the text of the appropriate section of Johannes and Holtz's major work (pp. 83–126). But even with this instruction I do not understand what would be the effect of increasing water concentration in a crystal suspension in which crystal interactions might well exert a more potent influence on the overall physical properties.

At the point of interlocking, the feldspar framework must be extraordinarily porous, and the above estimates of 35% remaining melt are amply confirmed by the open texture revealed by almost any large, polished slab of coarse granite. At this juncture the history of the magma is yet incomplete and we shall need to return later to a discussion of the final stage of consolidation and the role of the interstitial liquid in modifying the rock we see in outcrop.

Whatever the complexities and calculations I believe that we can reasonably expect the rheological condition of granitic magmas to change progressively during ascent and especially in the upper crust, though clearly magma in bulk batches, as in diapirs, will be slower to react than in extended sheets, as in dykes. Some may well begin life as Newtonian fluids, especially those hot, relatively dry magmas of marginal and island arcs. Others will not, especially those derived by crustal remelting, for they will already be crystal-bearing and so become increasingly viscous, perhaps elastoviscous, as they evolve. As crystallization advances, these magmas must thicken as the grains interact, stiffen as the crystals bond, then interlock, and set rigid. It is neither easy to integrate theoretically, nor to quantify, the changing effects of crystal content, crystal density, and differing growth rates, nor to take into account the effect of the decrease in pressure on ascent, and we will probably have to await the evidence of real-time experiments (but see Johannes and Holtz, 1996).

We may also contemplate what happens to the overall composition of crystal mushes during transport. If local solid–melt segregation is as common a feature in plutons as it is in dykes, then internal heterogeneities may arise merely from the changing proportion of crystals by flow differentiation. Indeed, such a mechanism has been proposed by both Sultan and Speer and their respective co-workers to explain just such local compositional variations in granitic plutons. This is not my personal experience, but the fact that no one bulk sample may ever represent what was once liquid presents a very real challenge to the geochemist. Certainly, if plutons are repeatedly fed by dyke systems, as I shall later propose, then we may need to return to the detail of single outcrops. At this point it is appropriate to turn in some detail to this matter of flow.

FLOW IN GRANITIC MAGMAS: A FIRST STATEMENT OF A PREJUDICE

FLOW IN THE MUSH: A PRELIMINARY COMMENT ON MAGMA DEFORMATION

If granitic magmas have a strength as just postulated, they will respond to the application of stress by pure shear, the resultant strain being recorded by permanent changes in shape, orientation and length within the crystal fabric. We can expect a progression of fabrics in response to the changing rheology with time as the complex flowage of magmatism gives way to the creep and dislocation of metamorphism. Paterson, Vernon and Tobisch envisage four stages: greater than 50% melt, magmatic flow; between 50 and 20% melt, submagmatic flow; less than 20% melt, high temperature solid-state flow; passing into low temperature solid-state flow. In this continuum what is regarded as flow as distinct from deformation is a question of semantics: viscous flow must always involve an element of pure shear and our concern is to discover when the overall strain will become indelible. If it is useful to make a distinction, then Paterson and co-workers define magmatic flow as deformation by dis-

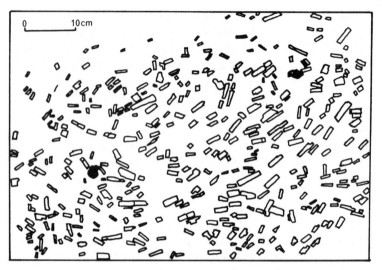

Figure 4.2 Swirling flow pattern picked out by the potassium feldspar megacrysts in the granite of the contact zone of the Land's End pluton, Port Ledden, Cornwall. Note disengagement of most of the megacrysts in this two-dimensional plan view of the outcrop.

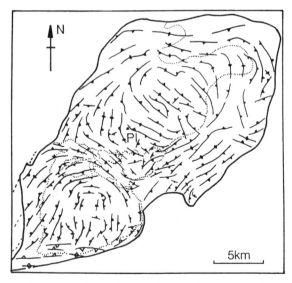

Figure 4.3 Mineral foliations in the Plouaret Complex, France. Note how foliations cross pulse boundaries (dotted lines). Adapted from Guillet et al. (1985).

placement of melt, implying a rigid body rotation of crystals, but without a sufficient interface between the crystals to induce the plastic deformation that follows crystal bonding.

Perhaps the nearest approach in nature to the fluidal patterns, such as those obtained by Nickel and co-workers in their experiments with intrusion fabrics, are the outcrop-sized swirls picked out by the potassium feldspar megacrysts within certain porphyritic granites such as those of the Serra da Estrela, Portugal, and Southwest England (Figure 4.2). On a larger scale we might include the meandering mineral alignments in the cores of some few plutons such as that of the Enchanted Rock, Texas, and of Plouaret, Brittany (Figure 4.3), but we need to be aware that such arcuate foliations could equally result from the diapiric upsurge of fresh magma into the core of the pluton.

Though not often so clearly expressed, some porphyritic granites show these local flowage alignments, thus demonstrating that fairly dense suspensions can be mobile, even turbulent. Many more plutons show much more linear, peripheral foliations of minerals and enclaves – foliations which are more obviously related to the overall shape of the pluton, most often subparallel to the external contacts. A number of workers continue to attribute these structures to wholesale flowage, even to indicate the form of convection currents. However, measurements of the associated strain suggest otherwise.

The strain is most easily measured from the degree of distortion of the enclaves, so that it becomes imperative to determine the origin of the latter – whether they represent blobs of contemporaneous magma of similar ductility to their host, or whether they are true xenoliths derived from the country rocks, when they may or may not have acquired this shared ductility. For the magma blobs the recorded strain will include increments imposed before and after the crystal interlocking stage of their host, and only if the planar foliation is shared by the enclave and host can we be sure that some part of the strain was acquired in the quasi-solid state. Furthermore, the true xenoliths have their own problems for the investigator, if only because they are much more likely to have an original schistosity: the latter will have controlled xenolith shape and consequently the orientation within a magma.

I will return to this question of strain interpretation, but as an example of the possibilities and pitfalls I refer to a notable example of this general problem, that of the interpretation of foliations within the Criffell–Dalbeattie pluton, Scotland, studied over a number of years by John Phillips. Within this zoned pluton Phillips distinguished between one arcuate planar structure picked out by the minerals and another, essentially parallel, measured by the disc-shape of what he considered to be true country rock xenoliths. He and his co-workers interpreted the min-

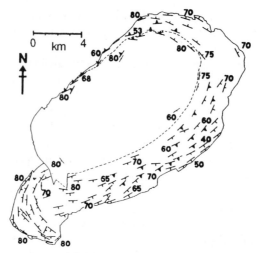

Figure 4.4 The Criffell–Dalbeattie pluton showing the foliation formed by the orientation of the xenoliths (dash symbol) and the deformation foliation (triangle). Granodiorite forms the incomplete outer shell and a rapidly transitional junction has been found between it and the porphyritic granite of the core. Reproduced from Phillips *et al.* (1981) with the permission of the Geological Society, London.

eral alignment as a penetrative deformation, following the original terminology of Balk (Figure 4.4), but the enclave orientation was considered to
represent a primary flow foliation produced by laminar flow of mushy,
convecting magma from which crystals were accreting on the walls. Later,
in a follow-up study in 1980, Stephens and Halliday showed that the core
of the pluton represents a separate and later pulse, and it was probably the
emplacement of the latter that caused the mineral foliation. Whether or
not the enclave orientation represents an earlier stage of flow is bedevilled
by doubts about its origin, whether igneous or metasedimentary, and
overall it seems that there is no convincing evidence for bulk convection.

Suffice it to say here that opinion on the interpretation of foliations will
vary with the example chosen, but I believe, along with Fernandez and
others, that many of the gross features of the fabrics of granites result from
a continuous history of incremental strain and rarely represent free flow
in a liquid. Indeed, long ago Hans Cloos, as reported by Robert Balk (1937,
p. 59), noted that the so-called 'flow-lines' correspond in direction to the
longest axis of the deformation ellipsoid, implying that the planar and
linear structures commonly seen rarely represent the original direction of
magma movement, a conclusion which is the thrust of this text.

In true liquids the explanation for the preferred orientation of crystals is
complex, yet, once within the realm of plastic or solid behaviour, the
apparent rotation can be easily attributed to a flattening strain. The lesson

is that in the petrofabric study of granites we must remain acutely aware of the transposing effect of deformation.

EVIDENCE FROM THE SINKING OF STOPED BLOCKS

The way in which stoped blocks of country rock sink into a magma ought to provide some clue as to its physical state at the time of arrival at its present level in the crust. However, at this late stage the addition of any large volume of relatively cool material would cause most granitic magmas to quench rapidly – except, perhaps, in the highly fluxed magmas of the alkalic granites. This rapid quenching is the reason why the blocks are frozen in the act of being stoped, and why their distribution within the periphery of a pluton sometimes reflects the original stratigraphic and structural order within the roof, so preserving a 'ghost stratigraphy', a phenomenon I discussed in 1970.

The observation that the core zones of granitic plutons are often free from stoped blocks has little to do with any easy free fall in the hotter interiors, but is due to the fact that piecemeal stoping plays but a secondary role in the emplacement process: the blocks rarely enter the magma chamber! If, on occasion, they do, then it seems likely that they will sink rapidly – that is, relative to the life of the pluton itself. Thus Carron finds that the rates of sinking of blocks in melted natural obsidian is of the order of several thousand metres per year, obviously rapid enough to clear any body of granite melt well within the likely time of congelation. Nevertheless, as I have already discussed, there are very strong reasons for believing that small xenoliths, of the order of 0.5 m or less, and also the feldspar megacrysts, remain suspended in all but the most highly fluxed magmas.

There is another lesson to be learned from this sinking of large blocks: that the latter fall straight down from the roof, unaffected by any bulk magma flow such as we might envisage in a convecting magma chamber. I have seen examples of this free fall in a wide variety of situations as varied as the Thorr and Fanad plutons in Donegal, the Pativilca pluton of the Coastal Batholith of Peru, and the outer ring dyke of the Ririwai Ring Complex of Nigeria. The fate of the blocks stoped from the roof of the Thorr pluton is particularly instructive. High in the pluton their attitude seems undisturbed from the original, but deeper down they are disoriented and seem to be tumbling as they begin to respond to the flow or strain pattern of the host. This free fall is one reason why I do not believe that large-scale convection operates at this arrival stage of a pluton's history.

This is an appropriate point to refer to a remarkable example, again from my own knowledge. It concerns the lower part of the pluton of Tumaray, Peru, which is marvellously well exposed in the canyon sides. A flat-lying layer, 300 m thick, of small, matrix-supported, densely packed,

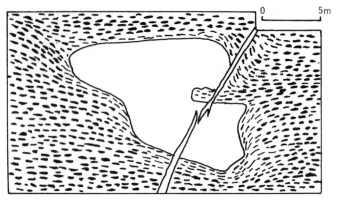

Figure 4.5 Distortion, as if by a dropstone, of an enclave layer around a xenolithic block of granite derived from an earlier intrusion forming the roof of the host pluton. Vertical section in the cliffside of Quebrada Quintay, Sayán, Peru. Note that this is a good example of the deformation of mafic magma blobs while immersed in a crystal mush.

mafic enclaves straddles the core of this pluton. This layer is seen to have been punctured by a few 10 m-sized granite blocks derived from the roof (Figure 4.5), a phenomenon so analogous to that of dropstones in the glacial environment, even to the wrap-around of the enclave alignments, that we can easily picture the blocks free falling through an upper 1000 m of magma, then to splash into a mushy floor of crystals mingled with viscid enclaves: if only we could measure the times we might even evoke Stokes' law! However, at least we can appreciate the nature of such a bottom layer and realize that crystal and crystal–enclave mushes are no figment of the imagination.

A SOURCE CONTROL OF MUSHINESS: PER MAGMA AD MIGMA

If natural magmas have these special properties we must expect, as I have already hinted, that their effects will vary in degree depending on the geological environment, magma source, generative process and mode of emplacement.

The wholly molten magmas of the fluxed and relatively high tempera-ture, alkali granites of the anorogenic environments represent, perhaps, the nearest approach to a true Newtonian fluid. In other environments the situation is different. Those mantle-derived, tonalitic magmas located within the high heat flow zone of the arcs may contain some suspended crystals, though rarely of the proportion representing critical crystallinity until late in their evolution. As a consequence they were able to rise high into the roofs of volcanic centres. However, in the case of the granodiorites intruded during major uplift, either at the continental margin or within the continental plate, the associated magmas were probably sufficiently cooled, during their rise through thick crust, to reach the point of critical crystallinity during their ascent. In total contrast we might expect those magmas generated by crustal remelting deep within a continental collision zone only to have become mobile at the very point of critical crystallinity. Thus they would remain loaded with crystals as they detached themselves from the source, so becoming likely candidates for the pseudoplastic condition.

Such crystal mushes represent the *anatexites* of Sederholm, the *migmas* of Reinhard, and the *parautochthonous granites* of Read. Their representative rocks should be easy to recognize, at least in the early stages of their detachment and migration. And so, I suppose, they often are, especially when in close space–time association with deep-seated, metamorphic complexes. A prime example is the Hanko Granite of southern Finland, which is loaded with restitic biotite clots and schlieren, all with garnet and cordierite. No wonder those famous professors of Helsinki University Sederholm and Eskola, viewed this and similar granites as migmatites, because the evidence is in every building stone in their city!

SCHLIEREN AND BANDING: EVIDENCE OF MAGMA FLUIDITY?

There are some clearly observable features of granitic rocks which have long been taken as evidence of a high degree of fluidity – that is, mineral layering in the form of schlieren, comb layering, orbicular structures and contrasted mineral banding.

SCHLIEREN: THE OMNIBUS TERM

Many granite outcrops show orientated streaks or layers of crystals, otherwise known as schlieren, a name borrowed from the early German glass workers. Broadly interpreted, such mineral schlieren are of four sorts: early crystal accretions, crystal precipitations, disrupted and streaked out xenolithic material representing the refractory residue of the source rocks, and smeared out blobs of coeval basic magma. Of these, it is perhaps relatively easy to understand how enclosed material can be streaked out by shear in a viscous host. Thus one of the most common features of granitic outcrops is a progression from dark ellipsoidal enclave to mineral schliere, with a parallel increase in the evidence of magma mixing (Figure 4.6). Perhaps it is the dramatic increase in the surface area that enhances mechanical and chemical interactions.

The recrystallization and partial assimilation of country rock, especially by the spalling-off of schistose lithologies, also creates streamers of biotitic and hornblendic schlieren (Figure 4.7), which presumably indicate some

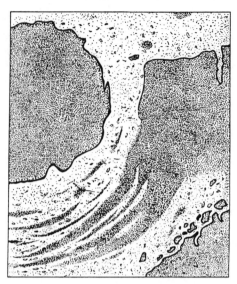

Figure 4.6 Formation of mafic schlieren by disruption of gabbroic enclaves in the Ronez Granite, Jersey. Reproduced from Wells and Wooldridge (1931) with the permission of the Geologists' Association.

kind of flow regime. However, what are we to make of the layering defined by alternating bands enriched or depleted in the constituent minerals? These are the true schlieren of Ernst Cloos, and represent the proper usage of the term according to workers in the Sierra Nevada, California.

There are really very few unequivocal examples of the gravitational separation of crystals – like that represented by the flat-lying, rhythmic layering within the Tigssulok Granite, southwest Greenland, so beautifully illustrated by Harry and Emeleus (Figure 4.8); this, however, occurs in a granite representing a highly mobile, fluorine-enriched, magma. According to Peter Brown, in 1994, even this example is equivocal and other examples are even more so. Dealing with a less extreme magma composition, Barrière describes curved laminae and graded layers in the subalkaline granite of the Ploumanac'h Granite, Brittany, attributing them to convection currents. Some of the granodiorites of the Sierra Nevada, California, are locally banded, with a grading and rhythmicity suggestive of a similar sedimentary origin.

There is little doubt that, until recently, a sedimentation model for at least some of these layered granitic rocks would have been acceptable, but the radical reinterpretation of igneous layering in general requires that all should be reviewed. Parsons and Butterfield have done just that for two layered syenites in South Greenland, concluding that, although sorting of locally derived crystals by currents accounted for one layered structure,

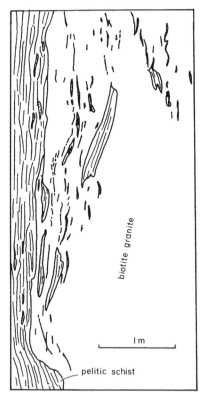

Figure 4.7 Formation of biotitic schlieren by the spalling-off of pelitic schist fragments from a disrupted septum of country rock within the Main Donegal Granite south of Lough Avarnis, Donegal. Plan view of outcrop.

the crystal supply in the other was controlled by order of nucleation and growth rate effects, although crystal settling still occurred: a classical, multifactorial, geological answer.

It seems that, for the most part, mineral layering in granitic rocks is not unequivocally of sedimentary origin. It is commonly a feature of contact zones, and correspondingly often dips steeply, even vertically, and in some roof situations it may have the form of a schlieren dome. Furthermore, any grading in the proportion of the mafic minerals generally faces away from a steep contact, so that gravity is most unlikely to be involved in the mineral separation. Rather it results from a physicochemical process, involving rhythmic precipitation or, following Wilshire, from the mechanical process of flow-sorting during accretionary, sidewall crystallization – a shear flow model not inconsistent with the non-Newtonian view of magmas. Thus it is of particular interest that both Hans Cloos and his pupil Robert Balk realized that a degree of synplutonic deformation may be involved in the formation of this schlieren layering. They ob-

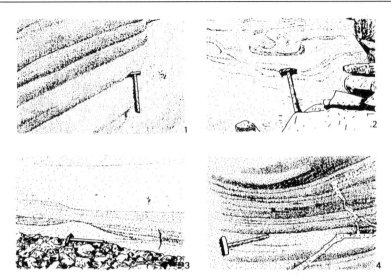

Figure 4.8 Different types of biotite-rich schlieren from the granitic intrusions of Tigssaluk and Alangorssuag, southwest Greenland. After Harry and Emeleus (1960) *Proceedings of the 21st International Geological Congress, Copenhagen*, Volume 14, pp. 172–181. (1) Schlieren with a sharply defined base but blurred upper parts, the whole reminiscent of rhythmic layering; (2) convoluted schlieren with forms reminiscent of slumping; (3) thin mafic layers merging upwards as if influenced by an underlying obstruction during their formation; and (4) parallel schlieren with sharply defined bases and blurred upper parts containing a number of large flat-lying crystals of feldspar. Reproduced with the permission of H. Emeleus.

served that not only were the mafic minerals aligned parallel to the layering surface, but the bands had been thinned by extension. Possibly the stretching itself might have provided a sufficient shear flow to promote sorting.

Within such compositional layering the crystals sometimes lie at high angles to the layer surface, as in the Willow Lake-type of Taubeneck and Poldevaart, otherwise called comb layering by Moore and Lockwood. In a well argued discussion of the origin of this comb layering in parts of the Sierra Nevada Batholith, California, the latter workers confirm the frequent association with contact zones, and also emphasize the close spatial relationship between schlieren layering, comb layering and the formation of orbicules.

From the analogous study of metals we learn from Worster (1990) that a phenomenon during the solidification of alloys is the formation of partially solidified mushy zones or layers. These take the form of a forest of dendritic crystals, oriented principally along the direction of strongest thermal gradient, with fluid filling and flowing through the interstices; even escaping via narrow, crystal-free chimneys. It is easy to envisage this same phenomenon in magmas, with the nucleation and growth of dendritic, whisker crystals on a boundary surface or within an individual layer. Indeed the bending of the whiskers may well indicate the sense of the shear flow. But why, then, is this not a common feature? A clue may lie in the rarity of foreign crystal inclusions within the whisker thickets of both the comb layers and orbs, which suggests, according to Vernon in 1985, that the boundary layer material was supercooled. On this basis and noting that disruption of the layering also occurs, Moore and Lockwood sought an explanation involving a fast flowing, low viscosity and crystal-free fluid, separate but derived from the main body of viscous crystal-bearing magma, and concentrated along structural traps in the walls. I believe that a more feasible cause of the disruption of the Sierran layering is that portions were sloughed off the walls as a result of seismic shocks. I have seen roof layers of pegmatite broken off and rotated in just this way.

As we need to find a satisfactory alternative explanation to that of sedimentation for much of this rhythmic layering, it may be useful to turn to the closely related problem of orb formation, especially as adjacent orbs and schlieren can show identical layering sequences.

THOSE MYSTERIOUS ORBS

It would probably be a mistake to claim that all orbicular structures have the same general origin, especially in view of a plethora of descriptions and hypotheses. However, of the many examples that I have seen in outcrop, all demand a purely magmatic explanation. Of these I illustrate, for historical interest, one of the earliest properly described, an example occurring in the Thorr Granodiorite of Donegal, Ireland (Figure 4.9).

It is usual that the crystal whiskers radiate from some kind of nucleating surface, be it an isolated megacryst, a xenolith or a specific contact; also that both the dark mineral banding within an orb and the orb's surface itself can be distorted by interference with an adjacent orb, and that the shells of dendritic whiskers do not, as noted above, enclose dispersed crystals like those of the host. As for the mode of occurrence, the orbicule bearing rocks, whether they are granite, granodiorite, diorite or gabbro, form but a tiny volume of any particular pluton, sporadically concentrated near contacts in the form of layers, pipes or composite aggregations of orbs.

In a general review of this problem in 1985 Vernon concluded that the virtual absence of nuclei again indicates a significant degree of super-cooling within a narrow boundary layer, a situation surely favouring dendritic growth. Appropriately, a number of workers, including Lofgren in 1980, Parsons and Butterfield in 1981, and Larsen and Sorensen in 1987, envisaged rhythmic supersaturation occurring alongside an inwardly

(a)

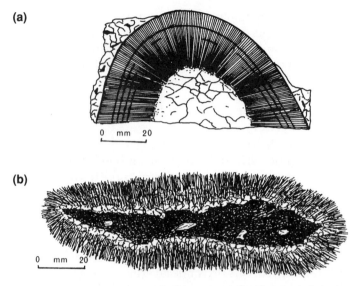

(b)

Figure 4.9 (a) Orbicular structure, Mullaghderg, Co. Donegal. Acicular oligoclase with biotitic shells. From an original drawing by Hatch (1881) *Quarterly Journal of the Geological Society of London*, **44**, plate XIV. (b) Orbicule nucleated on to a xenolith of biotite schist. Original drawing by Cole (1916) *Scientific Proceedings of the Royal Dublin Society*, **XV** (NS), 15, figure 3. Reproduced with the permission of the Royal Irish Academy. Scales added.

advancing, crystal–melt interface, a process requiring local supercooling, perhaps induced by a very local increase in water saturation. The required sequential precipitation of high and low temperature mineral phases must depend on nucleation density and the specific heats of both melt and crystals, coupled with a delicate balance of the rates of growth, diffusion and cooling. Certainly Swanson's experience in dealing experimentally with such systems is of the overriding importance of increasing supercooling, with large single crystals first giving way to coarse skeletal and then dendritic crystals, and finally to compact spherulites.

Overall, it is not at all surprising that comb layers and whiskered orbs are relatively uncommon features except in rather special contact and roof situations. It is here that a volatile-rich phase can collect in a narrow zone, and the situation is analogous to that of pegmatite selvedges on hanging walls where elongate crystals hang and branch out into the 'magma'. Of such features Stephenson provides a special insight based on his work on the layering in the felsitic granite of Hinchinbrook Island, North Queensland. The plain fact is that all these special features represent rather special conditions. In granitic magmas sedimentation processes are probably unusual and supercooling rare, and the structures we observe do not often tell us much about the general condition of the magmas.

ON BANDING DUE TO DEFORMATION

There is another form of banding in granites which has an even more complex history. Its formation illustrates well the fact that natural processes usually involve coupled mechanisms and only by the detailed study of specific examples of layering can we appreciate the relative importance of sedimentation, flow sorting, rhythmic precipitation and synplutonic deformation. I suspect that the latter is involved more often than presently accepted. Certainly many granites have suffered a self-generated, synplutonic deformation involving a measure of autometamorphism, as many of the textures signify. The most visible result is the development of a strong, often steeply dipping mineral layering, in effect a gneissosity which has often, and mistakenly, been identified as a relict bedding, the veritable 'grin of the Cheshire cat' of granitization!

The strong banding shown by the Main Donegal Granite pluton, Ireland, is a case in point (Figure 4.10). As described by myself and Anthony Berger this granite represents a composite intrusion of relatively simple sheets, but because it suffered a synplutonic deformation due to continued movement on its acceptor shear zone, it is now represented by a granite gneiss. According to Berger the banding and associated mineral alignments both result from this synplutonic deformation, the banding in particular forming by the segregation of the interstitial melt containing the components of potassium feldspar, yet with an average composition near to that of unbanded granite. Perhaps we have here a real example where a high rate of deformation reduced the effective viscosity. And no one can be surprised at the local migration of late stage, potassium-rich silicate fluids during the deformation (in this instance squeezing) of a crystal mush.

Reflecting on coupled mechanisms, this layering in the Main Donegal pluton is the result of at least three – that is to say, segregation, the smearing out of country rock material and dyking. Contemporaneous deformation transposed the resulting structures into a remarkable parallelism, to form a veritable composite gneiss graphically described in 1902 by the famous Irish geologist Grenville Cole, in a paper entitled, *The Composite Gneisses of Boylagh*.

Deformational layering or banding is certainly not confined to the Main Donegal pluton: another Irish example involves the Carna pluton of Galway, for which a similar origin has been proposed. On a broader canvas many of the granite gneisses of the world that I have seen represent such deformed intrusives, some deformed while still crystallizing from a magma, others deformed and recrystallized during the various stages of consolidation and cooling, and yet others during a later period of superposed metamorphism. Certainly banding in granites is rarely evidence of flow or fluidity.

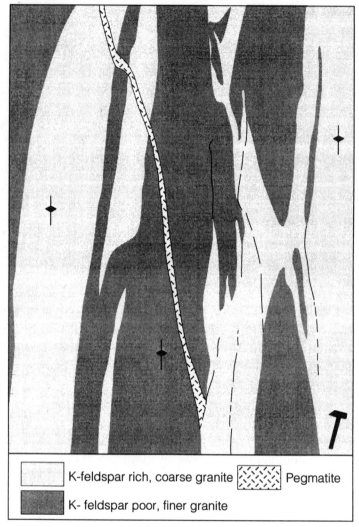

Figure 4.10 Banding in the Main Donegal Granite pluton, near Lough Attrive, Donegal. Mineral foliation (◆) and banding in the biotite granite is essentially vertical.

PER MAGMA AD MIGMA USQUE AD TEXTUS

In summary, it seems that we must expect a continuous change in the rheological condition of a pulse of granitic magma during its life history, and that this will often involve a stage when elastoviscous properties will prevail. Except when the magma has a high volatile content, the strength of such a fluid is sufficient to hold in suspension crystals and small enclaves. However, it is insufficient to prevent the free fall of any large stoped blocks that survive being frozen into the peripheral magma. The non-Newtonian properties ensure that many of the mineral orientations may not represent free flowage, but a permanent strain indelibly recording the effects of applied stress. Evidence of local turbulent flow sometimes survives, and may be some part of the early deformation of mafic blobs records a first streaking out in a fluidal host, but otherwise it is rare to find preserved any earlier, unmodified intrusion or convection flowage. The reader will have detected another prejudice!

When a crystal framework is eventually formed it is so very open, and so porous, that the interstitial, often volatile-rich, residuum is likely to remain highly mobile and capable of reacting with the framing matrix of crystals. This interstitial fluid may flow into transient shear zones or otherwise be expelled under gas pressure. There is nothing simple about the life history of a natural magma! The evidence for this later part of this history is recorded to a greater or lesser degree in the existing texture of a granite, and it is to that important subject that I now turn.

SELECTED REFERENCES

McBirney, A.R. and Murase, T. (1984) Rheological properties of magmas. *Annual Review of Earth and Planetary Sciences*, **12**, 337–357.

Pitcher, W.S. (1979) The nature, ascent and emplacement of granite magmas. *Journal of the Geological Society of London* (Presidential Address), **136**, 627–662.

Wickham, S.M. (1987) The segregation and emplacement of granitic magmas. *Journal of the Geological Society of London*, **144**, 281–298.

The evolution of the granitic texture: a continuum of crystal growth

5

... most granitic rocks exhibit textural and mineralogic features which are less related to the ultimate origin and early history than to subsequent subsolidus recrystallization.

W.C. Luth (1976) In: The Evolution of
the Crystalline Rocks, p.336.

INTRODUCTION

To understand the evolution of granitic textures is to begin to understand the genesis of granites. Central is the study of texture dynamics, which concerns the shapes and mutual disposition of matrix grains – that is, the morphology in the terms of a metallurgist. We need to distinguish carefully between what is seen of the geometry, what are the processes involved in its construction, and what are the driving forces. We must be alert to subsolidus reactions, which are especially active during the synplutonic deformation associated with the emplacement of many plutons. In what follows I deal with the more pristine textures largely attributable to crystallization from a melt, though almost inevitably I shall be tempted into consideration of the solid state, in particular certain reinterpretations of the groundmass texture of the granite porphyries.

ON GRANITIC TEXTURE IN THE MAGMATIC CONTEXT

TEXTURAL GEOMETRY

Until comparatively recently the geometric aspect dominated petrography, but the limitation imposed by two-dimensional observation, together with the lack of a basic understanding of the mechanisms of crystallization, led to a certain lack of confidence in the interpretation of granite textures. Moreover, generations of petrologists have sheltered behind the descriptive nomenclatures established long ago by Zirkel and Rosenbusch.

It is to Erdmannsdörffer and Drescher-Kaden that we owe the recognition of the complexities of granitic texture, in particular the importance of crystalloblastesis, with the need to distinguish secondary, metablastic, from primary, endoblastic textures. Unfortunately, the overenthusiastic interpretation of metablastic textures has led some petrologists to advocate a metamorphic origin for granites when, in truth, crystalloblastesis is the natural consequence of mineral readjustments during long cooling in the presence of water, and in response to both synplutonic deformation and burial metamorphism. Nevertheless, it is only fair to point out that some modern workers, such as Larry Collins, continue to emphasize the role of internal metasomatism in the origin of the compositional and textural variations in granitic rocks. Although I do not countenance this view, preferring to follow a more prosaic path, it has to be admitted that a mere turning of the pages of Augustithis' *Atlas of Textural Patterns of Granites, Gneisses and Associated Rocks* is illustration enough of the variety and complexity of textures, of which, as yet, we can only pretend to understand the simplest patterns.

What is clear is that this textural variety is intimately related to the magma-tectonic environment. Thus perhaps the most pristine of the melt–crystallization textures, approaching the tabular and adcumulate, are those of the granitic rocks in the volcano-plutonic arcs. In most other environments even apparently homogeneous granites in outcrop show some degree of blastesis. This varies from intergrain reactions, grain enlargements and the granulation of quartz in post-tectonic, high level plutons, to wholesale recrystallization and chemical readjustments in the deep-seated, syntectonic diapirs involved in the early stages of collisional orogenesis. Furthermore, the most thorough-going reconstructions occur in some A-type, alkali granites, where the concentration of volatiles in extra hot magmas promotes extensive subsolidus reaction and recrystallization. Finally, there are high level granites in a variety of global environments that suffer particularly rapid quenching in response to the quick release of concentrations of the volatile constituents, and these granites are prone to develop complex hiatal textures.

Textures rather clearly reflect the evolutionary histories of their host rocks, but we need to be aware that the external shape of a crystal and its interrelation with other crystals is ever changing during this history and, like the host granite itself, its frozen morphology is that of an arrival state. An awareness sharpened by studies such as those of Whitney in 1988 and Nekvasil in 1991 on the effect of decompression on an ascending and partly crystallized magma.

A FRESH STEREOLOGICAL APPROACH: STEREOLOGY

To advance in textural studies we need to carry out statistically controlled examinations of grain shape, size, distribution, and also the grain inter-relationship, including the grain contact frequences. This is very labour intensive work but it is here that the relatively new technique of three-dimensional analysis, using serial sections and image digitization, proves so valuable, particularly in revealing the roles of the porosity and intersti-tial melt in generating the compositional variation.

Bryon and his colleagues applied this technique, in 1994 and 1995, to the study of the textural development of a suite of arc-related rocks within the Coastal Batholith of Peru. These rocks were from the Linga suite and range from granodiorite through monzogranite to syenogranite, all showing the pristine preservation of their primary textures common to this Cordilleran environment. From their stereological examination it appeared that the textural variation, though progressive, could be conveniently described in terms of three stages as crystallization proceeds, namely (Figure 5.1):

1. Early uninhibited crystallization from magma with a high proportion of melt;
2. Development of a touching crystal framework;
3. Interstitial crystallization and porosity occlusion within interconnected pore spaces or isolated pore cavities.

In a specific example, plagioclase is seen to have been the first phase to start crystallizing when, with amphibole, it dominates this early stage. Biotite, quartz and alkali feldspar then appear as the melt evolves towards the cotectic (Figure 5.3, p. 75). The order and timing of the appearance of these latter phases gives rise to the textural variation. The longer the period of unaccompanied plagioclase crystallization the greater the chance that plagioclase crystals will form a framework by themselves, while the earlier the other phases appear the greater their contribution towards the framework. A model example of the various spatial relation-ships that plagioclase crystals are likely to adopt is shown in Figure 5.1, and is close to actuality. In the most mafic granodiorite the interstitial nature of the alkali feldspar indicates that the latter did not start growing until after the plagioclase framework had formed, the interconnections between neighbouring pockets implying that much of the growth oc-curred before the isolation of the pore spaces.

In the more evolved granodiorite of the suite the plagioclase crystals continue to form a framework except where they occur, in the form of isolated crystals or as small clusters, as inclusions within large alkali feldspar plates. This is interpreted as evidence for the onset of alkali feldspar growth while the melt still contained a crystal suspension and, consequently, its localized contribution to the framework.

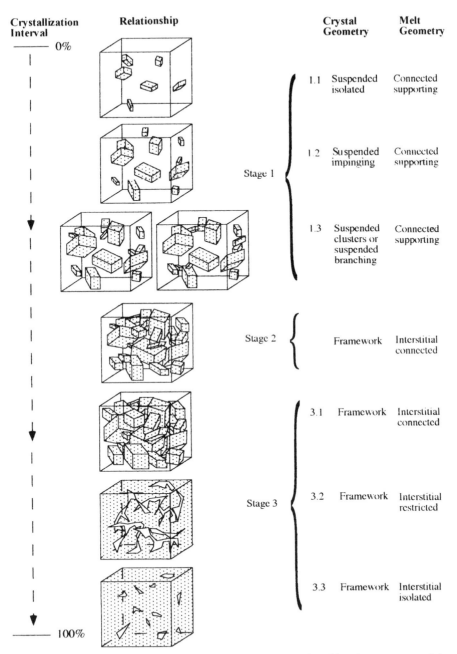

Crystallization Interval	Relationship		Crystal Geometry	Melt Geometry

Figure 5.1 A model of the changing geometrical relationships between crystals and melt during congelation. Explanation in the text. Reproduced with permission of Bryon *et al.* and the Mineralogical Society of Great Britain and Ireland.

There is no continuous framework in the monzogranite where the appearance of clusters of plagioclase crystals suspended in a mosaic of alkali feldspar and quartz implies that the crystallization of both the latter phases started before the plagioclase had time to form a framework, but not before its crystals had started to impinge and form localized clusters. A similar texture is shown by the most evolved rock of the suite, a syenogranite, but the isolation of the plagioclases and the absence of clustering points to an even earlier appearance of the plagioclase and quartz.

A particularly interesting facet of this study concerns the mode of interstitial crystallization. In the granodiorite, where plagioclase was the first to appear, the continuing crystallization of this phase within the interstices occurred by overgrowths on the existing framework crystals in

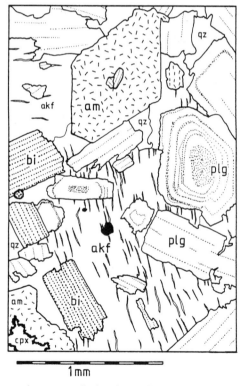

1mm

Figure 5.2 Texture of quartz diorite from Ballachulish, Scotland, drawn from horizontal thin section. The rock contains zoned plagioclase primocrysts, subhedral biotite and amphibole, with interstitial quartz and poikilitic alkali feldspar (flame perthite). The edenitic amphibole (am) evidently formed at the expense of augite (relict cpx, lower left). Reproduced from Weiss and Troll (1989) with the permission of the authors and Oxford University Press.

the form of thin shells that do not alter the euhedral margins. In contrast the first appearance of alkali feldspar varies throughout the suite: in the least evolved granodiorite it grew in the interstices as oikocrysts, that is, as relatively large crystals each filling several pore spaces and poikilitic to the surrounding framework (Figure 5.2, from another study, illustrates this particularly well). In further contrast the start of alkali feldspar crystallization at a relatively earlier stage in the more evolved rocks resulted in individual crystals contributing to the framework. Subsequently part of the post-framework crystallization of alkali feldspar occurred by the overgrowth of these crystals.

This and other crystallization histories are shown by the authors to be compatible with those deduced from melt pathways in the quaternary An–Ab–Or–Qz system. Thus almost for the first time we can begin to appreciate the three-dimensional geometry of a granitic rock in relation to its crystallization history. Futher advances in the textural understanding of grain boundary relationships can be expected from the use of fractal analysis as suggested by Petford and co-workers in 1993, though crystal growth interference boundaries do not seem to me to represent fractals in the sense of Mandlebrot.

A MEASURE OF POROSITY AS REVEALED BY TEXTURAL STUDIES

Studies of primary textures such as illustrated above show that a considerable volume of melt rested in the pore spaces within the framework even after the magma had stiffened. This raises the possibility that melt might move through and away from a porous crystal framework via interconnected pore spaces: a migration that would surely be enhanced if early microfractures were to develop in the framework. These are complex matters which involve pore interconnectivity, the role of overgrowths in reducing porosity and permeation, and the complex problems associated with melt extraction mechanisms. They may prove difficult to resolve, but certainly the extraction of rest melt at this late stage in the crystallization interval would explain many features of granitic rocks in outcrop, particularly the ubiquitous pegmatitic bodies segregating *in situ*, aplite dykes filling synplutonic fractures and, not least, the formation of felsic-mafic banding.

Such textural studies ought to provide some guide as to the nature of the critical melt fraction that we discussed earlier. By using the digitization techniques to strip away, so to speak, the overgrowths on the framework feldspars to the point of simple touching, it may be possible to estimate the volume of melt occluded at the very start of linkage. This has yet to be done, but even so this estimate of melt volume at the transition from suspension to framework would clearly be specific to the sample and probably differ for each compositional member of a particular rock suite.

Intuitively I would guess that this melt volume might be of the order of 40% in the example of the plagioclase-rich granodiorite quoted above. However, in the same example, Bryon and his colleagues, in 1996, were successful in determining that the pore network would have become disconnected at porosities below about 3–4%. Nevertheless, despite good evidence for interconnectivity over a wide range of porosity these authors are forced to conclude that composition convection is unlikely to have occurred during interstitial crystallization!

These questions of porosity will need to be discussed later at greater depth, but already we can better appreciate its likely importance in granites and so more confidently refer to modern views on porosity, melt segregation and melt migration already well established in studies of the mafic rocks, as usefully reviewed by Ribe in 1987, and also discussed later in this book.

THE BACKCLOTH OF THEORY AND EXPERIMENT

Because of the dominance of the feldspars and quartz in the felsic rocks the form of the experimental system An–Ab–Or–Qtz–H_2O has been used to model the precipitation histories of the mineral phases, so underlining the close relationship between textural development and phase chemistry. As an example all the compositions representative of the Linga suite discussed above lie within the plagioclase volume of the system (Figure 5.3a,b). Consequently the early crystallization of plagioclase in the granodiorites resulted in the melt evolving, via the plagioclase–quartz divariant surface, to three-phase saturation on the cotectic, the order of crystallization being therefore plagioclase, plagioclase + quartz, plagioclase + quartz + alkali feldspar, which is the case in fact.

Such a coincidence between experiment and reality allows us to appreciate the close control of pressure, temperature and water content on the nature and order of the primary mineral assemblage, and so explain the overlap in the individual mineral phases as a response to cotectic precipitation. Nevertheless, the involvement of the mafic components can hardly be trivial in the case of the abundant tonalites and granodiorites, especially as we know that the precipitation of hornblende will inevitably control melt composition in the early stages of evolution. Furthermore the synthetic granite tetrahedron developed from the primary studies of Tuttle and Bowen in 1958, and also by Carmichael in 1963, is but an idealized view of the phase relations in natural magmas and then only represents the liquidus relations at a constant H_2O content (Figure 5.3). In reality, the modifying effect of varying the H_2O content is so dramatic that, coupled with the changing water content of the rest melt throughout the crystallization history, the form of the system must also change. At this point it is worth reminding ourselves that most felsic magmas reach

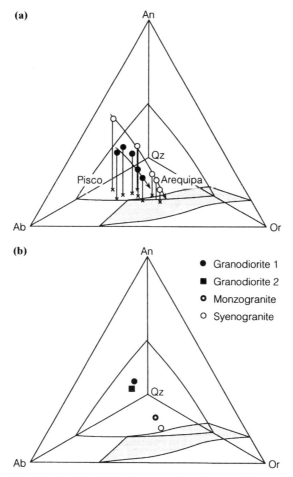

Figure 5.3 (a) An–Ab–Or–Qtz–H$_2$O diagram showing model crystallization paths determined for two separate plutons of the Linga superunit in the Coastal Batholith of Peru. (b) Plot of the compositions of the four rocks studied by Bryon and his colleagues from the Linga superunit. All four lie in the plagioclase volume and evolved by varying degrees of plagioclase fractionation to three-phase saturation on the cotectic. Determined by Atherton and Sanderson (1985) and Byron and his colleagues (1996).

saturation only during the late stages of their crystallization, if at all. The details of these effects were reported in 1988 and 1992 by Nekvasil, including the finding that increasing the water content at constant pressure leads to a contraction of the liquidus volume for quartz; also that the interrelations within the feldspar system, particularly the degree of resorption of early plagioclase in the sodic region of the system, are governed by this same activity of water, even to the extent that resorption may be followed by reprecipitation once the activity of water has increased due to its concentration during crystallization. Clearly we can begin to model many of the textural complexities of granites in terms of the changing compositions of their magmas, even the more complex scenario where precipitation of first one and then two feldspars is followed by partial or complete resorption of the plagioclase, and finally by the reprecipitation of two feldspars.

Even with this degree of similarity we have to accept that in real magmas the relative rates of crystal nucleation and growth are additional critical factors in determining just when and how precipitation occurs, and thus the overall morphology.

A COMMENT ON NUCLEATION AND GROWTH

From everything we have learned so far about granitic magmas – particularly the textural observation that the phenocrysts often enclose minute crystals – it is likely that these magmas carry seed crystals from the time of their birth. Thus nucleation is always likely to occur on a pre-existing substrate and, furthermore, it is even possible that a partial melt will retain a 'memory' of the feldspar structure. Consequently, nucleation rates may be subordinate to those of growth in determining the evolution of all but the simplest of textures: a wishful thought, perhaps, engendered by Cashman's revelation in 1990 that a predictive model for heterogeneous nucleation remains elusive, especially in multicomponental systems.

However the nuclei arise, it is their nature and density which is all important, as is shown dramatically by Lofgren's observation, in 1980, that by changing these two parameters he could duplicate experimentally the entire range of textures produced in basalts by variations in the cooling rate! Maybe this is less true of granitic magmas, in which the presence of pre-existing crystals obviates the need for nucleation of the liquidus phases, so that growth will proceed as soon as cooling begins. The result is that the shapes of the crystals are not so strongly a function of cooling rate as when nucleation must precede growth.

At this point it is particularly pertinent to recall that artificial melts have almost invariably been seeded so that growth models could be better constrained! And constrained they need be, because of the complexity of the growth process itself, which involves diffusion down a concentration gradient along pathways of very different character, such as through a fluid down to a grain surface, or along a disordered grain boundary, or through a defective crystal lattice. Perhaps some further simplification may arise from the extremely sluggish character of lattice diffusion as attested by the common survival of zoning.

This is not the place and neither am I qualified to discuss crystal growth theory and the surface chemistry involved, except to remind the reader that the three driving forces are: chemical potential, the surface energy resulting from the incomplete bonding of surface atoms, and deviatoric stress. It is particularly the variation in surface energies that explains many of the growth features of crystals – their shapes, their tendency to euhedralism and the magnitude of the dihedral angles of the grain contact corners. Wheeler offered a useful review of this complex field, particularly in emphasizing the role of stress in the development of the overall texture, and not simply of preferred mineral orientation. Further, A.G. McLellan examined in depth the thermodynamics and local equilibria in such stressed systems. I am simply content to draw particular attention to this dynamic aspect of the crystallization of granites.

From the experimental point of view the work of Lofgren, Fenn and Swanson on nucleation and growth in hydrous granitic melts provides some important guidelines in interpreting rock textures. Thus the simple finding that the growth rate of the feldspars far exceeds their rate of nucleation possibly explains why the feldspar-dominated textures of plutonic rocks are so generally of coarse grain: rapid growth from a few scattered nuclei overwhelms the competition! Yet another experimental finding is that a feldspar phenocryst need not represent an early stage of growth: megacryst and groundmass can grow simultaneously.

As always, the degree of water saturation plays a key part. In particular, the wetter the magma the lower the nucleation density and the faster the growth, an experimental observation entirely consonant with the common occurrence of feldspar megacrysts in those granites generated in hydrous environments, such as many S-types, or more generally in granitic contact facies wherever there is indication of a late stage concentration of water. Overall, the effect of the water content on the crystallization history of the feldspars, as reported by Nekvasil in 1992, is profound.

In outcrops of many granitic plutons there is a remarkable uniformity in grain size and grain distribution over large areas of a single magmatic pulse. This may be because of an even distribution of seed crystals, itself a result of localized turbulent flow at an early stage of magma evolution. It seems to me that it is only when flow becomes laminar, as in contact zones, that the nuclei and their overgrowths may sort into grain density layers, which are reflected in differences of grain size. Of course there are other examples of plutons where close examination reveals subtle modal and textural heterogeneities which I would suspect are due to the mixing of different draughts of essentially the same magma during transport (p. 83).

I have no such ready explanation for the remarkable fact that mineral size and shape, even the textural interrelationship, are often specific to either one particular magmatic unit or a particular suite of several units. Perhaps it suggests an underlying compositional control inherited from the source, and one likely to involve the very nature of the volatiles.

PARAGENETIC SEQUENCES

It has been traditional to rely on crystal idiomorphism, the 'inclusion principle', and apparent replacement features to establish paragenetic sequences in granitic rocks, but this kind of microscopical evidence can be equivocal and was seriously questioned by Flood and Vernon in 1988. Nevertheless it is difficult not to accept that, as established long ago by Rosenbusch, there is a general order of first appearance, with distinct, though overlapping, growth phases. The primary accessories are first to appear (though sometimes they may represent inherited restite) and are followed, as we have seen, by the mafic minerals and plagioclase, together forming the framework matrix within which crystallize quartz and potassium feldspar. The final, irregular grain boundaries between the latter and plagioclase attest to much co-precipitation and overlap, the plagioclase evidently continuing to enlarge by adcumulus growth. The general result is the hypidiomorphic texture of Rosenbusch. This simple view of growth sequence was sufficiently adequate to enable Stephen Nockolds and his colleagues, in a series of seminal papers on the Scottish granites (1946 and 1956), to integrate the textural history and the geochemical data and thus construct a liquid line of descent modelled by an observed sequence of crystal fractionation.

Such modelling is now commonplace and is particularly successful in the almost ideal case of arc-related tonalites and granodiorites. However, in the granites of other tectonomagmatic environments the strict application of the textural rules often reveals metablastic textures of a considerable complexity, almost certainly a result of varying degrees of unmixing, regrowth, recrystallization and entirely new growth (Figure 5.7, p. 86). These changes usually follow the formation of the crystal framework and then continue into the completely solid state, representing the transition between the magmatic and the metasomatic or pegmatic stages, in which the mineral replacements often result from the circulation of hot fluids.

This complexity is well illustrated by reference to Schermerhorn's elegant description of textural sequence and paragenesis in the crustal-derived granites of northern Portugal (Figure 5.4). We do not have to agree with every detail to appreciate the complexity of this particular textural story with its overlapping and polyphasal growth histories of the constituent minerals. But we need to read such histories if we are to understand the commonly occurring granites of the world.

From this general introduction I turn now to a brief discussion of certain aspects of the growth histories of individual mineral phases, starting with

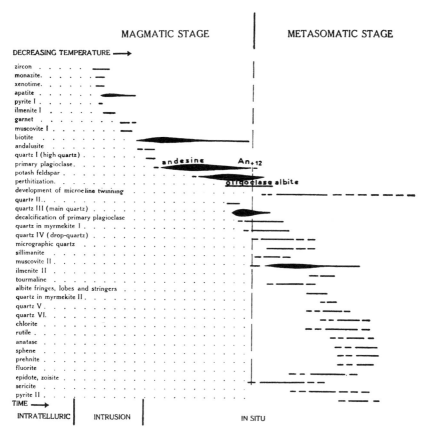

Figure 5.4 General paragenesis of the granites of the Viseu, northern Portugal, according to Schermerhorn (1956). Reproduced with the permission of the author.

the accessory minerals, which have an importance far outweighing their modal proportions.

INDIVIDUAL CASE HISTORIES OF MINERALS AS A GUIDE TO PROCESSES

THE ACCESSORY MINERALS: A SOURCE OF VITAL INFORMATION

The accessory minerals such as apatite, zircon, monazite, xenotime and allanite certainly do not escape this complexity of growth, and their individual histories are especially crucial to the determination of U-Pb and Th ages, the correct interpretation of the REE budget in partition studies and, not least, to understanding the early history of the magma. Zircon is a case in point, with an inherited crystal core overgrown by one or more phases of magmatic precipitation: a core which, along with monazite, provides the only certainly identifiable restitic minerals – and a powerful way of determining granite provenance.

There is much to learn, both of the necessary analytical techniques and the complexities of interpretation, in a group of papers published separately in 1992 by Williams, also Paterson and his colleagues, and Pidgeon and Compston. All are concerned with analysing and dating the core and overgrowths of the zircons in various granitic hosts, with important results. Not only can several generations of zircon coexist within one population, but zircons within the same host may sometimes have different histories. It is as if, as Pidgeon and Compston point out, the crystals grew in different parts of an early, inhomogeneous melt, only later to be brought together into the existing host: a scenario that would fit well with a crustal melting model. A further finding is that there is often an abrupt change in composition between each of a sequence of overgrowths of the same general age, suggesting that each shell represents a distinct and separate phase of magmatic precipitation, again possibly the result of the upward pulsing of the host magma.

There is even more to be learned about the ambient physical conditions in the magma from the relationship between these conditions and the habit of the zircon crystal, as described in the works of Pupin (1980, 1985 and 1988), and his colleague, Caironi (1985). Clearly the so-called accessory minerals are quite the reverse in terms of understanding the evolution of granitic magmas.

In general, the tiny crystals of these accessory minerals are concentrated within the phenocrysts and aggregates of the major mineral phases, where they are the hosts for large fractions of the trace elements of intermediate and felsic igneous rocks. Following a study by Green and Watson on the growth of apatite, Bacon offered an attractive general explanation involving the local saturation of the incompatible elements within the boundary layer adjacent to a growing crystal. This arises because the latter rejects the incompatible elements, which cannot then escape into the body of the magma on account of slow diffusion. The accessories thus crystallize as attachments or inclusions within the early growth phases of the essential minerals, a thesis that carries the implication, according to Bacon and Druitt, that the rates of growth of the host crystals were large relative to the probable lifetimes of magma chambers; yet another indication that the overall crystallization is quickly accomplished.

A FOOTNOTE ON EPIDOTE

As distinct from early allanite, epidote is yet another mineral of granites that commonly shows several phases of growth, giving rise to a certain scepticism about a primary magmatic origin. Thus, although I agree with Zen and Hammarstrom that epidote is commonly euhedral against biotite in tonalites and granodiorites, I do not share their confidence in its primary growth, or readily accept their interpretation of this primary status as establishing, from certain stability data, the crustal depth of the host magma at the time of its crystallization.

MAFIC MINERALS AND THE FUGACITY OF OXYGEN AND WATER

The mafic minerals of granites are good indicators of the intensive variables governing the formation of the assemblages in which they occur, in particular, the temperature, and the water and oxygen fugacities. Good examples are provided by the mineralogical studies of our Japanese colleagues, particularly Ishihara, in 1977, and Nozawa, Kanisawa, and Kuroda and others, all in 1983. They showed that, in essence, the granitic rocks of Japan group into a magnetite series and an ilmenite series which relate to a relatively oxidized or reduced status of the solidifying magmas, the f_{O_2} differing by two or three orders of magnitude. This is reflected not only in the proportions of magnetite and ilmenite, but also in the compositions of the biotite and hornblende: in the magnetite series the latter minerals are magnesium-rich and aluminium-poor, whereas in the ilmenite series they are iron- and aluminium-rich. An immediate consequence of this contrast in oxidation, as reflected by the iron oxide mineralogy, is that granitic rocks show great variations in magnetic susceptibility. For example, in the Sierra Nevada its measurement by Bateman, Dodge and Kistler revealed gross regional differences, and these translate into a regionality of oxidation ratios which may well image that of the source rocks.

An example of the influence of water activity, and incidentally the application of textural criteria, is the frequent reversal in the order of stabilization of augite, hornblende and biotite. With increasing fractionation and water enrichment, the conventional view is that hornblende will be stabilized before biotite in calc-alkaline melts. However, Wones and Gilbert opined that biotite may often appear immediately after augite and before hornblende in granodiorites, and Naney showed that this particular order is controlled by the water content. On the basis of this kind of mineralogical evidence Weiss and Troll claimed that the monzodiorite and granodiorite magmas of the Ballachulish Complex, Scotland, were undersaturated with water, to the extent that it was initially only of the order of 1 ± 0.5 wt.%, a claim supported by the metastable preservation of potassium feldspar in the structural state of orthoclase and intermediate microcline, and also by the total lack of hydrothermal effects. I believe this to be a general finding.

THE PROBLEM WITH MUSCOVITE

In the paragenetic scheme presented by Schermerhorn (Figure 5.4) several growth stages of white mica are identified, which is indeed a common finding. Miller and his colleagues, in 1981, reviewed the textural evidence for distinguishing between that muscovite which appears to be of magmatic growth and that which appears to have resulted from subsolidus reactions. Unfortunately, as already hinted in the foregoing, their criteria – which include coarse grain comparable to associated magmatic phases, clean crystal terminations especially when euhedral, and enclosure in a clearly magmatic phase – can be challenged. Euhedralism is particularly equivocal as is well illustrated by the complex growth histories recorded by the muscovite in many granites. Even in the same specimen well-formed flakes are seen to be in euhedral parallel intergrowth with biotite, while other flakes, often as well formed, morphologically replace feldspar, and yet others, usually of more ragged outline, form separated aggregates or coat microfractures. Only the first mode of occurrence is likely to be of primary magmatic origin, but even so the presence of zoning, such as reported by Roycroft from the Leinster Granite, is a useful confirmation. Like the tourmaline with which it is so often associated, much of the muscovite growth belongs to the post-consolidation, pegmatitic phase. Certainly the presence of muscovite is far from being an infallible indication of peraluminicity.

The two-mica granites of northern Portugal provide a particular insight into these complexities of polyphase muscovite growth, and from their study Neiva has been able to distinguish several phases, not so much by textural features as by the recognition of a systematic variation in composition, which correlates well with a decrease in estimated temperature of precipitation and replacement.

From the stability data provided by Yoder and Eugster we should expect muscovite to appear as a stable phase in both magmatic and subsolidus environments, but only in magmas at a relatively high pressure. However, fluorine is a likely component in such a system and would significantly lower the pressure required; otherwise, translated into crustal depth, the implication is that primary muscovite-bearing granites can never have risen high into the crust – a conclusion with which I am most unhappy.

A COMMENT ON THE SIGNIFICANCE OF ZONING

Mineral zoning is a common feature in granites and it ought to be possible to read from these inorganic 'tree rings' a record of the physical and chemical conditions pertaining during intrusion and consolidation of the host magma. The problem is to relate specific patterns of zonation to specific genetic processes.

The zoning of plagioclase is a case in point. Although the oscillatory pattern is almost certainly produced by crystallization kinetics which are local to each crystal, the abrupt changes in composition represented by the truncated zoning appear to correlate with changes affecting the magma body as a whole. There are a number of possibilities, such as large-scale convective overturn, rapid intrusion of magma pulses involving changes of pressure or temperature, and even the effect of the co-precipitation of biotite and amphibole in reducing the water content of the melt. The explanation favoured by Vance in 1965 is that the truncated zoning sheathing a calcic core indicates resorption due to pressure reduction during the rapid ascent of the host magma, a thesis supported by the thermal cracking often apparent within the calcic core. Whether such cores represent crystals residual from a remelting process or simply an early phase of crystallization is not easy to assess, though the former alternative fits well with the anatectic model now much in favour.

My own view is that such truncated zoning is more often the result of the mixing of magmas during the multiple recharging of plutons, a process which turns out to be ubiquitous and universal. The calcic plagioclases may be derived directly from synplutonic invasions by basic magmas, and these, and any early formed calcic plagioclase crystals, may well be resorbed within an ascending, granitic pulse, only to provide nuclei for reprecipitation when their host mixes with the cooler, standing magma into which it is intruded. Such multiple recharge may well explain the two- or three-fold repetition of truncation reported by Mason in 1985.

There are many variants of the subsequent history depicted in the peripheral zoning. The sequence described by Loomis and Welber from the Rocky Hill Pluton of California is of particular interest in that it shows a broad compositional 'plateau' in which the oscillatory zoning is best developed, and they envisage this as due to the slow growth of crystals in a host magma continuously agitated and mixed by convection and turbulence, that is, until anchored by the formation of a crystalline matrix.

The results of cathodoluminescence and back-scattered electron imaging of feldspar phenocrysts reveal more detail of the their growth history in a dynamic magma chamber and, futhermore, reinforce the mixing model. As an example, the zonation characteristics of the alkali feldspar megacrysts of the Shap Granite, northern England, have been studied in this way by Cox and others (in 1996). Both growth and dissolution events are revealed by the zoning, in which the fine-scale zones are thought to represent as many as 20–50 mixing events! An even more important finding is that these megacrysts are zoned with respect to $^{87}Sr/^{86}Sr$, demonstrating that magmas from different sources were involved.

ON BIG CRYSTALS OF POTASSIUM FELDSPAR

The significance of euhedralism is again called into account in any discussion of the status of the potassium feldspar megacrysts, a theme that has aroused intense debate in granite studies. In particular, potassium feldspar megacrysts in granitic plutons have been variously interpreted as representing either phenocrysts or porphyroblasts. As is well known, the metasomatic growth of 'les dents de cheval' formed the cornerstone of the granitization hypothesis. The arguments have been endlessly rehearsed and the reader is referred to leading accounts by Drescher-Kaden (1982), Stone and Austin (1961), Mehnert and Büsch (1985), and particularly Vernon (1986). What follows is a multifactorial compromise with which I remain content.

The evidence called upon to support crystallization from a melt includes euhedralism, conformity of size, and a zoning marked by inclusions and chemistry, whereas a porphyroblastic origin is favoured by invasive texture, the way in which megacrysts lie across boundaries of aplitic veins or within or across the margins of mafic enclaves, or occur within the adjacent country rock hornfels. However, although it is true that the K-feldspar megacrysts commonly show the euhedral shapes expected of growth from a melt, such gross euhedralism is not, in itself, convincing evidence of such an origin. The crystal boundaries often show a microscopic invasion of the groundmass, the minerals of which appear to be partly resorbed, even replaced, forming an outer zone of extension

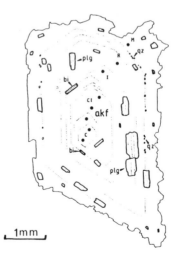

Figure 5.5 Zoned potassium feldspar with laths of plagioclase accreted on growth surfaces. Ballachulish Granite (pink granite facies). After Weiss (1986), *Dissertation München*, **60**, p. 228, with the permission of the author.

that has been interpreted as a replacement front, although, equally, it may simply represent a change from free to impeded growth as the groundmass crystallizes. Less equivocal is the evidence of the internal zones of tiny inclusions of plagioclase, biotite and quartz, which have accreted on to crystal faces during still-stands in the growth of the host (Figure 5.5). That this represents precipitation from a melt is supported by a compositional zoning revealed by fluctuations in the barium content, the higher concentration of this element in the core according with the geochemical expectation.

The meticulous work of Mehnert and Büsch, involving textural and chemical data, provides a satisfying model for the formation of potassium feldspar megacrysts in many granites. An early formed, relatively small potassium feldspar crystal, either singly or accreted with others, forms a nucleus for the rhythmic precipitation of the bulk of a single large feldspar. Such a precipitation ends with the growth of an invasive rim in which there are great fluctuations in barium reminiscent of the situation in the pegmatitic phase of many granites. Essentially the model again involves the multistage growth of a phenocryst from a melt, ending with blastic growth in the near solid.

Turning to the situation whereby potassium feldspar megacrysts penetrate or lie across aplitic veins, examples of which I have myself described (Figure 5.6), I now favour the view that such megacrysts were split, after which growth continued on the truncated surfaces by the preferential precipitation of feldspar components derived from the aplitic fluid. This healing of the fractured feldspar gives the appearance of a primary crystal. As for the growth of megacrysts within or across the boundaries of enclaves, I will later argue that most examples best fit a model of incorporation of early formed crystals into a mafic magma blob. Possibly there remain examples of the growth of 'dents de cheval' in contact hornfels which are difficult to dismiss as mere porphyroclasts. Nevertheless, I am convinced that most potassium feldspar megacrysts represent crystallization from a melt with some extension of growth in the near solid state.

However, within the body of a pluton there are many examples of complex interactive textures which seem to show potassium feldspar replacing plagioclase or even the reverse (Figure 5.7), an interpretation that I am loath to abandon. Thus I think it impossible to deny metasomatic growth where there is supporting evidence of late stage potassium metasomatism. Certainly few would dispute the reality of the microclinization associated with the later stages of the evolution of A-type granites, particularly as exemplified in the Nigerian examples described by Bowden and Kinnaird in 1984 (p. 263).

I end this particular discussion with a mere note on a famous subject of debate: the causes of rimming of potassium feldspar by sodium-rich

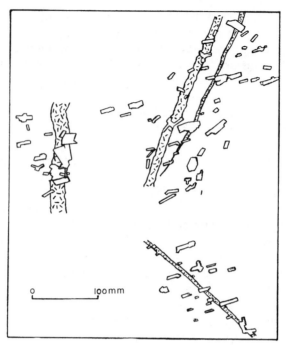

Figure 5.6 Granodiorite (blank) of the Thorr pluton cut by aplo-pegmatite veins (stippled). Potash feldspar megacrysts, showing a crude alignment, apparently overgrow both vein and host. Horizontal outcrop, 800 m east of Meenatotan, Donegal, Ireland. Reproduced from Pitcher and Berger (1972) with the permission of John Wiley and Sons.

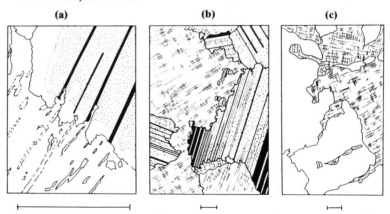

Figure 5.7 Microcline–plagioclase relationships in the Thorr pluton. (a) Interdigitation interpreted as microcline microperthite invading weakly zoned plagioclase; (b) microcline interpreted as replacing and enclosing relicts of plagioclase; and (c) microcline–quartz relationship – one interpretation is that quartz is invading microcline. Scale bars 0.25 mm. Drawings by E.H.T. Whitten in Pitcher and Berger (1972).

plagioclase, often referred to as the rapakivi texture. In the light of theoretical and petrological studies by Hibbard in 1981, Bussy in 1990, Wark and Stimac in 1991, and Nekvasil in 1992, it would seem that a model involving decompression of an ascending, crystal-bearing magma that is changing composition by mixing with a mafic magma, offers an adequate explanation, certainly to me because it recognizes the importance of mixing in the dynamic environment of magma upwelling.

MORE ON MYRMEKITE

This shifting debate on the origin of potassium feldspar megacrysts is paralleled by that concerning the origin of myrmekite. The rims and warty lobes of intergrowths of plagioclase and quartz that invade potassium feldspar from its grain boundaries with matrix plagioclase have puzzled petrologists since their description as myrmekite by Michel-Lévy in 1874. By the turn of the century this particular debate had divided into the usual bipartisan line with Becke favouring a metasomatic origin, while Schwantke moved towards an exsolution model based on the presumed presence of a hypothetical molecule, $CaAl_2Si_6O_{16}$, held in the potassium feldspar. A consensus view has now emerged favouring replacement, though another contrary hypothesis, involving direct crystallization from a residual fluid, was advocated by Hibbard in 1979. Of course, we might again suspect that the process is multifactorial, especially as several phases of myrmekite formation are often discernible in the same rock section. However, it is clear that myrmekite is related to grain boundaries or to fractures within grains, with the implication that the feldspar matrix was largely established before its growth. The textural evidence of replacement is convincing, with the cauliflower-like lobes representing an advance into the potassium feldspar. Furthermore, the lobes are largely independent of exsolution lamellae.

It has long been known that the plagioclase of the intergrowth and of the contiguous matrix are in optical continuity, and a study by Hopson and Ramseyer, involving cathodoluminescence microscopy and microprobe analysis of one good example, confirmed that the plagioclases are indeed continuous and identical. This is surely good evidence of heterogeneous nucleation on an existing surface.

The volume for volume replacement with its co-precipitation of quartz cannot wholly be explained by exsolution from the host potassium feldspar, as it requires a net gain of sodium and calcium with a loss of potassium from the system. However, such a localized metasomatism is not difficult to envisage in the latest stages of crystallization, and is in accord with the obvious enhancement of myrmekite growth during the increased hydrothermal activity associated with ore mineral deposition.

With respect to the geometry of myrmekitic intergrowths, we can note that Shelly in 1970 likened the structure to that of rod eutectics in metals. He proposed that it results from the need to conserve energy by favouring low surface areas when quartz is constricted during its growth within plagioclase: rods are superior in this respect to planes. Interestingly, the vermiform rods can often be seen to end at the growth front in the form of quartz drops within the potassium feldspar host, suggesting that the rods advance by coalescence of quartz spheres, the most energy conserving shapes of all.

Deformation also favours the formation of this intergrowth, as is indicated by its association with the mortared structure of deformed grain boundaries. In this connection Simpson and Wintsch explained the likely role of strain energy, suggesting that the high concentrations of tangled dislocations at grain boundaries localize myrmekite growth: perhaps these are the 'smekal defects' mentioned by Drescher-Kaden. Simpson and Wintsch opined that although the strains involved in the growth of the matrix crystals may be sufficient in themselves to disturb the local chemical equilibria, any applied stress would greatly enhance this effect.

However, on the assumption that the delicate interconnecting quartz vermicules could hardly survive much strain, Vernon in 1991 considered that the role of deformation is only indirect, simply facilitating the access of fluids to the growth front and so changing the local chemical environment. He envisaged the growth of myrmekite as occurring ahead of an advancing front of deformation, but even so the association with deformation remains, and it is clear to me that applied stress stimulates this and a variety of other solid-state interactions, not least the reordering of potassium feldspar with the conversion of orthoclase to microcline.

RIMS AND SWAPPED RIMS

The contacts between the feldspar grains of granites are often rimmed by albite, the formation of which presents a similar genetic problem to that of myrmekite, with which these rims are often associated, that is, whether the material is exsolved or is the result of introduction along grain boundaries. The observation that such rims are often in optical continuity with, and pass into, perthitic lamellae, supports the exsolution thesis, but again I suspect that both processes are involved, with intergrain, pressure-induced solution playing an important part.

Between the grains of potassium feldspar and plagioclase the albitic rim extends into the former and is in virtual continuity with the latter. However, where potassium feldspars ajoin, albitic rims encroach and lobe into both grains from the common boundary. Each rim is in optical continuity with the exsolution lamellae of the *opposing* feldspar grain, hence the term 'swapped rims'. The cathodoluminescence technique shows up such textures to perfection as has been illustrated by Rae and Chambers (1988, figure 4). It seems that unmixing of the alkali feldspars is preferentially located along the grain boundaries, being aided by the percolation of late-stage fluids along the latter, when each albitic rim nucleates and grows from the surface of the external grain.

It would be surprising indeed if granite textures did not respond to the long periods of cooling down to the ambient country rock temperature, during which they must have been often permeated by hot waters. As Wickman, Levi and Åberg pointed out in 1981, many of the textures described here provide good evidence of the low grade, burial metamorphism which is more apparent in the adjacent country rocks.

Having succumbed to the temptation to review these all too familiar late stage textures of granites at some length, I turn finally to just one other which has now assumed a new importance – that is, the fine-grained groundmasses of porphyritic microgranites.

SOME SPECIAL FEATURES OF MICROGRANITES

MICROGRANITIC TEXTURES

Porphyritic microgranite marginal and roof phases are not an uncommon feature of granitic plutons the world over, especially where there is an attendant mineralization and a retinue of porphyry dykes, aplites and pegmatites. Such fine- to medium-grained textures are most easily explained by a reduction in water pressure elevating the liquidus when, as a result of the ensuing crystallization, vapour is suddenly vented into fracture systems of the uppermost crust.

This is the likely explanation of most fine-grained contact facies: they do not represent actual chilling. The way in which large potassium feldspars and sheaves of muscovite hang from and grow away from hanging walls into an aplitic microgranite, exactly as in aplo-pegmatite dykes, implies that volatiles have concentrated on the hanging wall, and in re-entrants in that wall, leaving a relatively exhausted melt in the immediate area. The effect is to promote growth at the fringe and nucleation inside it.

Fine-grained variants are not restricted to contact zones, however. Considerable bodies of potassium feldspar megacrystic granites are often texturally heterogeneous, with groundmasses consisting of a fine- to medium-grained equigranular mosaic of quartz, microcline, albitic plagioclase and biotite flakes, compositionally representing a more highly evolved magma than the corroded and partially resorbed feldspar and quartz megacrysts.

Such a bimodal, disequilibrium texture is especially characteristic of the tin-bearing granites of Southeast Asia, where the studies of Pitfield and Cobbing and their colleagues in 1990 revealed a continuous spectrum from a single phase, and presumably primary textured granite, through a two-phase variant, approximating to a granite porphyry, to a microgranite. On the basis of recognizing crystal bending and fracture, microbrecciation, microveining, corrosion and replacement, these workers interpret the progression in terms of a transformation of a largely consolidated coarse-grained granite, the megacrysts representing xenocrysts residual from the explosive disintegration of the primary texture. Although it is true that the groundmass texture is identical to the fabric produced in the quench experiments of Swanson (1977) and other workers by the sudden reduction of water pressures, nevertheless the evidence of the disaggregation of a primary crystal framework and its partial incorporation into an evolving residual melt convinced Cobbing and his co-workers that the process is more complex, involving the formation of fluidized mixtures. If so, the progression may well represent a change in the rate of flow of penetrating gases, a fixed fluidized bed evolving into a turbulent fluidized bed. Subsequent recrystallization and re-equilibration is suggested by the regrowth of the xenocrysts by encroachment of the polygonized groundmass.

Such a mechanism was early proposed, in 1954, by Doris Reynolds at the time that fluidized bed technology was first being developed, when she introduced the term 'tuffisite' for the resulting rock. Fluidization processes would naturally result from the pressure build-up on the concentration of late stage fluids and gases sealed within the carapaces of many high level plutons, and we can expect to find evidence of even more dramatic changes as the fragments and volatiles vigorously interact. Indeed, Martin and Bonin believe that the late stage concentration of water may sometimes induce a near complete refusion. Furthermore, because the vapours involved in these disruptions scavenge and concentrate the metals in such highly energetic systems, these textures are an important indicator of attendant mineralization.

IMPORTANCE OF PETROGRAPHY

I have only touched fleetingly on the textural aspects of granitic rocks, though perhaps sufficiently to demonstrate that we avoid the complexities of texture at our peril. Petrographic examination has been an essential element of many of the granite studies that I now turn to describe but, regrettably, its central role is not everywhere appreciated. It is, after all, the modal and textural changes which provide the primary evidence for the diversity of granitic rocks, the causes of which will be the continuing theme of this book.

SELECTED REFERENCES

Bryon, D.N., Atherton, M.P. and Hunter, R.H. (1995) The interpretation of granitic textures from serial thin sectioning, image analysis and three-dimensional reconstruction. *Mineralogical Magazine*, **59**, 203–211.

Dowty, E. (1980) Crystal growth and nucleation theory and the numerical simulation of igneous crystallization. In: Hargraves, R. (ed.), *Physics of Magmatic Processes*. Princeton University Press, Princeton, NJ.

Pitfield, P.E.J., Teoh, L.H. and Cobbing, E.J. (1990) Textural variation and tin mineralization in granites from the Main Range Province of Southeast Asian Tin Belt. *Geological Journal*, **25**, 419–430.

Vernon, R.H. (1986) K-feldspar megacrysts in granites – phenocrysts, not porphyroblasts. *Earth Science Reviews*, **23**, 1–63.

Vernon, R.H. (1991) Questions about myrmekite in deformed rocks. *Journal of Structural Geology*, **10**, 979–985.

Differentiation in granitic magmas: zoning as an example of multifactorial processes at work   6

The question, to which American geologists especially were, in general, somewhat insensitive, was whether this magmatic differentiation model outlined an actual sequence of events which really operated in nature to produce granite.

Matt Walton (1965) Science, **131**, 635.

ASPECTS OF DIFFERENTIATION IN GRANITIC MAGMAS: A MULTIFACTORIAL PROCESS

It is likely that a considerable number of processes are involved in the diversification of magmas, the most important of which are fractional melting, fractional crystallization, restite fractionation, magma mixing and assimilation. Only minor roles can be expected of immiscibility, thermogravitational diffusion and volatile mass transfer. None is exclusive but nevertheless we may expect that each will take the lead at a particular stage of magma evolution. Thus fractional melting may well dominate the generative process, with assimilation, mixing and fractional crystallization variously controlling the path of the subsequent differentiation, leaving diffusion and immiscibility relegated to the late stage. Nevertheless, one might hope to identify the predominant process from differences in the behaviour of the minor, if not the major, elements in such different physical chemical systems. Of prime importance is the ability to distinguish between differential partial melting and crystal fractionation, and between either one of these and magma mixing.

Appropriately, advances in the understanding of trace element behaviour in melt–crystal systems have taken the form of mathematically defined, model processes which can be used quantitatively to test the merits of competing hypotheses. Thus, early on in the development of such schemes, a test for fractional crystallization was proposed by Allègre and Minster (1978), partial melting by Shaw (1965), assimilation combined with fractional crystallization by DePaolo (1981a, b), mixing by Langmuir and others (1977), and cumulate formation by McCarthy and Hasty (1976). The search has continued for such tests as illustrated by the studies of Maaløe and Johnson (1986), who suppose it possible, on the basis of the relative concentration of the incompatible elements, to distinguish between magmas derived by extraction from migmatites or otherwise by batch melting or accumulative processes. However, it is salutary to recall that many of the necessary manipulations depend on assumptions regarding the values assigned to the partition coefficients, which are known to vary with the physical parameters such as oxygen fugacity, temperature and concentration. Wayne Sawka's analysis of the strongly zoned McMurry Meadows pluton, California, illustrates well the difficulties encountered in such modelling in the face of large variations in the accessory minerals – the main repositories of the REEs, changes in the paragenetic sequence, and also in the values of the partition coefficients. Nevertheless if the repositories of the REE can be identified, the relative abundances, properly plotted on the so-called spider diagrams, are capable of discriminating between possible generative processes in simple cases. For example we are led to expect that a strong depletion in the heavy REE signals either the important involvement of garnet in the fractionation process or its presence as a residue from melting in the source region. Another example, and one of particular importance in granite studies, concerns the measure of the europium anomaly as an indication of the importance of both plagioclase fractionation and the oxidation conditions in the source region. I shall have cause to refer to such usages as this account unfolds: the detail of the methods is the stuff of text-books.

It is, of course, unlikely that geological systems are sufficiently singular or sufficiently closed to be simply modelled. For this reason and in appreciation of the complexities, Langmuir proposed a dynamic model of the melting process in which a proportion of melt at any stage is retained mechanically trapped within the source material, to mix then with the next-formed liquid. Similarly O'Hara treated the crystallizing process as an open system, envisaging melt being periodically drawn off as a contribution to volcanism, with the magma chamber replenished by new, hot, unfractionated liquid. Combinations of such dynamic systems offer powerful and flexible models, but we need to be aware that the various processes may or may not have sufficient time to run to completion. Hildreth and Moorbath appreciated this time factor in their composite model involving melting, assimilation, storage and homogenization, conveniently designated by the acronym, MASH.

SEPARATING OUT OF CRYSTALS FROM LIQUID OR LIQUID FROM CRYSTALS

With the exception of a few special situations pertaining to fluorine, boron or water-enriched magmas, I have never seen convincing field evidence for the bulk settling out of crops of crystals in granites, though I cannot pretend that this is everyone's experience. On the other hand, the frequency of steeply dipping compositional zones on all scales is good evidence for sidewall accretionary crystallization.

I have already stated my belief that most granitic magmas rising into the crust develop a sufficient yield strength to prevent the settling out under gravity of both crystals and small enclaves (p. 51), the physical reason being that they become suspensions of interacting crystals; also perhaps the highly siliceous, polymerized melts have themselves non-Newtonian properties. Furthermore if such magmas undergo a multitude of small-scale turbulences, as suggested by Sparks, Huppert and Turner, then crystals might well be maintained in agitated suspension. As might be expected there is some debate about this, as noted by Martin and Nokes, and the expected swirling patterns are only occasionally preserved in outcrop – but then spectacularly as in the Cornish granites. However, I remain convinced that gravity-aided settling is an inadequate mechanism in those granitic magmas bearing substantial concentrations of suspended crystals.

From their modelling of magma chambers Sparks and his colleagues have proposed an alternative explanation for the separation of fluid from crystals. They found that the local compositional variation consequent on crystal growth reduces the density of the immediately adjacent melt sufficiently for this melt to rise away and exchange for an aliquot of the bulk melt. The overall effect is to produce a compositional grading in the liquid during crystallization outboard from all boundary surfaces, not least the sidewalls, whereby the continuous exchange of elements occurs across the boundary layer between the intercrystalline pore fluid and the main body of magma, that is, the solidification front. Moreover, such an exchange is more effective in the transfer of material than diffusion, the very low diffusivities of silicate melts restricting the latter to the crystal–liquid interface. Here then is an adequate explanation of compositional zoning without appealing to the heavy precipitation of crops of crystals strewn over crystal floors, and it is one which equally applies to sidewall accretion.

It is a concept that has been part of petrological thought for a century, being enunciated in various guises by Becker in 1897, Pirsson in 1905, and Grout in 1918, although it has only recently come to the fore with the advent of a new understanding of fluid dynamics, the borrowing of ideas from the study of metals and the contribution of simulation experiments

(Marsh, 1996, p. 22). This explanation carries with it an additional degree of freedom in that it is not only the early phenocrysts that then control chemical evolution, but the entire crystallate. The various laboratory experiments that led to the present proposals also showed how complex are the convection dynamics of a multicomponent system involving saturated aqueous solutions of salts, and we can hardly guess how much more or less complicated are multicomponental siliceous melts with their much greater viscosities, especially when loaded with crystals entrapping a melt which is continually interreacting with the matrix.

In an extension of the saturated aqueous solute approach, Tait and Jaupart were able to increase the viscosities to nearer those of silicate melts and so demonstrate that not only did compositional convection still operate, but that it involved a 'field of plumes' rising from the crystallizing region. An even greater realism has been achieved by Seedman and Donaldson, who observed compositional convection in experiments involving a basaltic melt seeded with olivine crystals. Moreover, they were able to describe the actual detachment process, which, in one set of conditions, was marked by low viscosity plumes arising away from the surfaces of olivine crystals!

Compositional convection is clearly a viable physical process. Nevertheless Marsh, in his Hallimond lecture of 1996, remained cautious in accepting the evidence as it stood as an accurate model for real magmas, pointing out that the solidification fronts in magmas are fundamentally different from those of metal alloys and aqueous solutes, with the dendritic growth in the latter contrasting with the more complex three-dimensional clusters, ganglia or strings in magmatic systems. Clearly there still remain many problems in determining the precise role of compositional convection in the evolution of a real magma chamber.

In attempting to circumvent the difficulties inherent in the experimental approach many workers have resorted to the theoretical modelling of the processes operating within this mushy layer at the solidification front – a layer likely to include those crystallinities between 25 and 50% (Marsh, 1996, p. 8). Concerning the form of this layer we have to ask difficult questions such as how deeply does the interstitial melt circulate within it in the process of replacing the nutrient-rich melt which sustains growth in the mush, and at what stage in the increasing crystallinity will circulation be halted by closure of the interstitial channelways. If circulation is envisaged as penetrating deeply, perhaps down to the linked, rigid matrix of >50% crystallinity, then we need to differentiate carefully between compositional convection in the boundary layer and this porous medium convection inboard of the boundary. In the former it is the viscosity of the melt which is important, while both the viscosities of the melt and the matrix need to be considered in the latter. Such an approach immediately focuses on the possibility that the rate of extraction of the melt is controlled more by the viscosity of the matrix than the melt itself, especially when compaction or shear motivates melt migration.

We then need to understand the fluid mechanics of the percolative flow of melt through a viscously deformable, permeable matrix, and to be able to integrate the density difference – which expresses buoyancy – with permeability, the viscosities of both the fluid and matrix, the thermal diffusibility and, not least, the thickness of the crystal mush. In essence this integration can be expressed in the form of a ratio between the destabilizing buoyancy and the stabilizing forces that oppose motion, and is represented in formulations as the dimensionless Rayleigh number, or numbers when both melt and matrix are involved. Such a calculation, together with simulated experiments, does provide a threshold *matrix* R number below which percolative flow is unlikely, so that by utilizing such a formulation and feeding in physical parameters determined from the textural and compositional studies of a specific natural example, together with appropriate experimental data, it ought to be possible to assess the likelihood that fluid flow could have operated. For example, utilizing the granodiorite of the Lima superunit mentioned previously (p. 70), Cheadle (in Bryon *et al.*, 1996), generated *matrix* Rayleigh numbers 1.5 orders of magnitude lower than the theoretical threshold and was thus led to reject melt migration through the matrix lattice as an explanation for adcumulus growth. Be this as it may interstitial growth often demands refreshment of the melt, and either the formulations are inadequate or accessory channelways are involved independent of the interstices themselves, perhaps represented by the 'chimneys' reported in experiments.

Of recent years this modelling has reached a high degree of sophistication, as, for example, in the works of Sleep (1974), McKenzie (1984), Ribe (1987), Worster (1991) and Spiegelman (1993), with the last three providing useful reviews. These are not matters I easily understand, especially when I learn that different authors define the boundary conditions differently to suit selected conditions. I am certainly at a loss when such authorities as Huppert and Turner (1991) and Marsh (see discussion and reply, 1991) disagree so profoundly, especially as their very different views on the physical conditions within a magma chamber are particularly pertinent to the present discussion. Whereas Huppert and Turner envisage vigorous, whole chamber convection with crystallization occurring simultaneously everywhere in the system, Marsh considers that heat transfer is so efficient within the chamber that initial temperature differences are quickly erased and any wholesale convection is short lived. Crystallization is then largely restricted to the advancing solidification front.

Almost intuitively I incline to many of Marsh's views, particularly on the issues of convection and convective flow. The simple fact that felsic magmas are not superheated, coupled with the excellence of their insulation within rapidly dehydrated aureoles deep in the crust, ensures that magma chambers will lack the initial temperature difference required of Rayleigh–Bernard convection (see also Bergantz, 1991, p.33). However, we need to remain aware that any simple model is likely to founder before the reality of a continuing recharge by fresh, hot magma!

There are many other relevant matters discussed in Marsh's Hallimond lecture (1996), including the nature of the solidification front and the way in which this is envisaged as capturing any suspended phenocrysts (p.11). But I am not sure that any of the theorists can adequately explain the remarkable homogeneity of crystal size and texture that is so often a feature of individual pulses within granitic plutons, though the concept of the steady inward advance of a solidification front, across which there are only minute changes in temperature on the liquidus, offers a possible explanation.

I especially warm to Marsh's contention that the key to protracted differentiation through fractional crystallization is *not* crystallization in stationary, closed chambers, but the repeated transport and chambering of magma, or the periodic resupply to chambers of phenocryst-rich magma. As we shall see granite plutons are certainly multipulsal, with much evidence of replenishment even in the separate pulses, but a suffi- cient residence is required to permit the winnowing of either the early formed crystals from pristine magmas or the restite from crustal remelts. Perhaps this is the most important role of thick continental crust in the whole of this complex process. As Marsh suggests, magmas ascending through a stack of interconnecting channel-like chambers may hesitate or even come to rest, when strong fractionation would be most likely, with a resulting 'gravel bed of crystals' through which fresh or separated magma would need to percolate, the whole functioning as a fluidized bed (Marsh, 1996, p.28). These are ideas and problems that we will need to return to, especially in seeking to account for the space required for such a stack of chambers in a crust rarely more than two or three times the expected thickness of a single pluton. Perhaps these conduits are thin and sheet-like and not at all like the great plutons we have been led to expect.

CONVECTIVE MOTION AND MINERAL ORIENTATION

If convective motion in granitic magma chambers were a major process it ought to be recorded in the orientation of minerals and schlieren. Unfor- tunately the significance of the latter is far from clear cut, if not wholly equivocal. However, despite my earlier remarks it would be unfair not to present the rather contrary view of Schermerhorn, who stresses the sig- nificance of schlieren, claiming that because they represent primocryst accumulation they provide 'abundant evidence that internal flow does indeed occur in granitic magmas' (1987, p.617). On this basis he vigor- ously advocates a leading role for large-scale convection in the accumulation of crystals and their separation from the melt. Schermerhorn cannot be wrong in reminding us that such a mechanism could operate in nature, but it is my view that most schlieren are neither of simple cumulative origin nor sufficiently common in granites as to assume a lead role in the overall interpretation.

As for the simple arcuate patterns picked out by mineral foliations, especially in the peripheries of plutons, I later substantiate my view that these more often represent the traces of fairly late stage synplutonic deformation than the early pluton-scale flowage associated with major convective overturn. However, the localized swirls picked out by potassium feldspar megacrysts in the cores of a few plutons, such as, for example, those of Dartmoor in southwest England and Serra da Estrela in Portugal, may represent the small-scale turbulent flow that the studies of Sparks and co-workers lead us to expect. It is the aggregate effect of this much smaller scale convective flow coupled with compositional convection that provides the real alternative to crystal settling, with the melt moving away from the crystals, not the reverse.

CO-LATERAL PROCESSES: THE RECHARGE MODEL

Although the separation of liquid from crystals must provide a central explanation of differentiation, other mechanisms are also clearly involved. Thus the strong evidence for magma mixing at all stages in the evolution of certain types of plutons strongly supports the O'Hara model involving the continuous recharge by magma – quite often basaltic – fed into the base of a magma chamber through swarms of dykes. Such a continuous addition of hotter magma will maintain the temperature in balance and thus provide sufficient time for adequate mixing with the rest magma deriving from fractional crystallization.

CO-LATERAL PROCESSES: ASSIMILATION

However, although this recharge model can explain many variations within rock suites, there is often compositional evidence for the assimilation of recycled crustal rocks. This may result from the direct addition of country rock material to a magma, but perhaps the more fundamental process is the partial melting of such crustal material at depth. There is here the possibility that a sequence of varying melt compositions will result from the progressive unmixing of melt from solid restitic material. Such partial melts may constitute magmas in their own right or form mixtures with other magmas.

The possibility of direct assimilation of country rock, especially in the low thermal energy environments of the upper and middle crust, is now often subordinated to a mere complication. I must admit to agonizing over its relative importance for many years, as any examination of my work will show, but in the event it seems obvious that the addition of any significant amount of foreign material will quickly freeze the contact zones of most granitic magmas despite the release of heat of crystallization, a fact most clearly demonstrated by innumerable examples of the retention of unmelted xenoliths in wall and roof situations. This is not to say that the melting of wall rocks and xenoliths never occurs at high levels in the crust; it does in some subvolcanic magma chambers and associated diatremes, but it is not the rule. Even when there is good field evidence of a limited assimilation of solid material we are told on the good authority of Bowen in 1928 and McBirney in 1979 that the addition of most crustal materials will not drastically change the composition of a fractionating system and that smooth variation diagrams will accordingly result.

Overall, it seems that the energy requirements are such that significant contamination by crustal materials can occur only near the source of remelting. Halliday, in 1985, even argued that the evidence of crustal contribution, as implied by a progressive increase in the appropriate isotope ratios in some zoned plutons, is a good indication of the deep source of their magmas. Conversely, the lack of such a variation is one possible sign of differentiation *in situ*!

LATE STAGES IN THE DIFFERENTIATION PROCESSES

The models just discussed concern differentiation before or during the main phase of crystallization, but it is clear in the field that the separation process is somehow continued to a very late stage in the consolidation history. Often felsic melt can be seen to have segregated along the walls and roof of a pluton, and also into flat-lying sheets in its upper part, showing that it must have separated out after a sufficiently coherent crystal matrix had formed to sustain brittle fracture, at least over short timescales.

We learn from the experiments of Jurewiez and Watson that interstitial melt does not easily collect and segregate, so that such an extraction requires a mechanism such as crystal compaction or squeezing – that is, the filter-pressing so often invoked by the early writers. A more modern version, suggested by Mahood and Cornejo from their studies of the La Gloria pluton of Central Chile, envisages flowage of the interstitial rest melt into tension fractures that are continuously forming and rehealing as the crystal matrix is deformed synplutonically. Another possibility at this late stage of congelation is that contraction on cooling, when coupled with the tensional forces consequent on central subsidence, will open voids into which melt is effectively sucked. Furthermore, the developing high vapour pressure will assist in expelling fluid. Indeed, with such a variety of possible mechanisms it is surprising that such a segregation is not more widely observed, though Mahood and Cornejo have pointed to the possibility that a back-mixing process may often resorb the liquid.

The merits of a variety of these proposals were discussed by Miller and co-workers, Petford, and Wickham (in 1988, 1993 and 1987, respectively), with the conclusion that all require special circumstances, or are physically or geologically unrealistic. I do not necessarily agree, but I am attracted to the Miller group's proposed alternative. Learning from studies of liquid phase sintering, they envisaged that the first-formed, self-supporting framework of crystal grains will compact under gravity to the extent of expelling some portion of the melt fraction. The efficacy of this gravity compaction model is uncertain but at least neither it nor any other of the extractive models contradict the mechanism proposed by Sparks, so that we can now pass on to consider the causes of compositional changes in real plutons.

GENETIC SIGNIFICANCE OF ZONING IN PLUTONS

I agree with Shimizu and Gastil in their contention that most granitic plutons are compositionally zoned, even though in some the available data may not be of sufficient sensitivity to detect the variation. Perhaps surprisingly, it is the measurement of magnetic anisotropy that turns out to be particularly useful in this respect, as reported by Bouchez and Gleizes in their re-examination of the zoned plutons of the Pyrenees. With the caution that it represents just one magma-tectonic environment, a useful estimation of the frequency and degree of zonation is provided by Nozawa and Tainasho's survey of 71 granitic plutons in Japan. Setting aside the eight of these that are clearly nested ring complexes, and four others that are reversely zoned, 30 of the remaining are strongly zoned in the normal sense, the rest only weakly. In the remaining four it is possible to record an upward zoning.

It would seem that such zoning can arise by a variety of processes: sidewall accretion, recharges by mafic magma, or multipulse injection of melts either fractionated elsewhere or derived by successive remeltings. Rarely is marginal assimilation a significant factor. Although the marginal accretionary process is probably the most important cause of zonation, I suspect that a combination of magma mixing and multipulse injection often offers a viable model.

If zoning is viewed from the point of view of magma-tectonic environment, a clear interrelationship emerges whereby the rheological differences between the magma types and the availability of contrasted sources provide crucial controls of the process of variation. Thus zoning is particularly evident in I-type granitic suites, presumably as a result of the fluidal nature and the dual input of material from mantle and crustal sources. In contrast, zoning is relatively weak in those S-type suites involving mushy magmas derived as partial melts from metasedimentary crustal rocks; what zoning there is here is likely to be due to differing compositional draughts from the remelting of the source rocks. Finally, in view of the superior mobility of fluidal A-type granitic magmas, it is hardly surprising that the zoning here is most often expressed by multiple intrusion representing repeated draughts from an evolving magma source at depth.

PATTERN OF ZONING

The general sequence of lateral zoning inwards from the periphery most generally involves a change from quartz diorite, through granodiorite, to granite, with gabbro occasionally forming an outer rim. Where basic enclaves occur they are almost invariably concentrated in the outermost granitic unit. This mafic to felsic inward variation can be expected to result from accretionary processes, whether occurring at emplacement level or in the form of draughts from an evolving, deep supply chamber. Furthermore, because magma mixing is likely to be most effective in the early stages of magma generation it will also produce a rock series in this 'normal' order. But reversed zoning does sometimes occur and is best modelled as the type of multipulse filling of a pluton where each magma pulse, though derived from the same compositionally layered magma body at depth, arrives in the reverse order, possibly as a result of subsequent remelling from bottom to top, of the layered body by the intrusion of fresh, hot magma into its base. Allen, in her 1992 paper, usefully discusses this and other models in the light of her studies of the reversely zoned, yet multipulse, Turtle Pluton of southeastern California.

Of the detail of the zoning, which varies between plutons, I can do no better than report Nozawa and Tainasho's summary statement concerning the mineralogy and chemistry of Japanese examples as being generally applicable. The main inward changes concern an increase in the modal proportion of quartz and potassium feldspar and a decrease in the mafic minerals, both of which are strongly reflected in the abundances of the major elements. The proportion of plagioclase remains little changed, though the An content generally decreases as might be expected of a process dependent on fractional crystallization. These mineral changes are clearly reflected in the chemical changes such as in the europium anomaly and the iron:magnesium ratio, the former resulting from the early precipitation of calcium plagioclase, the latter from a change in hornblende composition, itself a consequence of a falling temperature gradient towards the core of a pluton. Furthermore, it is hardly surprising that the incompatible elements increase with the differentiation index while the compatibles decrease.

When in the 1950s it first became possible to analyse serial traverses of zoned plutons, the Thorr pluton of Donegal presented myself and my colleagues Whitten and Mercy with an ideal example for modal and geochemical studies (Pitcher and Berger, 1972, p.96 *et seq*). In the event I was over-impressed by the outcrop evidence for the incorporation of country rock material and mistakenly interpreted my colleague's analytical results as evidence of the overriding importance of marginal assimilation. At about the same time, in 1967, but far away in Oregon, Taubeneck presented a particularly good study of the Cornucopia stock which enabled him to dispense with the then current marginal assimilation model because the zoning within the Cornucopia pluton showed no relation whatsoever to the country rock envelope; neither was there any evidence of mixing with basic magma. According to Taubeneck the pattern of modal and chemical zoning, coupled with the changes in hornblende mineralogy and texture, strongly supported a thesis involving fractional crystallization within a volatile-rich magma that crystallized inward from the margins. Taubeneck searched for a process 'that can draw off residual liquid after partial crystallisation', concluding that the compositional gradient originates largely by fractional crystallization, with material exchanging between the interstitial liquid and the core magma by diffusion through a progressively water-enriched medium. This model was, at about the same time, enthusiastically advocated by Vance of New Zealand in 1961, and if we now substitute convective fractionation for diffusion, we can see how this model converges with the current view.

A SPECIFIC EXAMPLE OF SIDEWALL BOUNDARY LAYER DIFFERENTIATION

A particularly good example of the current model for the accretionary process is provided by the study by Sawka, Chappell and Kistler (1990) of the Tinemaha pluton in the Sierra Nevada. This pluton is both horizontally and vertically zoned over some 1000 m and all the compositional variations are directly related to its sides. It is envisaged that sidewall crystallization created a boundary layer of fractionated liquid outboard of an advancing wall of crystallate in the manner discussed in the foregoing. Less dense than the bulk magma, this buoyantly moved upwards to be continually replaced by liquid from the main bulk of convecting magma. The vertical zoning is thought to be due to this light liquid collecting in the roof zone of the pluton.

The convective interchange between crystallate and melt, and between layers of melt of different density, was probably rather complex, involving differing rates of diffusive heat exchange and compositional diffusion, but I like to believe that the interchange was effected through operation of a myriad of relatively small and ever-changing convection cells rather than a pluton-scale convection system.

Although I find this overall model attractive, I am not convinced by the authors' claim that certain compositional peaks within the zonation of the Tinemaha pluton, and even a distinct compositional gap in the zonation of the intersecting McMurry Meadows pluton, can be accommodated within the internal accretionary model. I suspect that such hiatuses are more likely to result by multipulse intrusion, a general prejudice I defend below.

SOME OF THE COMPLEXITIES OF ZONATION

Concerning the form of the zonation we have to accept that opportunities are limited for observation in three dimensions, though it seems that zonal boundaries, especially where they are marked by actual contacts, most often dip steeply in parallel with the outermost contact. Occasionally, however, favourable topographic relief reveals a flat-lying zonation in which the texture and composition are particularly well-marked with, for example, a coarse, hornblende–biotite granodiorite passing up into a finer-grained, biotite leucogranite containing abundant miarolitic cavities. At this high level, such a pluton is often intersected by flat-lying sheets of aplo-pegmatite which are completely without feeders, and which have obviously segregated *in situ*. Of such vertically zoned plutons, that of Tumaray in Peru provides a splendid example from my own experience (Figure 6.1).

That a particular zonal pattern was formed *in situ* is not easily proved. A diligent search often reveals internal contacts of various types and of varying significance. Although some mineralogical and textural changes may simply denote a mineral stability change during cooling and crystal-

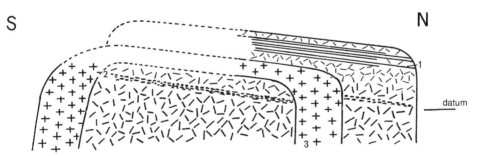

Figure 6.1 Controlled sketch section through the Puscao pluton of the Huaura Centred Complex, Peru. Coarsening of stipple represents increase in colour index in a downward transition from granite to granodiorite. Depicted are: (1) the aplogranite sheets in the roof zone of the pluton; (2) a 300 m thick layer of mafic enclaves; and (3) a later intrusion of the Sayán Granite. Vertical and horizontal scale about 10 mm = 1 km. The whole pluton has apparently been tilted about 15° to the north.

lization, others represent either rapid transitions or zones of magma inter-mingling, neither of which are likely to represent long time-intervals between the magma pulses. Perhaps the most tantalizing of such internal contacts is marked simply by an abrupt ending of a swarm of synplutonic

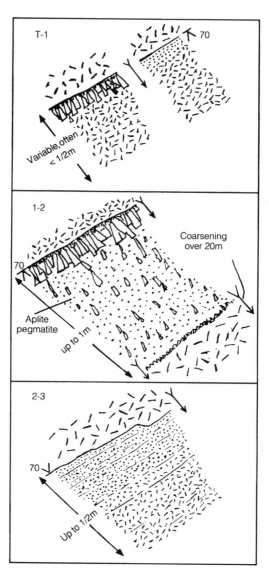

Figure 6.2 Types of internal contacts within the Rosses pluton, Donegal. T-1, outer contact; 1–2, contact between pulses 1 and 2; 2–3, contact between pulses 2 and 3. Stipple depicts changes of grain size. Triangle-like shapes depict potassium feldspar crystals; arrows indicate younging direction.

dykes entering from outside a pluton. However, even the vague, discontinuous, will-o'-the-wisp contacts that so often plague the conscientious surveyor have a story to tell, in this instance of small-scale magma surges which we would expect in the dynamic system presented by the models of both Sparks and O'Hara and their respective co-workers. Many other more obvious boundaries seem to represent a significant flow past of magma in that they are accompanied by a compositional and textural hiatus of some magnitude, and these are most often represented in outcrop by sharp contacts marked by the nucleation of minerals inboard of the interface (Figure 6.2), or by changes in granularity. When the compositional variation between the pulses is much greater than within any single pulse, this may well represent a fresh influx of magma and the establishment of an entirely new magma cell. Indeed, even without such an immediate contact, and with a transition revealed only by a step in a trend surface, Stephens and Halliday in 1980 recognized the existence of such an internal magma cell within the Criffell pluton of southern Scotland. It is from such evidence, repeated world wide, that I have come to appreciate that only small differences in rheology must separate fluidity and rigidity in granitic magmas.

Fresh influxes of magma are most likely where the mechanism of emplacement involves the stoping out and foundering of a central block or blocks, with the consequent upwelling of magma. This is probably the explanation for that extreme form of zoning whereby gabbro forms the periphery, forming an example, as Ayrton suggested in 1988, of an early, 'failed' ring dyke, whereby basaltic magma is pumped from below and intermingles with the granitic pulse that immediately follows on; this is but a gross example of the recharge process. A subsidence mechanism also provides a further explanation for reverse zoning, where the most mafic facies is centrally located, either as a consequence of the foundering of a mafic carapace or from the pumping up into the central position of new, unfractionated magma.

I would hardly regard these multipulsal plutons as truly zoned, but any field survey quickly reveals that, in the formation of zonal patterns, there is every transition between multiple injection and the single pulse, *in situ* process of differentiation. I have been privileged to work on a number of zoned plutons and each revealed some different aspect of emplacement history.

SOME MORE EXAMPLES OF ZONED PLUTONS

In Donegal, Ireland, the four magma pulses of the nested pluton of the Rosses are separated by sharp contacts, and although the inward order of intrusion neatly parallels a change in composition, the separate pulses each appear to be homogeneous, as if representing instant samples of a differentiating deep source, each mobilized by a subsidence event (Figure 6.3a).

In contrast, the inner, second pulse of the adjacent Ardara Diapir represents an independent, zoned unit, in effect an inner and separate melt cell (Figure 6.3b, G_{2A} and G_{2B}). Nevertheless, the zonation within this inner cell is so widely eccentric in its pattern of distribution that it is most unlikely to have developed *in situ,* and may actually represent the continuing magma infilling of a ballooning diapir by magmas in the process of mixing.

The zoned Thorr pluton of Donegal shows a much more continuous zoning, though again eccentric (Figure 6.3c). At first sight the neat rela-

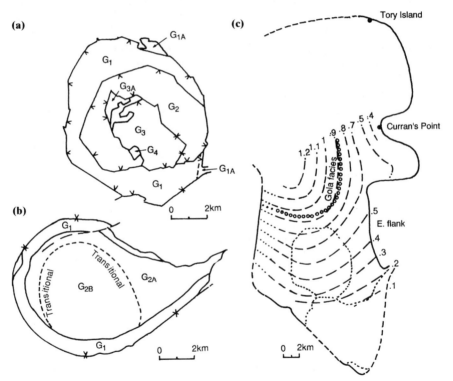

Figure 6.3 Examples of zoned plutons in Donegal, Ireland. (a) Rosses: magmatic pulses G_1, G_2, G_3, G_4 separated by internal contacts. (b) Ardara: outer pulse, G_1, separate from transitionally zoned inner pulse, G_2. (c) Thorr: apparently a continuous compositional change inwards, but zonation eccentric to outer contact, as is well shown by a third degree polynomial trend surface for modal attribute. The most granitic facies is that of Gola. (---) potassium feldspar/plagioclase isopleths; (oooo) transitional boundary of Gola facies; (. . .) boundaries of later granites. (c) Partly after Whitten (1966), reported in Pitcher and Berger (1972). See also p. 130.

tionship between the density of enclosed country rock fragments and the colour index of the host granodiorite suggested that the zoning had resulted from marginal contamination. After much research it would appear otherwise, with crystal fractionation now identified as the most important process, and with assimilation and magma mixing playing but a minor role. However, in view of this loading of the outermost mafic facies with locally derived xenoliths in greatly varying amounts, it seems most unlikely that the eccentric zonation of the Thorr pluton was established *in situ* (Figure 6.3c), so that multipulse infilling of a magma differentiating elsewhere is again envisaged.

There are many examples of eccentric zonation, often with the core facies intruding across the outer boundaries and into the country rock envelope. Thus within the Coastal Batholith of Peru the vast, lozenge-shaped pluton of Santa Rosa is continuously zoned, yet with minor compositional steps marked by mappable, 'flow-past' contacts. Here, too, the outcrop of the penultimate pulse can be mapped as breaking out into the country rock envelope. These discontinuities might simply reflect relatively trivial upwellings of still mobile core magma and, accepting this view, Atherton in 1981 considered that the zoning at Santa Rosa represents *in situ* differentiation, and certainly a large variation within one particular pulse is not accompanied by discernible internal contacts. I am not so convinced, believing that the evidence more truly supports multipulse infilling with a form of compositional variation nearer to that depicted in Figure 6.4b. Indeed there are a substantial number of studies in which the compositions, particularly in terms of the trace elements, indicate that the different pulses are not directly related to a single parent magma (Paterson and Vernon provided a list in their 1995 review).

Of course, there are some single pulse plutons representing a single melt cell. One such is the Cañas pluton of Peru, which William Taylor showed in 1985 to be a compositional sandwich (Figure 6.5) – a particularly elegant example of the use of regression surface techniques. Apparently structurally homogeneous and devoid of enclosures, it is the perfect candidate for modelling as a case of *in situ* differentiation by some kind of convective fractionation. Furthermore, from its emplacement within a ring fracture, it is easy to deduce that the pluton is of pastille-like form, about 6 km in diameter by about 2 km in thickness, so that here, at last, we have a model magma chamber to delight the theorist!

This catalogue of variation and complexity could easily be extended, but the point has been sufficiently made that compositional zoning often results from multiple injection, albeit ranging from mere mobilization of a pluton core to the multiple intrusion of nested plutons. We are rarely privileged to observe an unequivocal example *in situ* differentiation. However, isotopic evidence can redress this deficiency in large measure.

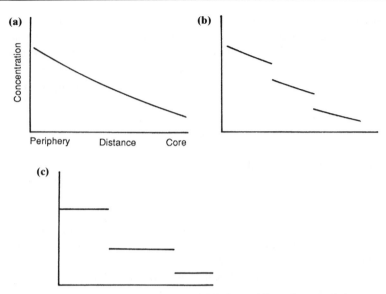

Figure 6.4 Schematic form of the compositional trend lines from periphery to core of a zoned pluton. (a) Continuous, likely to have been established *in situ*; (b) stepped discontinuities produced by periodic upwelling of differentiating core magma; (c) totally discontinuous as a result of multipulse infills.

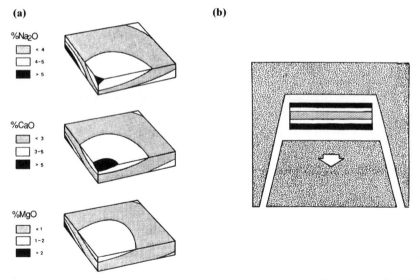

Figure 6.5 (a) Block diagram showing some of the oxide distribution trends within the Cañas pluton. The dimensions of the block are 6000 × 6000 m with the upper face at 1220 m and the lower one at 610 m. After Taylor (1985). (b) Explanation of the oxide distributions in terms of the positions of the sample block within the cauldron subsidence.

ISOTOPE RATIOS AND ZONATION

Perfit and his colleagues' accounts of the 13 Ma-old Captains Bay pluton from the Aleutian island arc are particularly relevant to this discussion, because there is no question here of any input of continental material and, indeed, the chemistry is exactly that of the volcanic country rocks. This particular high-level pluton is zoned inwards from a narrow gabbroic rim through quartz diorite to a granodiorite core. The modal and chemical changes are most consistent with a model of *in situ* fractionation involving a progressive changeover from the accretionary crystallization of calcic plagioclase, lesser pyroxene, olivine and Fe–Ti oxides, to that of sodic plagioclase and hornblende. Perhaps remarkably, in such a situation, the Sr_i ratios do vary by as much as 0.70299–0.70377, but unsystematically, the average value being comparable with those of the arc volcanics, clearly indicating an initial mantle source for the magmas.

Turning from the island arc environment to that of a marginal continental arc we begin to see isotopic variation of a greater degree. As we may expect, such an isotopic zoning does not occur within those plutons within the ambit of the primitive arc with its single source, but only where magmatism has moved into the continental lip with the opportunity for mixed sources or progressive contamination. However, there is a revealing study by Hill and co-workers in 1985 of three zoned plutons from the Peninsular Ranges Batholith of California. Within these plutons a marginal, mafic tonalite changes systematically inwards to a felsitic granodiorite, the abundance of dark enclaves in the former strongly suggesting that basaltic magma mixed with a differentiating magma. In common with the rock suites of circum-Pacific batholiths the suites of these three plutons are identical between themselves in all chemical and petrographical features including the $\delta^{18}O$, but with an important exception that the Sr_i values differ significantly between identical rock facies in each pluton. Hill and co-workers considered that this is incompatible with solidification from a large, closed-system magma chamber, and so were persuaded to adopt the recharge–fractionation model in which a differentiating body of magma intercepted and incorporated all liquids rising into its base from a deeper source region in which the Sr_i was geographically variable.

From the very different environment of the marginal accretionary clastic apron representing the Lower Palaeozoic of southern Scotland, Halliday, Stephens and Harmon, in 1980, examined three zoned plutons emplaced into a thick sequence of greywackes (Figure 6.6a). All three show the characteristic zoning of calc-alkaline rock suites, though variously extended towards the most felsic end-member. As is so often the case microdioritic enclaves crowd the mafic border facies, and these probably represent disrupted igneous material. In the Fleet pluton in particular

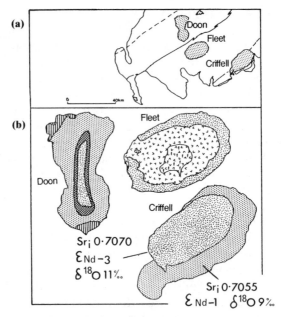

Figure 6.6 (a) Location and (b) petrology of the three main Galloway plutons, southern Scotland, with average Sr_i, ε_{nd} and $\delta^{18}O$ values for Criffell inserted. Diorite (⦀); granodiorite (░); biotite granite (▒); biotite muscovite granite with fine-grained facies (∴); 'hybrids' (▓). After Stephens and Halliday (1979) and Halliday, Stephens and Harmon (1980).

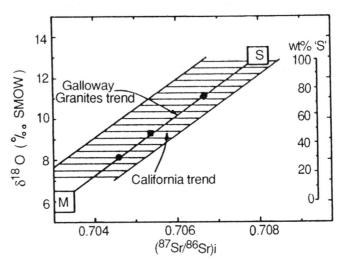

Figure 6.7 Averaged $^{87}Sr/^{86}Sr$ and $\delta^{18}O$ values for diorites (●), granodiorites (■), granites (◆) and sediments of the Galloway area. S = average metasediment; M = inferred mantle. After Halliday, Stephens and Harmon (1980). The scale on the right is a crude calibration of the model mixing line and the barred zone represents the trend established by Taylor and Silver (1978).

there is also the familiar change from metaluminous to peraluminous chemistry, but of particular significance is a zonation in the values of both Sr_i and $\delta^{18}O$ in all three plutons which correlates well with the lithological zoning (Figures 6.6 and 6.7).

On the basis of such findings these workers seek a common petrogenetic process which takes into account the lack of any field and petrographic evidence for local contamination of the magma by the country rock at the level of exposure. Again, it is the isotopic evidence that guides their choice (Figure 6.7), the values from all three plutons defining a linear array which lies between the fields defined by primitive mantle (or lowermost metaigneous crust) and a sedimentary crust of the type represented by the greywacke lithologies of the accretionary apron. This suggests that a sedimentary component was progressively involved in the generation of the magmas, if only because the whole range in Sr_i values, from 0.70414 to 0.71092, is far in excess of any variation to be expected for a purely mantle source.

This contamination can only have occurred deep in the crust, and the possible processes involved have been widely discussed. Taking account of Halliday's latest studies of coupled Sm-Nd and U-Pb systematics, Stephens and his colleagues then provided a balanced review of all these possibilities in particular regard to the Criffell pluton, finally embracing the combined assimilation-fractional crystallization model. Subsequently, in 1992, from a follow-up study of a detailed traverse, Stephens reported mean values for the outer granodiorite of Sr_i 0.7055 and $\delta^{18}O$ 9‰ at 64% SiO_2, and for the inner granite of Sr_i 0.7070 and $\delta^{18}O$ 11‰ at 70% SiO_2. By examining the zonation in terms of the composition surfaces of the bulk parameters, such as SiO_2 and Rb/Sr (Figure 6.8), he showed that the variation is best modelled in terms of multipulse intrusion, in accord with the field studies. Furthermore a greater variation was found to exist between the pulses than within a single pulse. Stephens then showed that in order to satisfy the isotopic data the Criffell pluton can have evolved only by a complex interplay of processes in open and closed systems, these being distinguished on the basis of changing or fixed Nd and Sr ratios, respectively. The openness of the system is revealed both by the isotopic evidence of derivation of the granodiorite and granite from different, isotopically distinct, sources, and by the input of a mafic component to the former, as represented by the observed mafic enclaves; its closure is revealed by the internally consistent evidence for restite separation and fractional crystallization in the actual differentiation processes.

Of the sources, it is hypothesized that one, largely metaigneous, yielded the marginal, I-type granodiorite, whereas another, largely metasedimentary, yielded the internal member with S-type granite characteristics. Stephens envisaged the I-type magma as evolving largely by restite separation in equilibrium with the mafic magma which triggered the whole melting episode. In contrast, the second pulse differentiated in

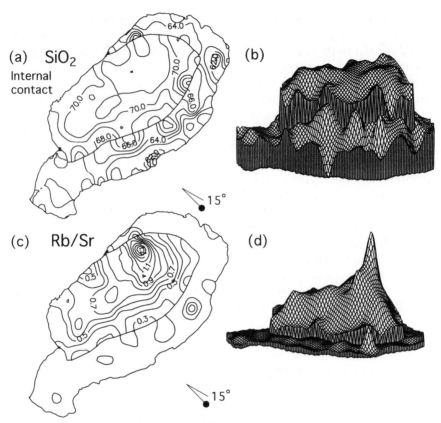

Figure 6.8 Maps of whole-rock variations within the Criffell pluton: (a) Two-pulse contour model for SiO_2, with pulses separated by the internal contact which isolates each pulse in the contouring routine. (b) Isometric view of the surface of (a), taken from southeastern corner at an angle of 15° from the horizontal, as indicated. (c) Rb/Sr map for two-pulse model. (d) Isometric view of the surface of (c), taken from southeastern corner at an angle of 15° from the horizontal, as indicated. Modified from Stephens (1992) with the author's permission.

a closed system by fractional crystallization. At last we can begin to appreciate the true realities of Nature's complexity!

There is then the problem of explaining how two magmas born together remain separate. It is of particular interest that Stephens sought a rheological explanation based on the possibility that magmas of different viscosity will respond differently to shear stress while flowing up a conduit, and will thus segregate, a process of viscosity segregation advocated by Carrigan and Eichelberger which we shall discuss in some detail later.

As for the multifactorial model of magma generation itself, this may be of general application, and one that could easily apply to the Ardara pluton mentioned earlier. Rarely is a single model sufficient to explain the zonal pattern of plutons, though the zonation process within the plutons of the arcs most often conforms to this ideal model. More often, especially in the cratons, it seems that, variously, mafic melts, and one or more felsic melts separately derived by remelting–unmixing processes from different sources, may then mix or just mingle together, both during ascent and also during fractionation. All this takes place in a highly dynamic environment where there is every likelihood of a magma chamber being continually recharged, even refilled. In a sense such complications are revealed in the complexities such as the zoning and resorption of plagioclase crystals. Furthermore, once a sizeable body of granitic magma forms in the crust it will offer a density and thermal barrier to upwelling mafic magmas, and the interface between the two will become the generating surface of a variety of hybrid magmas.

Only now are we beginning to appreciate the real complexity of so-called magma chambers, especially those deep in the crust – that they represent mush columns with extensive solidification fronts, with the resultant magma diversity being critically dependent on the relative volume of active mush and crystal-free magma. Furthermore, as Marsh tells us, 'The ideas that magmatic systems are periodically resupplied by magmas carrying important amounts of crystals that undergo extensive sedimentological sorting, reworking and recrystallization are concepts central to understanding magma' (1996, p.37). And, I would add, that are periodically disturbed by earth movements.

Clearly each natural example requires a separate assessment of the roles of the several different processes through time, though I would hazard an opinion that *in situ* liquid or crystal fractionation will be found to have played a lesser part in the differentiation of granitic magmas than has been so often envisaged in the past. But before turning to discuss the other processes involved I digress a little to consider how far the expulsion of melt from a pluton makes a contribution to volcanism, and how far the nature of this extract can reveal the pattern of evolution within the supply pluton.

SELECTED REFERENCES

Atherton, M.P. (1981) Horizontal and vertical zoning in the Peruvian Coastal Batholith. *Journal of the Geological Society of London*, **138**, 343–349.

O'Hara, M.J. and Mathews, R.E. (1981) Geochemical evolution in an advancing, periodically replenished, periodically tapped, continuously fractionated magma chamber. *Journal of the Geological Society of London*, **138**, 237–277.

Sawka, W.N., Chappell, B.W. and Kistler, R.W. (1990) Granitoid compositional zoning by side-wall boundary layer differentiation; evidence from the Pallisade Creek intrusive suite, central Sierra Nevada, California. *Journal of Petrology*, **31**, 519–553.

Sparks, S.R., Huppert, H.E. and Turner, J.S. (1984) The fluid dynamics of evolving magma chambers. *Philosophical Transactions of the Royal Society of London*, **A310**, 511–534.

Stephens, W.E. (1992) Spatial, compositional and rheological constraints on the origin of zoning in the Criffell pluton, Scotland. *Transactions of the Royal Society of Edinburgh: Earth Sciences*, **83**, 191–199.

In particular the process of liquid fractionation, otherwise compositional convection, is discussed by:

McBirney, A.R., Baker, B.H. and Nilson, R.H. (1985) Liquid fractionation. Part 1: Basic principles and experimental simulations. *Journal of Volcanological and Geochemical Research*, **24**, 1–24.

The volcano–plutonic interface: not Read's hiatus

7

Personally I do not share the author's (H.H. Read) disapproval of Judd's Denudation Series, uniting active volcano with dissected pluton.

Edward Bailey (1958) Transactions of
the Geological Society of Glasgow, *23*, 45.

INTRODUCTION

In his teachings, Herbert Read was wont to echo Reginald Daly's dictum that, volumetrically considered, the igneous rocks were of two different types, granite and basalt, the former almost exclusively intrusive, the latter extrusive. This view underpinned Read's strongly held opinion that granite and basalt were generated by very different processes. Of course, this was an overemphasis purposely designed to counter the then current monogenetic view, but, as Edward Bailey once wryly remarked, Read's *The Granite Controversy* should have been plainly marked 'for adults only' on its cover, with the admonition that geological adolescents should not, without preparation, be so instructed!

NATURE OF THE INTERFACE

Doubtless there are some granites of wholly crustal origin that are divorced from the basaltic and volcanic association, but this is not the case in most global tectonic environments. Even though basaltic magmas may be rheologically more disposed to extrude than felsic magmas, the latter, in fact, are very well represented by vast outpourings of dacitic and rhyolitic ash-flow tuffs, so that the volumetric contrast is largely redressed. Indeed, the fallacy that silicic magmas seldom erupt was firmly laid to rest by Smith in his 1960 review of ash flows. Silicic magmas can reside high in the crust and vent rhyolite and dacite, sometimes directly as lavas, but much more often as fragmented glass and crystals, fluidized and entrained in a gaseous medium.

As the intrusion of granite is so often genetically associated with uplift and erosion, the crustal roofs of plutons tend to be removed, with the result that the evidence of the upward passage of granite into rhyolite does not often survive. Nevertheless, there remain sufficient examples to demonstrate the intimacy of the connection, although it is on two levels: direct, with a pluton supplying volcanic products, and indirect, with plutonism and volcanism generally related in time and space.

The general case is easily substantiated. In the island arcs, as for example in the Aleutians and the Kuriles, and also in the volcanic islands, as represented by Kerguelen, felsic rocks form an integral, if only minor, part of the volcanic centres. In the relatively young, complex arc of Japan, the shallowness of erosion allows us to observe the direct relationship between the felsic volcanism and comagmatic plutonism, which is particularly well displayed within the volcano–plutonic cauldrons. Then, in the continental margin arcs, with their immense granitic batholiths intruding thick volcanic sequences, the connection remains perfectly clear whenever erosion has not been too severe. For example, the Sierra Madre Occidental Batholith of Mexico is capped by an estimated 350 partly eroded calderas, while the more deeply dissected Coastal Batholith of Peru carries only the relics of three such calderas (Figure 7.1), even though innumerable nests of plutons clearly represent the understories of many others. In contrast, only traces of caldera infills are left to represent the volcanic edifice that once must have overlain the Sierra Nevada Batholith.

Quite apart from these volcano–plutonic arcs the outcrop structures of the granites of the late-stage uplifts often reveal various levels in the classic denudation series: caldera, ring complex and nested pluton – the Newer Granites of the Caledonian of Scotland provide a good example, especially the Lough Etive and Ben Nevis multipulse plutons with their close time, space and composition relation with the eroded caldera of Glen Coe.

Of all the environmental types, the alkali granites of the cratons show particularly well this denudation series between felsitic pluton and ignimbrite-filled caldera, and it is here that we see best how the effusive and plutonic histories run in parallel. A typical history is of collapse on a ring fault, the latter providing the site of an arcuate system of volcanoes venting their products into a central caldera. A felsic ring dyke intrudes the ring fault, but whether or not it represents the direct conduit and supply to the volcanoes is often equivocal. One or more centred plutons then stope into the caldera fill, but note that this is late in the evolutionary history. Examples are legion and of all ages, though particularly well known are those of Oslo, Niger–Nigeria, trans-Baikal, Okhotsk, New Hampshire and the volcanic belt of the western USA. The latter provides

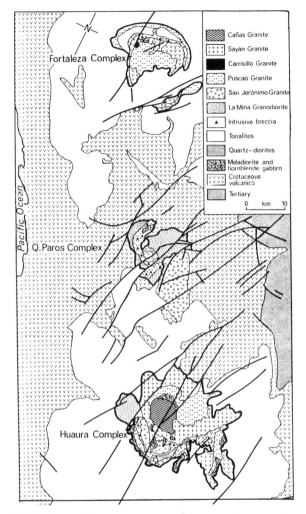

Figure 7.1 Distribution of the spaced centred complexes within the Coastal Batholith of Peru between the rivers Huaura and Fortaleza. After Bussell, Pitcher and Wilson (1976).

some particularly illuminating case histories involving the more direct connection.

One of these concerns New Mexico, where good topographic relief, combined with tilting along the eastern margin of the Rio Grande Rift, reveals a 3–5 km section through the Latir Volcanic Field, the associated Questa Caldera and an underlying pluton. As this extended section was interpreted by Lipman as showing a caldera capping a contemporary feeder batholith, the details are of particular interest. The caldera-fill consists of andesitic and dacitic lavas overlain by a great thickness of silicic ash-flow tuff and intruded by small granitic plutons. From the gravity data the latter were interpreted as coalescing at depth into a continuous body, about 30 km long and 15 km wide, and this, in turn, is taken to represent a pluton of porphyritic granodiorite, the roof of which is exposed by deep erosion beyond the outcrop of the volcanic field. Lipman envisaged the differentiated top of this pluton as having been erupted pyroclastically, its place then being taken by a resurgence of granodioritic magma from below. As a consequence of this plutonic activity the caldera infill was arched and faulted. Then, in response to extension in the Rio Grande Rift, the fault blocks were rotated to display the underpinnings of the caldera floor. The whole process took a few million years and ended with an important mineralizing event.

It is more common to find that subvolcanic plutons have risen, resurgently, into the roofs of calderas, and do not, therefore, represent the volcanic supply chambers. In the deeply incised roof zone of the Coastal Batholith of Peru this is particularly obvious when the whole caldera–pluton connection is visible in single canyon sections. There it is also possible to see in detail another relationship – that is, that the associated swarms of synplutonic dykes do not necessarily arise from the immediate subcaldera plutons but represent new magma from much deeper levels in the crust. Basaltic andesite magma was evidently able to squirt through still unconsolidated granitic magma to supply surface volcanoes. In this sense there is a hiatus of sorts.

Nevertheless, from the particular to the general, it is clear that no fundamental hiatus exists between plutonism and volcanism: they are connected processes with a common cause. Of course differences exist in style and relative abundance, because these reflect the intensity and duration of the heat source and the rheological differences between the magmas of the different granite types. However, I believe that those granites born in the process of continental collision are the exception. They are generally the products of minimum melting of crustal rocks and their relatively cool, 'wet', peraluminous S-type mushes would hardly be expected to erupt as volcanic lavas and pyroclastics. And the lack of evidence for associated S-type volcanic rocks cannot be wholly attributed to deep erosion. This generality might have to be tempered in the knowledge that there are S-type, cordierite-bearing lavas in association with the S-type granites of the Lachlan zone of southeastern Australia (p. 326), and also cordierite-bearing ignimbrites along with the two-mica granites in the Sakhalin magmatic belt of the Russian Far East, except that I do not recognize either of these particular granite assemblages as resulting from continental collision. More on this matter of S-type volcanic rocks in the sequel!

Even though so much evidence supports the essential continuity of plutonism and volcanism we have to expect that, following eruption, the continuing evolution of the residual magma, coupled with late stage subsolidus interactions, will probably result in a final compositional mismatch between otherwise closely related intrusive and extrusive rocks. Indeed, as we shall see, the processes operating in the magma chamber, as revealed by studies of the ejecta, are so varied and so polyphasal as to suggest strongly that many apparently simple granitic plutons, as now revealed by deep erosion, either represent a mere residue frozen in the process of decline, or were never supply chambers at any time in their history. We need to be aware that, as Señora Beatrix Levi once told me – and it takes on added emphasis in Spanish, 'the spew does not represent what remains in the stomach'.

EVOLUTION OF SILICIC MAGMA IN SUBVOLCANIC MAGMA CHAMBERS

The ejecta do, however, tell us something of what went on during the vital early stages of a pluton in cases where it actually represented a supply magma chamber. Also, a stratigraphic succession of ash-flow tuffs surely reflects the ever-changing history of that chamber. Indeed, as Druitt and Bacon pointed out, there are considerable advantages in this approach because the quenching of magma during ignimbrite eruptions can provide pristine samples of liquid and crystals. Furthermore, a single eruption can discharge so vast a volume of magma that levels may be tapped as deep as 2 km into a chamber. Equally important is the ability of an eruption to sample the magmatic body at its stage of maximum development.

In addition to such compositional information the geophysical mapping of these young volcanic fields provides convincing models of the three-dimensional shape of the underlying chamber itself. And, not least, useful estimates of the longevity of magma chambers can be obtained by dating the cyclic volcanic sequences standing above them.

EXAMPLES FROM WESTERN NORTH AMERICA

A particularly rich library of such data is available from ongoing studies of the ash-flow tuffs of the volcanic fields of the USA and Mexico. One important interim conclusion is that many of the source magma chambers must be vertically zoned in composition, temperature and density. Another equally important finding, derived from seismic P-wave delay studies of recent calderas, is that a thermally disturbed zone exists down to well below the Moho, perhaps for 200 km or more beneath the Pleistocene Yellowstone Caldera, for example. This is a salutary reminder of the immensity of the heat plumes that accompany even high level felsitic magmatism.

In detail, the ejecta from these eruptions tell us a great deal about the early history of silicic magma chambers. An excellent example, documented by Druitt and Bacon, concerns the climactic eruption of the Mount Mazama magma chamber, Crater Lake, Oregon, where approximately $50\,km^2$ of magma was discharged, largely in the form of rhyodacite pumice. A striking feature is the compositional uniformity of this pumice, and the narrow range of temperature, 880–886°C, recorded by the use of the Fe–Ti oxide geothermometer. Evidently this rhyodacite was erupted from a chemically and thermally homogeneous body of melt.

In contrast, there are late erupted scoria clasts which are compositionally, texturally and mineralogically diverse, but distinguishable as the products of at least two contrasted types of andesite, one high in strontium, the other low, with corresponding differences in isotopic composition. Some of these scoriae are particularly crystal-rich and represent cumulates with a high abundance of individual, equant, and not obviously zoned crystals. It seems that before eruption, the scoriae formed part of dense crystal mushes produced either by crystal settling or by the compaction which accompanies the extraction of an intercumulus liquid.

Druitt and Bacon calculated the compositions of viable parents for the cumulates of the two scoriae types, discovering that these approximate to two compositional varieties of andesite, a result consonant with the occurrence of quenched globules of these within pre-climactic rhyodacitic lavas. Cleverly, these workers estimated the cumulate compositions by assuming a loss of an intergranular liquid amounting to 20–40% of the original mass. On the basis of this kind of data and taking due account of the order of intrusion, the form of the original layering can be reconstructed as a homogeneous rhyodacite liquid, underlain by layers of the high and low strontium-bearing crystal mushes (Figure 7.2) – a model consonant with expected densities. Convincing arguments were advanced for a layered structure with a sharp density interface between largely liquid rhyodacite and an andesitic to basaltic crystal mush, marked by a step of 16% in crystal content.

Presumably an early rhyodacitic magma chamber was repeatedly recharged by andesitic magmas, even occasionally by basalt on the evidence of scanty basaltic scoriae. These fresh influxes ponded at the base, and some recharges may well have triggered the eruption of the overlying silicic magma. The recharge magmas are considered to have cooled and crystallized with the separation of the residuum by convective fractionation, the successive recharge events building up, layer by layer, a pile of crystal mush, the residual liquid continuously feeding upwards into a growing silicic layer. Although Druitt and Bacon did not dismiss the

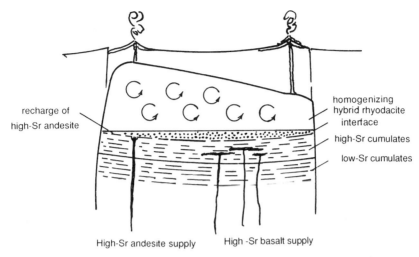

Figure 7.2 Schematic diagram illustrating the model for the evolution of the Mount Mazama magma chamber. Adapted from Druitt and Bacon (1989) with the permission of the authors.

possibility of sidewall crystallization, they favoured the dominance of horizontal deposition.

This is a fascinating story which is supported by other similar studies, particularly by that of the Bishop Tuff of California carried out by Hildreth in 1981. Examination of this sequence of ash-flow tuffs not only convinced Hildreth of the existence of a layered magma chamber; the high degree of equilibration between crystal and glass led him to argue that a compositional gradient existed in the parent magma before the phenocrysts formed. He envisaged this being established early by a diffusion mechanism accelerated by convective transfer of the melt towards the cooling surfaces. This radical hypothesis, essentially of thermogravimetric diffusion, has not found general favour, but nevertheless it inspired a great deal of further study, and even the possibility reminds us not to reject diffusion, even if only as an ancillary process.

The evidence provided by the ejecta, coupled with the descriptions of actual layered, subvolcanic plutons, and supported by the experiments of Sparks, Huppert and Turner, provides a good model for the evolution of some felsitic plutons, especially those involving the more highly fluxed magmas. However, as will be clear from the various reservations already expressed, I am not persuaded that all plutons, or indeed very many, have followed this evolutionary path, leaving behind but a mushy residual magma to represent the declining stage of their history.

Furthermore, a good number of zoned granitic plutons were constructed from magma pulses generated elsewhere, so it is not surprising
that there are very few unequivocal examples of cumulate, mafic, basal
layers. Of course, it is possible to dismiss this latter finding as due to an
obvious lack of topographic opportunity, and to point to the few examples
of plutons that have allegedly been sufficiently tilted to reveal a dioritic
base. Flood and Shaw, in 1979, referred to such a pluton in New England,
and Barnes and co-workers, in 1990, to other examples from the Klamath
Mountains, though in all but one I remain unconvinced of the structural
evidence for tilting in the face of certain circular arguments.

So I conclude, not altogether rationally, that although volcanism and
plutonism are generally inseparable, not many plutons represent true
supply chambers that have differentiated in this highly contrasted way,
venting their silicic tops to leave just a dying relic of themselves for the
field geologist to argue over. In other words, there is a kind of hiatus,
though hardly as fundamental as envisaged by Read. But here the reader
will sense yet another personal prejudice!

RHYOLITES AND GRANITES: ARE THEY THE SAME?

Siliceous volcanics rocks in any considerable volume are generally associated with thickened regions of the crust, being especially characteristic of
the continental marginal arcs. In this environment, as is so well demonstrated in the Andes, rhyolites, dacites and andesites are closely connected
in space and time in the form of compositional sequences which closely
parallel those of their plutonic analogues. Although there are some differences it is reasonable to conclude that these volcanic rocks evolved from
the same sources and by similar differentiation processes. As we have
seen, the relationship can be as direct as a granitic magma chamber supplying rhyolitic tephra as in the classic model. Nevertheless, again as
hinted above, regional studies often seem to deny this direct connection,
leaving me with the suspicion that the coeval volcanic magmas are often
derived independently from very deep in the crust (e.g. Pitcher and
Cobbing, 1985, p. 287; Brown, 1982, p. 446).

The reason for this independence is quite fundamental yet poorly understood, though we can suspect that differences in the water content of
the magma, its temperature in relation to the relevant solidus, and the rate
of its ascent will be the deciding factors. And I would add the tectonic
accident that permits magma to hurry to the surface. Whatever the complications there is no serious difficulty in modelling the origin of the
Cordilleran dacitic and rhyolitic ignimbrites in terms of an extended process of fractional crystallization.

There is, however, a second magmatectonic environment that presents
a problem particularly germane to any discussion of the origin of granites.

This is the tholeiite–rhyolite association of the rifted volcanic plateaux, the stark bimodality in which has intrigued petrologists since the time of Bunsen. In particular, Reginald Daly puzzled over the problem of explaining this compositional gap in his classic studies, in 1925, of the volcanic rocks of Ascension Island, and in 1973 Yoder discussed its significance in terms of a range of possibilities: immiscibility, extreme crystal fractionation, the melting of granitic country rock and the partial melting of a mafic progenitor. The problem is particularly acute when, as in Iceland, there is no continental crust to call on to provide a silicic partial melt, and where rhyolite and basalt can be products of the same volcano and even form a kind of emulsion in the same lava flow (Walker, 1963).

The Icelandic rhyolites have been interpreted, variously, as differentiates of basaltic magmas, partial melts of a lower part of the lava pile (O'Nions and Grönvold, 1973), or of oceanic plagiogranite (Sigurdsson and Sparks, 1981), though recent researches favour the first of these models, the starting point being a basalt derived by partial melting of mantle material within the garnet lherzolite, stability field (MacDonald et al., 1990).

According to MacDonald and his colleagues, within a particular volcanic complex, such as that at Torfajökull in Iceland, there is no significant compositional hiatus between basalt and rhyolite, which are geochemically consanguineous, the intermediate terms being represented by icelandite, which, somewhat paradoxically, is only represented by inclusions in the lavas! Thus a gap remains in volumetric terms, implying that the differentiation must have been rather perfectly accomplished within a very narrow zone, and, furthermore, geochemical modelling has indicated that, for the rhyolites, it took place at a crystallization percentage greater than 90%.

It is a matter of particular interest that although rhyolites only form less than 3% overall of Icelandic volcanic rocks, they are more abundant locally among the products of the central volcanoes such as that of Torfajökull. Taking into account that there were a number of supply chambers, it is likely that in some of the latter the separation of the rhyolitic liquid would have been nearly perfectly accomplished – a remarkable situation. I can only understand this in terms of a highly dynamic system involving a continuous process whereby mafic magma, ascending as a diapiric column, is continually replenished with fresh basaltic magma as proposed by George Walker in 1975. It is in such an environment that Marsh, in 1996, modelled granophyre production in terms of the separation of tiny blobs of the differentiated siliceous liquid ahead of an unstable solidification front – blobs which rise away to collect and coalesce at the head of the column. Are these not the granophyric pools we so often see in the groundmass of tholeiitic basalts?

So it is possible to extract rhyolitic liquids from basaltic magmas by extreme crystal fractionation, but only in this very special environment. These volcanic rhyolites would never become the granites of the orogens but, nevertheless, extreme differentiation processes cannot be wholly dismissed from discussions on the origins of the rather special granites of the centred complexes of the rifts.

I now turn to those processes other than fractionational crystallization which can produce a variable suite of interrelated granitic rocks, dealing first with the possibility that felsic liquids may separate from a solid residuum or restite, either wholly or partly, during crustal remelting – a process representing a kind of unmixing.

SELECTED REFERENCES

Druitt, T.H. and Bacon, C.R. (1989) Petrology of the zoned calcalkaline magma chamber of Mount Mazama, Crater Lake, Oregon. *Contributions to Mineralogy and Petrology*, **101**, 245–259.

Lipman, P.W. (1988) Evolution of silicic magma in the upper crust: the mid-Tertiary Latir volcanic field and its cogenetic granite batholith, northern New Mexico, U.S.A. *Transactions of the Royal Society of Edinburgh: Earth Sciences*, **79**, 265–288.

MacDonald, R., McGarvie, D.W., Pinkerton, H., Smith, R.L. and Palacz, Z.A. (1990) Petrogenetic evolution of the Torfajökull Volcanic Complex, Iceland: I. Relationship between the magma types. *Journal of Petrology*, **31**, 429–459.

The evidence for restite: unmixing as an alternative hypothesis

8

The diversity of granitic rocks is caused by sedimentary processes.
Tom Barth (1962) Theoretical Petrology, *p. 379.*

A STATEMENT IN FAVOUR

Ever since the beginning of the serious study of granite–country rock interaction by Lacroix, the facts that the mafic minerals so often form tiny aggregates, and enclaves so often copy the mineralogy of their host have been matters for comment and debate. I think that we now have to accept that each of the explanations that have been proposed can hold true for particular examples; that is to say, such enclosures may variously represent the natural agglutination of growing crystals, the dis- aggregation of early precipitates, the dismemberment of synplutonic in- trusions, the disintegration of surmicaceous country rock xenoliths, the resisters or residues that escape complete granitization, or residual mate- rials inherited from the partial melting of the source rocks.

The decision in any specific example is often difficult, especially when account is taken of the continuous recrystallization and reciprocal ex- change of material which must often take place during transport within the host magma or migma, processes so eloquently described by Stephen Nockolds in the 1930s, and detailed by Didier and Barbarin in 1991. Increasingly, however, petrologists have come to suspect that granitic magmas often represent partial melts, and are thus likely to carry a refrac- tory residue with them during their upwelling. The use of the term *restite* by Mehnert in 1953 (1968, p. 298), comments by Bateman and co-workers in 1963, and also by van Moort in 1966 attest to this growing awareness. The problem then arises as to the degree of retention of this residue or restite. On the assumption that it is often carried up in substantial amounts as an integral part of the magma, White and Chappell have elevated the restite hypothesis to a central position in the discussion of the causes of chemical variation in granitic suites (e.g. 1988, 1992). I would remind the reader that this is essentially a discussion of process, and separate from considerations of granite typology.

The restite hypothesis is not wholly dependent on the retention of primary restite. In essence it rules that the variation in the composition of granitic suites results from the different degrees of unmixing of SiO_2-poor restite and SiO_2-rich liquid during and after melting. Such an unmixing provides a neat explanation of both the linear trends of many Harker-type variation diagrams and the identity of the mineralogy of aggregates, enclaves and host. It carries the important implication that the granite and restite together directly image the source rock, so that by a simple calculation the composition of the source material can be determined; a truly remarkable proposition.

The rationale behind this restite model has been presented by Chappell and White in the following form (1991, p. 376). 'We take the view that most granite magmas result from partial melting of the crust. A mechanism of partial melting implies that a silicate melt must, at least initially, coexist with residual unmelted material or restite. Debate about the role of restite in granite genesis must revolve around the extent to which that silicate melt is completely removed from its restite at an early stage in the evolution of the magma. It is our contention that all degrees of separation are possible, so that there is a spectrum of restite involvement. At one extreme is the classical situation in which the granites formed from a largely or completely liquid magma. The opposite case is that in which the magmas retained varying amounts of solid residual source material within a low-temperature silicate melt.'

It is difficult to refute so rational a statement of the restite model. It accords well with the experimental findings of Winkler, Wyllie, and others, which show that rock melting can hardly be avoided at temperatures and pressures likely in the lower crust. More specifically, the experiments of Vielzeuf and Holloway on the fluid-absent melting of pelite (p. 44) demonstrate that a melt proportion as large as 40% is produced, at lower crustal pressures, at temperatures within the range 850–875°C, the narrow range being due to the buffering effect of the biotite breakdown reaction. Furthermore we have already learned from Wyllie's experiments that the mafic and felsic components behave as separate entities over a wide range of temperature and pressure.

From Presnall's geometric analysis of the process of partial fusion and his application, with Bateman, of this analysis to the problems of magmatism in the Sierra Nevada, we learn that the phase relationships are compatible with the separation of either completely or partially liquid magmas from lower crustal rocks. Partial melting, however, seems the most likely in that substantially lower temperatures are required, and also that the driving forces of buoyancy and tectonic squeezing should ensure mobilization long before the completion of melting. Thus the fact that granitic magmas might often carry restite seems undeniable and, indeed, I have already opined that such magmas avoid undercooling and primary nucleation simply because they carry seed crystals from their very beginnings. But how often and in what proportion?

A central question is how easily can liquid separate out from crystal mush. It is not a new enquiry, for the theoretical possibility of compaction and filter pressing has long been with us, and concerns liquid separation from crystal cumulates as much as the unmixing of partial remelts. I will return repeatedly to this matter throughout this text. At this juncture I merely report that Wickham discussed the relevant physical controls, reaching a conclusion in line with a consensus view that the rheological nature of low melt fraction crystal mushes will inhibit the separation of liquid. However, I still think it is possible that the continuous deformation of a crystal mush will promote extraction. In the case of bulk melting Wickham predicted that, as soon as a critical fraction of melt is exceeded, the effective viscosity will fall sufficiently low to bring about the for-

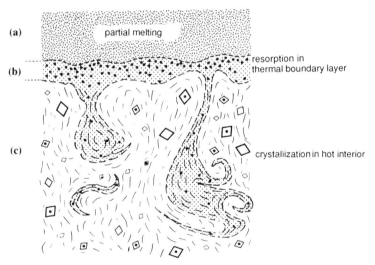

Figure 8.1 Schematic diagram illustrating thermal and fluid dynamic structure of melting roof. Partial melting occurs in a roof zone (a). The thermal boundary (b) is a region of heating and strong thermal gradients in which crystals (restite) are resorbed. The thermal boundary randomly detaches to form plumes which move downwards and mix into the hot interior. The convecting interior (c) is a region of small temperature variations where crystallization occurs due to mixing in of the cooler partial melt magma from above. Reproduced from Huppert and Sparks (1988) *Journal of Petrology*, **29**, pp. 599–624, figure 7, with the permission of Oxford University Press.

mation of suspensions. Localized convection can then be expected to occur, which will effectively mix the disintegrating restite with the liquid (Figure 8.1).

THE DEBATES

The debate between Chappell, White and Wyborn on the one hand, and Wall, Clemens and Clarke on the other, on whether rock suites result from differential separation of restite or crystal fractionation, illustrates well the difficulty of arriving at a general acceptance or rejection of the restite hypothesis. The discussion first focuses on whether restite can survive in a form that can be unequivocally identified within the mineral assemblages of granites. With the notable exception of the datable ancient cores within zircon crystals, I think not. I would hesitate to identify with any certainty the ragged aggregates of biotite, hornblende and apatite as primary restite, but concede that they may well represent material modified by interaction with a melt of ever-changing composition. Then again, although the iron oxide-sprinkled cores of hornblendes suggest a reaction series with pyroxene, the latter is as likely to represent an early precipitate as a survivor from the source. Moreover, both Vernon in 1983 and Wall and colleagues in 1987 cautioned against accepting the contrasted calcic cores within zoned plagioclase crystal as being of restitic origin, pointing out that there are orthodox petrological explanations for this texture. Thus I conclude that the textural evidence is often equivocal, and also that the presence of restitic seeds need not imply that there was ever an important restitic component. There is also good evidence that garnet, cordierite and andalusite can precipitate from peraluminous melts, but it is necessary to entertain fibrolite, in its form of sheaves within quartz grains, as being truly restitic in origin (Clarke *et al.*, 1976; D'Amico *et al.*, 1981).

Perhaps an acceptable solution of this problem of restite survival lies in Huppert and Sparks' theoretical studies of the physics involved in the generation of granitic melt by the intrusion of basaltic magma into continental crust. Assuming convection within the boundary layer of remelt, it seems that there would be sufficient heat to remelt a significant fraction of the refractory components, though these might soon reprecipitate as the new magma is cooled by the mixing or by moving away (Figure 8.1). As a result many crystals will be of genuine igneous origin though nucleated on to unresorbed residua, and from the geochemist's point of view they will represent the refractory component as made-over restite. Huppert and Sparks are bold enough to calculate the possible changes in the proportion of restite to precipitate as this changes with temperature and time, and predict that the precipitate will greatly predominate for most of the life of the crystal mush.

Turning to the significance of the form of Harker-type compositional plots in discriminating between unmixing and fractional crystallization, I regard the linearity of such diagrams to be inconclusive, and I have some sympathy with the view of Wall and co-workers that such trends say little about the actual petrogenic process involved.

Finally, there is much play not only with the question of the efficacy of liquid separation, but also with the possibility or impossibility of the rise of diapirs of restite-rich mushes. Although I urge the reader to follow the several threads of this informed debate with a certain respect for both points of view, I am inclined to turn first to the rock outcrops themselves to seek an answer.

AN OUTCROP EXAMPLE OF REALITY

As we shall see, the granites of the various global tectonic environments provide different answers. Those generated in the continental crust are often particularly enigmatic, their textures defying any simple explanation in terms of restite retention. An example is provided by the marginal zone of the Thorr pluton in Donegal, Ireland, where the outcrops of granodiorite show abundant mafic mineral aggregates accompanied by two types of enclave, one of which is microgranular, dioritic in composition and commonly fringed by coarse flakes of greenish biotite (Figure

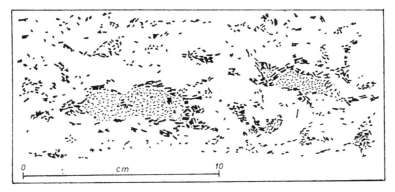

Figure 8.2 A surface of the Thorr Granodiorite showing the distribution of the dark minerals, largely biotite and hornblende. Originally described in Pitcher and Berger (1972, figure 5.14) as showing a complete gradation between obvious xenoliths and mineral aggregates, with the coarsening of biotite on the peripheries of the xenolithic patches. However, it is more likely that the microgranular enclaves of whatever origin – true restite, disrupted country rock or mingled magma – simply provided nucleating surfaces for the growth of biotite or hornblende. But what story will this same outcrop tell after another 30 years? Shore WSW of Dungloe, Co. Donegal.

8.2), whereas the other is coarser grained, surmicaceous and fringed by flakes of reddish-brown biotite. That we are seeing here the simultaneous operation of several processes is confirmed by reference to two contrasted contact situations of the Thorr pluton (see Figure 6.3c). At a northern locality, near Curran's Point, an appinitic basic body is seen frozen in the act of mingling with the Thorr magma, flooding the latter with microdioritic enclaves which were not wholly solidified at the time of their engulfment. These enclaves have provided surfaces for the nucleation and growth of coarse biotite and hornblende crystals which are not due to the disintegration of the enclaves themselves.

In other outcrops on the eastern flank of the pluton, particularly adjacent to large stoped blocks of coarse-grained, pelitic hornfelses, the surmicaceous xenoliths can be seen to have spalled off from the blocks, leaving no doubt as to their metasedimentary origin. However, it is important to record that the latter mingle with the microgranular mafic enclaves to produce a decidedly mixed assemblage.

Very clearly each of the processes of country rock dismemberment, magma mingling, crystal nucleation and crystal accretion has contributed to this particular magmatic pot pourri, and this, I believe, is often the case. Such processes probably operate simultaneously during the whole period of the upwelling of magma, so that we can expect a continuum which may well involve the survival of some primary restite. Indeed, in the example of the Thorr pluton old zircon has been recognized. This, then, is the reality, with the central problem being the decision as to proportionality.

To examine possible ways of deciding the relative importance of unmixing as a cause of rock variation, I turn to a discussion of Chappell and White's original investigations in the Lachlan region of southeastern Australia.

THE LACHLAN EXPERIENCE: SUITES AND SUPERSUITES: GRANITE TYPES AND THEIR SOURCE ROCKS

In southeastern Australia that part of the Tasman Orogen in the Lachlan region consists largely of a tightly folded slate belt (p. 326). During Silurian and Devonian times this was the site of extensive igneous activity (Figure 8.3), and in the area discussed by Chappell and White in a long series of publications (e.g. 1992) there are more than 800 separate plutons forming steep walled intrusions with simple contact aureoles. Many of these plutons form contiguous arrays – batholiths – many with outcrops in excess of $500\,km^2$; that of Strathbogie, for example, is of the order of $1500\,km^2$. The constituent rock types group into well-defined mappable suites which, because they appear in several separate plutons or groups of plutons, can be referred to as supersuites. Granites assigned to a single suite or to a supersuite share distinct textural, modal and chemical features, although Chappell has always held that the chemical criteria, especially the isotopic parameters, form the best test of the identity of suites.

The 8600 km^2 Bega Batholith (BE, Figure 8.3) provides a key example with the recognition of seven supersuites distributed between more than 150 separate plutons. The compositional individuality is well illustrated by the strontium versus SiO$_2$ plot of Figure 8.4, with its clear separation of a low strontium Glenbog suite, involving 14 plutons, from a high strontium Bemboka suite, involving just two plutons. Although the composition of the supersuites of the Bega Batholith are dominantly metaluminous, those of the Kosciusko Batholith (K, Figure 8.3) involve both metaluminous and peraluminous granites, the former represented by a Jindabyne suite distributed within eight plutons, the latter by an Ingebyrah suite consisting of five plutons.

Such findings about the regional coherence of rock suites is wholly in agreement with that of Cobbing and myself (1972) in our work on the Coastal Batholith of Peru, where we introduced the term 'superunit' to mean much the same as the 'supersuite' of Chappell and White, or the 'sequence' of Bateman and Dodge. Clearly the recognition of such

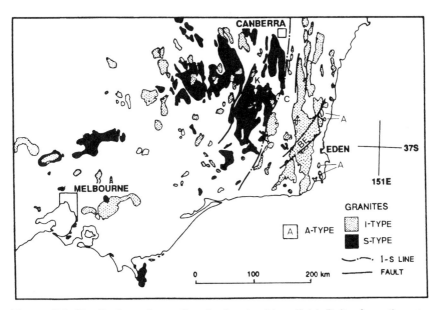

Figure 8.3 Distribution of granites in the Lachlan Fold Belt of southeastern Australia. Particular locations are Cooma pluton (C), Kosciusko Batholith (K), and Bega Batholith (BE).

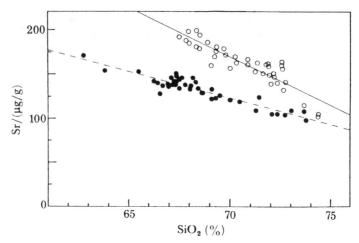

Figure 8.4 Harker diagram showing strontium variation in the Bemboka suite (open circles) and Glenbog suite (solid circles) of the Bega Batholith. After Chappell (1984), figure 2.

supersuites is crucial to granite studies because it allows the identification and mapping of granites that represent discrete batches of magma, each of which is likely to be unique in both origin and subsequent evolution. The Lachlan study adds yet another dimension: such suites and supersuites are equally valid in granites of very different source rock compositions.

As already noted the restite hypothesis carries with it the implication that the source rocks can be identified: that granites image their source. This was demonstrated in the Lachlan Belt by finding that the peraluminous and metaluminous granite types are largely separate in outcrop distribution. Fortunately, they are highly contrasted, the peraluminous type consisting of biotite granites bearing abundant cordierite; the other, metaluminous type being represented by sphene-bearing biotite-hornblende granites or granodiorites. The differences reflect the high aluminium content relative to alkalis and calcium in the former and the converse in the latter type. Although Chappell and White emphasize this contrast, there is such a span of composition that it would be difficult to deny a compositional convergence.

A significant feature is that the differences for many of the elements are not great – the most significant contrast concerns the lower content of calcium, sodium and strontium in the peraluminous rocks – but, nevertheless, this must be of fundamental significance as it correlates with the removal of these three elements during the breakdown of feldspars in the sedimentary cycle. Thus Chappell and White interpret this contrast as being inherited from the source, the peraluminous granites resulting from the partial melting of metasediments: these are the S-types. On the other hand, metaluminous granites derive from originally igneous material, again, it is thought, by remelting: these are the I-types.

Of course, the original chemical fractionation of source rocks is not the only explanation for peraluminosity, if only because fractionational crystallization in magmatic systems naturally leads to increasingly peraluminous compositions, especially where hornblende is a precipitate. However, for such a closely knit group of rocks, in both time and space, as represented by the S-types of Lachlan, where whole rock suites are positively peraluminous, and in much greater relative volume than might be expected of a mere fractionate, the sedimentary source-rock explanation is to be strongly preferred.

There are some other chemical contrasts that fit the model. Thus S-types have a lower ferric to ferrous iron ratio and are relatively reduced compared with I-types, possibly due to the presence of the reducing agent, graphite, in the sedimentary source rocks, as suggested by Flood and Shaw in 1975. This contrast in magma redox potential controls the behaviour of metals such as copper, molybdenum, tungsten, tin and gold, because the partitioning of these elements between melt, oxide and sulphide is redox dependent according to Candela and Bouton. Then again, according to O'Neil and Chappell, whether the oxygen isotope ratio $^{18}O/^{16}O$ (expressed in the standard form of $\delta^{18}O$) is less than or greater than 10 also provides a significant discriminant, because the isotopic signature of oxygen in sediments determines its value in the derived melts.

Since the early designation of S- and I-types in the Lachlan Belt a substantial body of new compositional data has strengthened the reality of the division, though each of the discriminants is more complexly variable than was at first envisaged (Table 8.1), not an unexpected complication considering the multifactorial nature of geological processes. Thus although a relatively high initial strontium isotope ratio remains a general feature of S-types, there is a range within the I-types at Lachlan from 0.704 to 0.712, leading to an overlap in this particular measurement which Compston and Chappell ascribed to the differing age of mantle derivation of the I-type source rocks. Furthermore, and again concerning the oxygen isotopes, Matsuhisa and co-workers showed that exchanges between magma and country rock greatly complicate the simple source-rock model. Indeed, exchanges of this kind, involving the most mobile of materials, silica and water, can lead to such profound changes in the redox potential of an I-type magma that hornblende fails to precipitate, ilmenite appears in place of magnetite, and red, titanium-rich, biotite appears in place of green biotite, all features characteristic of an S-type granite; Oglethorpe and also Simpson, Plant and co-workers provide examples (1987 and 1982, respectively).

Table 8.1 Summary of the distinctive features of the S- and I-type granites of the Lachlan Fold Belt originally proposed by Chappell and White (1974) and extended in later papers

I-types	S-types
(i) Metaluminous mineralogy; hornblende common and more abundant than biotite in mafic samples; accessory sphene common	Peraluminous mineralogy: biotite and muscovite predominate; no hornblende; some cordierite and/or aluminosilicates; monazite may be accessory
(ii) Hornblende-rich igneous-appearing xenoliths	Pelitic or quartzose metasedimentary xenoliths
(iii) Relatively high Na_2O	Relatively low Na_2O
(iv) Molecular $Al_2O_3/(Na_2O + K_2O + CaO) < 1.1$	Molecular $Al_2O_3/(Na_2O + K_2O + CaO) > 1.1$
(v) Normative diopside or small amounts of normative corundum	Normative corundum $>1\%$
(vi) Broad spectrum of compositions from mafic to felsic	Narrow range of more felsic rocks
(vii) Regular inter-element variations within plutons; linear or near linear variation diagrams	More irregular variation diagrams
(viii) Initial $^{87}Sr/^{86}Sr$ 0.704–0.706	Initial $^{87}Sr/^{86}Sr > 0.708$
(ix) Usually unfoliated; contacts strongly discordant; well-developed contact aureoles	Often foliated; sometimes surrounded by high grade metamorphic rocks

Despite all these possible reservations I believe that many petrologists would accept the reality of the supersuites, and also the general thesis that the contrast between S- and I-types at Lachlan is due to a fundamental difference in the composition of the source rocks. Further, there might well be a wide measure of agreement that the S-type magmas derive from a supra-crustal source, and I-type magmas from an infra-crustal source, in the sense that the latter was itself the product of an earlier, mantle-derived, crustal underplate. This holds whether they approve of the labels or not. Disagreement arises over the efficacy of the restite model, by involving differential unmixing, to explain the variations within the suites. The plot thickens when Compston and Chappell move on to calculate the composition of the source rocks on the assumption that the compositions of the derived granites of each suite will lie on the line passing through the granite compositions joining melt to restite: a line which should then pass through the compositional point representing the source rock. It follows that the composition of the most mafic sample can be taken as approaching that of the model source.

I think that we must accept Chappell's contention that the two proc-esses which involve the removal of crystals – in one case carried from the source, in the other precipitated from the melt – are complementary. One or the other may dominate or they may act together. Concerning the Lachlan experience, Chappell posits that the restite mechanism predominated in both I- and S-types, though fractional crystallization sometimes followed the clearing of all the restite from the magma; indeed there is one very clear example where restite separation is considered to have been dominant from 66% to 69% SiO_2, afterwards followed by an extended stage of precipitate separation; and another where the latter process was wholly dominant (Chappell, 1996, p. 167). But how is it pos-sible to establish this proportionality?

The essential tool is the geochemistry. Particularly significant are the variations in Sr, Ba and Rb, which according to Chappell often show patterns precluding origin by fractional crystallization, such as no change in element ratios. The modelling is somewhat involved, depend-ing on estimates of partition coefficients, and I leave the reader to follow it through in detail (1996, p. 163); in essence it involves calculating the weight percentages of the minerals that would remove Sr, Ba and Rb from a melt by crystallization and, finding that they do not simulate reality, at least in the Lachlan example, turning to a restite model to obtain a better fit of the variations at the various stages of magma evolution. I am, probably quite unreasonably, not wholly confident of calculations that rely on estimates of partition coefficients, but Chappell makes the telling point that in the case of Cordilleran granites this same calculation fits well with fractional crystallization being the dominant process, as nearly everyone supposes!

Further support for this restite model is provided by Burnham's quan-titative assesment, in 1992, of the likely feldspar–quartz liquidus relations using the analytical data from selected Lachlan suites. Calculation echoes reality as, for example, when the calculated plagioclase liquidus and residual plagioclase compositions turn out to be the same as those of the actual plagioclase cores, which strongly suggests that the residue was of a fixed composition compatible with a restitic origin.

Concerning the S-types of Lachlan I believe that the evidence support-ing the restite model is compelling, especially when account is taken of the increase in peraluminosity with mafic content. Furthermore, in compari-son with the I-types, these S-types are relatively heterogeneous, have a shorter range of composition, especially in silica, and show poorer geochemical correlations. There is even the added bonus in that the calcu-lation of the Nd model ages for the source rocks yields geologically reasonable results in harmony with the extended history of the protolith as determined by Williams, in 1992, from zircon age studies.

However, with the exception of the zircon cores, I still have difficulty in identifying with certainty restitic minerals in the rocks themselves, and therefore in making any estimate of the survival of the unmelted residue, yet I appreciate Chappell's insistence that the calcic cores of plagioclases ought to be given the same restite status as those of zircon, and I am confident that future research will decide this matter. I am certainly not happy with the view that the metasedimentary enclaves that I have seen represent the protolith: these is no sign of melting, only mechanical disruption, and the isolated metamorphic garnets are true xenocrysts, so that these enclaves appear to me to represent true xenoliths derived from the wall rocks: they are 'resisters' not 'restites'. Of course such personal opinions are based on a very limited experience with the Lachlan rocks and I could easily be wrong!

I also have to admit having residual doubts as to the success of the pure restite model in explaining the evolution of the Lachlan I-types because I find nothing particularly revealing in the chemical composition, chemical variation and the textures of these rather homogeneous, biotite-hornblende granodiorites, certainly nothing that I recognize as restite. Indeed the complex nature of the dioritic, microgranular enclaves that I have seen in these rocks is in keeping with a degree of magma mixing and mingling, and, moreover, there is the expected association with synplutonic basic dykes and basic lavas, though all this may simply reveal that a relatively minor ancillary process was also involved (compare Gray, 1984, 1990; Collins 1997; and the riposte by Chappell, 1997).

In very recent publications (1996 and 1997) Chappell and his colleagues vigorously defend their position concerning the restite hypothesis, especially in respect of the I-types of Lachlan within which they recognize sets of features that may indicate whether the variation in an I-type granite suite resulted from fractional crystallization or restite fractionation (Table 8.2). The resulting contrast echoes the earlier observations of Wones in 1980, and myself in 1982, that there is an important contrast between the I-type granite suites of the American Cordillera and those of Acadia, southeastern Australia and the British Isles, though I hasten to add that the Chappell group do not make this connection with tectonic site and context central to their thesis. I leave my reader to consider the genetic implications of Table 8.2, while I promise to return later to the significance of the geological context.

The best evidence that restitic crystals are entrained in ascending S-type magmas would be the occurrence of demonstrably metamorphic crystals in volcanic rocks. According to Wyborn and Chappell almost all of the phenocrysts present in the peraluminous Hawkins volcanics of Lachlan represent restite. They envisaged a continuum in the growth of metamorphic minerals whereby crystal growth was finally taking place in the presence of melt, thus explaining the general euhedralism of the crystals

Table 8.2 Features that may indicate whether the variation in an I-type granite suite resulted from fractional crystallization or restite fractionation

| | Compositional variation results from | |
	Fractional crystallization	Restite crystal fractionation
Field criteria and associated rocks		
Dominant regional granite type	Tonalite, low-K granodiorite	Granodiorite, monzogranite
Associated gabbros	May be present	Minor to rare
Associated volcanic rocks	Complementary felsic rocks	Matching compositions
Zoned plutons	Common	Uncommon
Mafic enclaves	Relatively rare	Common in more mafic rocks
Hand sample	Even textured	Heterogeneous
Mineralization	More likely	Most unlikely
Mineralogical and petrographic criteria		
Plagioclase cores	Absent or rare	Common
Plagioclase zoning	Normal	Complex
Mafic minerals	Often well-shaped	Clots common
Zircon ages	Inheritance rare	Inheritance common
Chemical and isotopic criteria		
$SiO_2 < 54\%$	In cumulates	Never
SiO_2 54–60%	More common	Uncommon
Trace elements	Trends may show inflexions	Regular variations
Cs, Rb (Li, F, Sn, W) mafic to felsic	Increase strongly	Moderate increase
Sr, Nd isotopes	Generally primitive	Generally evolved

Note that the above criteria are not intended to be individually definitive, and also differ in significance. Also, suites formed dominantly by restite fractionation may change to a fractional crystallization mode after all or most of the restite has been separated from the felsic melt. From Chappell *et al.* (1997).

of cordierite and also of weakly zoned plagioclase, both of which were thought to be largely of restitic origin. Along with both Vernon and Clemens (1986 and 1989, respectively) I find myself unhappy with these textural interpretations which I might well construe to mean quite the opposite: that these crystals initially grew from liquids!

Whatever the general validity of the restite model the fact that the distinction between the S- and I-types at Lachlan is so remarkably clear strongly indicates a difference of source and, I also suggest, a difference in genetic process. However, in other comparable studies, such as those of the Scottish Caledonides which I touch on later, the differences are not so clear cut, as we might well suspect in view of the greater possible range in source rock materials and the greater maturity of source-rock sediments in the Scottish example. How far the restite model holds generally and how valuable is the distinction implied by the specific labels will long be discussed.

At this point the reader will doubtless remember that I have already argued that many magmas are in fact crystal-bearing mushes for much of their evolutionary history, a view that might be expected to favour the restite hypothesis. But do they develop their mushiness intratellurically or inherit it? Again I do not think there is a general answer, and whether a particular granitic pluton represents the intrusion of a restitic or a crystallized mush has to be argued for in each example, as in the following.

POSSIBLE S-TYPES IN THE EUROPEAN HERCYNIAN: ITALY IN PARTICULAR

Peraluminous granites are widespread in the Hercynian of Europe and a particularly instructive example has been described by D'Amico, Rottura, Maccarrone and Puglisi from the Calabria–Peloritani Arc of southern Italy (D'Amico et al., 1981; Rottura et al., 1993), where plutonism followed the emplacement of a complex pile of nappes. While the predominant granitic suite of this region is calc-alkaline and metaluminous, with tonalites and granodiorites predominating, there is also a contrasting minor suite of peraluminous monzogranites and leucogranites. The latter occurs as relatively small, unzoned plutons from a few square km to $150 \, \text{km}^2$ in area, sharply cut out of a variety of lithologies, in the more pelitic of which they produce simple, static thermal aureoles.

Some of these granites are particularly peraluminous, with accessory muscovite and sillimanite, with or without cordierite and/or andalusite. Fine to medium in grain, rather leucocratic, and everywhere associated with pegmatite aplite dykes and veins, they present a strong contrast with their Lachlan analogues, which are generally less evolved and remarkably free of such volatile-rich phases. D'Amico and co-workers discussed at length the origin of the muscovite, cordierite and aluminosilicates, concluding from their textural studies that andalusite largely crystallized from the melt, but sillimanite and cordierite represent relics, although the several generations of muscovite include restite, crystallizate and a post-consolidation growth phase – a common finding for the growth history of muscovite in granites (Figure 8.5).

In view of our present discussion, the most significant textural feature is the ubiquitous occurrence of tiny metamorphic aggregates of either muscovite with fibrolite ± cordierite ± quartz ± plagioclase ± apatite and opaques, or otherwise with cordierite and biotite, or mineral clusters of muscovite and biotite with relics of fibrolite. For many years these have rightly been interpreted by Italian petrologists as representing relics from anatexis, and such evidence, coupled with the typical S-type

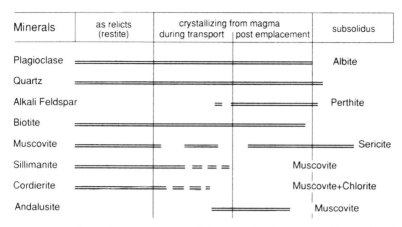

Minerals	as relicts (restite)	crystallizing from magma		subsolidus
		during transport	post emplacement	
Plagioclase	═══════════	══════════════	═════════	Albite
Quartz	═══════════	══════════════	═════════	
Alkali Feldspar		=	════════════	Perthite
Biotite	═══════════	══════════════	═════	
Muscovite	═══════════ ═	═══════	═══════════	Sericite
Sillimanite	═══════════	= = =		Muscovite
Cordierite	═══════════	= ═ ═		Muscovite+Chlorite
Andalusite			═══════════	Muscovite

Figure 8.5 Crystallization sequence of the peraluminous granitic rocks of the Calabria–Peloritani region of southern Italy. After D'Amico *et al.* (1981), but see Rottura *et al.* (1993).

geochemistry, convincingly argues for the partial melting of a metamorphic assemblage involving fibrolite ± cordierite paragneiss and muscovite schist, rocks which are present in outcrop within the Calabria–Peloritani zone.

Rottura and his colleagues have continued the study of these strongly peraluminous granites (1993), focusing on their origin in the belief that they would provide an excellent test of whether or not they represent independent melts of crustal origin. Most revealing is the fact that, despite the lack of a systematic correlation between elemental abundance and sample location, the geochemical studies showed strong inter-variable compositional correlations, and hyperbolic trends on element ratio *vs.* element concentration plots. Furthermore a multivariant statistical analysis indicated that a single dominant process was responsible, while the geochemical trends were compatible with either a two-component mixing or an unmixing melt–restite model. It would seem to me that the fact of spatial heterogeneity is most consistent with the latter alternative.

These authors then undertook an exhaustive programme of numerical modelling which took into account all the possible processes, finding that a particularly good data fit could be obtained with a model crystal–liquid unmixing process. This had the advantage of allowing estimates to be made of the possible compositions of the 'ideal' solid and liquid, and even of the likely proportions of the early-formed crystals. Even so a certain variability in the Sr_i and εNd_i values was thought to suggest that some degree of assimilation had continued throughout the evolution of the magma, which though very likely, might just as well result from the involvement of a heterogeneous crustal source.

Turning to the textural and outcrop evidence, none was found to support the mixing or mingling of contrasting magmas and, revising certain earlier opinions, the authors argued against a restite origin for any part of the biotite, plagioclase and quartz – the dominant mineral components – while continuing to accept sillimanite and cordierite as residua from partial melting. Rottura and his colleagues were not persuaded, in this 1993 paper, that the metasedimentary aggregates were of restitic origin, accepting them as true xenoliths, but the earlier suggestion to the contrary is, I believe, nearer the truth. From a welter of data and discussion the authors opted for a model involving crystallization from a liquid early separated from a crystal mush. But what was the origin of this mush?

It seems that the most important characteristics to be taken into account are: the relatively high Sr content and the decoupling of Ba and Rb; trace element patterns similar to the coeval I-type granodiorites; highly fractionated REE patterns, depleted in HREE; crustal Sr and Nd isotopic signatures. The REE evidence suggested derivation of a melt in equilibrium with a garnet-bearing residue, which would seem to accord with the finding that the estimated starting composition of the magma is near that of the experimental melts from pelites. However, the authors detected compositional discrepancies, and were unable to accommodate the Ba–Rb decoupling with the single hypothesis involving remelting of a wholly pelitic source. Neither did the Sr_i data support an exclusive derivation from any of the rock types presently in the Calabrian region, being more akin to the values reported for the associated I-type granodiorites, with which they share other geochemical features. Yet despite these affinities a direct derivation of the peraluminous suite from its calc-alkaline associate was ruled out in the face of a lack of field data and the incompatibility of the geochemical trends as a whole. Faced with this apparently contradictory evidence a rather complex explanation was envisaged involving a mixture of both crustal- and mantle-derived components, a possible parent being a distinct batch of calc-alkaline magma which had mixed with melts deriving from the lower crust, resulting in a mush of mixed melts and restitic solids.

I have reported this particular investigation at some length because it well illustrates the intricacies of formulistic modelling. Wholly rational though this is, I would not expect any formulation to model exactly the natural process or processes involved, and would settle for a rather simpler explanation involving the melting of a composite source, modelled as high grade pelites intruded by I-type granites. Indeed, such a thesis of anatectic melting had been adopted in the earlier publication of D'Amico and his colleagues, who had even addressed the problem of assessing the relative proportions of the melt and restite, using a formula established experimentally by Wyllie that showed that, to a first approximation, the amount of melt is about three times the percentage of alkali feldspar, which in this example is 60% for the monzogranites. Though they avoided discriminating between restite and precipitate, the total proportion of crystals is consonant with their interpretation of the mineral growth histories (Figure 8.5). It is the sort of approximate geological answer we might well expect knowing that it must vary throughout the whole period of unmixing, segregation and ascent. Furthermore, it is wholly in accord with the calculations of Huppert and Sparks discussed earlier. If only a relatively small proportion of the restite was entrained, the residue must have remained *in situ*, perhaps to be identified in the granulitic-kinzigitic rocks of the Calabrian basement, introducing a general thesis of the formation of deep crustal rocks by degranitization – that is, the extraction of granitic melt – which I will return to later.

On a more prosaic theme I draw attention to a further finding of the Italian team that the strongly peraluminous, S-type granites occur along with a somewhat less peraluminous, but more abundant, 'mesaluminous' suite of primary muscovite-bearing, biotite granodiorites and granites, characterized by the absence of aluminosilicates, a paucity of hornblende, and a lack of clearly identifiable restite – a situation common to many other segments of the European Hercynian. It seems, therefore, as confirmed by Barbarin in 1996, that coeval peraluminous granites fall naturally into two groups, with a rather leucocratic two-mica granite carrying magmatic muscovite and often associated with pegmatite, also poor in enclaves of any sort, contrasting with a more mafic biotite-rich granodiorite or granite, often rich in cordierite but poor in primary muscovite, lacking pegmatite but carrying restitic material. According to Barbarin the difference can be explained by reference to the experimental finding that a higher initial water fugacity is necessary for the precipitation of muscovite relative to that needed for cordierite: in short 'wet melting' yields a two-mica granite, 'dry melting' a cordierite granite. Barbarin also noted that the difference also tallies with the observation that the two-muscovite granites as characteristic of ductile shear zones – in which fluids might preferentially circulate – the cordierite-biotite granites showing no such relationship. It follows, he contended, that the physico-chemical conditions control which granite type results from the same source rock.

Depth in the crust might also be a factor, but I find most intriguing the possibility that differences in the protolith are also involved, noting that D'Amico and co-workers have suggested that existing igneous rocks may have formed the particular protolith of the mesaluminous magmas. Their view that granites themselves can be remelted and recycled is attractive, perhaps leading to a general thesis of repeated refining that might well explain the granite series, whereby true anatexites, migmatitic granites, restitic granites and evolved granites are but parts of a gradual refining of an ancient metamorphic and granitic crust.

S-TYPE MAGMAS

In view of this discussion we need to search for natural examples of partial melts in which liquid has separated out almost completely at source, leaving an impoverished residuum.

One of these is the generation of a thin sill of S-type microgranite in the aureole of a large body of a coeval gabbro at Cashel, western Ireland. According to Ahmed-Said and Leake, the changes in the composition of the aureole pelites imply a loss of material almost exactly equivalent to that of the rather leucocratic microgranite of the sill. Thus it seems likely that the latter represents magma produced *in situ* by melting, and segregated into a structural zone; this despite its compositional contrast with the leucosomes in the adjacent hornfelses, which is not an uncommon situation in migmatitic terrains. If so, it is easy to deduce that the elements were fractionated into the melt in an order almost exactly opposite to that in magmatic crystallization, even to the extent of the development of a negative europium anomaly in the segregated microgranite!

In a contrasted example, especially in terms of scale, magma with only a few suspended crystals was required in immense volume during the intrusion of the great leucocratic sheets of the Higher Himalayas. From the studies of Le Fort and co-workers there seems no doubt that these highly evolved, tourmaline-rich, and sometimes andalusite- and cordierite-bearing, two-mica granites, are of crustal origin, yet they are devoid of the xenoliths or mineral aggregates that might be construed as restite. There is good evidence supporting a model of extraction of homogeneous melt, by partial melting, from an underthrust slab of Tibetan gneisses (Chapter 18).

In terms of origin, Manaslu and Cashel provide two very different aspects of crustal melting, one resulting from the thermal blanketing effect of thrust piling, the other from the intrusion into the crust of voluminous basic magma. Both, however, are examples of the production of a granitic melt largely devoid of restite. This emphasizes the point made by Chappell and White that we must expect all degrees of separation, leading to restite-poor or restite-rich, primary magmas. I like to believe that the obvious contrast between the Lachlan granites and those of Manaslu – the absence in the former of aplogranite and pegmatite, and their abundance in the latter – is due to the simple fact that a high concentration of volatiles will promote liquidity and ease of extraction. I also like to believe that it is these liquid, mobile magmas that are preferentially segregated into structural traps such as are provided by intersecting faults and thrusts.

Although the multifactorial nature of geological processes such as the remelting of crustal rocks implies that the evidence can often mean all things to all men, my prejudice is that granitic magmas in general are first generated as liquids with but a modicum of suspended crystals, restite or precipitate, with crystal separation then proceeding apace as soon as the magmas move away from the source region. I will later return to these important matters, particularly this problem of sorting solid from liquid, in further discussions of partial melting in the production of migmatites, and bulk melting in the genesis of batholiths.

SELECTED REFERENCES

Chappell, B.W. (1997) Compositional variation within granitic suites of the Lachlan Fold Belt: its causes and implications for the physical state of granite magma. *Transactions of the Royal Society of Edinburgh: Earth Sciences*, **87**, 159–170.

Chappell, B.W. and White, A.J.R. (1992) I- and S-type granites in the Lachlan Fold Belt. *Transactions of the Royal Society of Edinburgh: Earth Sciences*, **83**, 1–26.

Chappell, B.W., White, A.J.R. and Wyborn, D. (1986) The importance of residual source material (restite) in granite petrogenesis. *Journal of Petrology*, **28**, 1111–1138.

Clemens, J.D. and Wall, V.J. (1981) Crystallization and origin of some peraluminous (S-type) granitic magmas. *Canadian Mineralogist*, **19**, 111–132.

Collins, W.J. (1996) Lachlan Fold Belt granitoids: products of three-component mixing. *Transactions of the Royal Society of Edinburgh: Earth Sciences*, **87**, 171–181.

Wyborn, D. and Chappell, B.W. (1986) The petrogenetic significance of chemically related plutonic and volcanic rock units. *Geological Magazine*, **123**, 619–628.

The mingling and mixing of granite with basalt: a third term in a multiple hypothesis

9

The nature of the mutual reactions which have taken place among the several rocks indicates that they were intruded in somewhat rapid succession and even in certain places that one was not completely solidified before it was invaded by the other.

Alfred Harker (1904) Memoir of the
Geological Survey of Scotland, *p. 181.*

PREAMBLE: MIXING AND MINGLING

The nature of the dykes and xenoliths associated with a granitic pluton provides important clues as to the rheological history of its magmas, its internal compositional variations and its mode of emplacement. Yet even to begin to discuss this vital evidence it is necessary to resolve a certain confusion in the nomenclature, for I can well understand Didier's exasperation when in 1964 he wrote: 'Les géologues de langue anglais désignent depuis longtemps les enclaves par un série des noms souvent pittoresque mais imprécis.' However, I fear that the alternative international terminology provides little improvement, introducing a rather artificial genetic division of the enclosures – of the enclaves, to accept the French equivalent.

There are, after all, just five classes of enclave within granites, representing: (1) restite, the solid residue from the partial melting of source rock; (2) fragments collected from the country rocks during the ascent of the host magma; (3) cognate material representing early consolidated variants of the host; (4) blobs of quasi-liquid magma derived from comagmatic mafic intrusions; and (5) accretions of intratelluric crystals. Didier and Barbarin provided a full discussion in comprehensive reviews in 1991.

Whatever the terminology, it is commonplace that the mafic enclaves of metaluminous granites often bear a close petrographic resemblance to the associated mafic intrusions. This is just one indication of the common coevality of basaltic and granitic magmas which has been noted from the earliest studies: as long ago as 1857 Durocher proposed that hybrids might be produced by the commingling of these two different magmas. Although the extent of the mixing of such rheologically contrasted melts is a matter for discussion, their coexistence is an observable fact, particularly within subvolcanic, centred complexes.

In 1904 Alfred Harker, in a masterly study of the Tertiary centre on Skye, succinctly stated what is the current model in the form quoted above. Since then many workers have taken the view that microdioritic (or microgranitoid) enclaves come from the same stock as their host – that they are cognate. My reading of the ideas current during the first half of the 20th century convinces me that a number of authors had also realized that the enclave was not solid at the time of its inclusion. Pabst, in his classic 1928 study of the mafic enclaves in the Californian batholiths, was certainly aware of liquid in liquid features, but the most telling proof came from Wager and Bailey's seminal observation, in 1953, that the mafic globules within the felsic ring dykes of the central complexes of St Kilda, Scotland, and Slieve Gullion, Ireland, had chilled against their felsic host. More recently Vernon, in 1984, has provided an updated statement of this thesis and it seems undeniable that mafic magma can inject felsic magma in a manner analogous to the formation of pillow lavas in water.

In describing such two-magma interactions it is indeed fortunate that the English language clearly differentiates between mixing – having the sense of combining – and mingling – meaning to move among while retaining an identity, a distinction wholly appropriate in these discussions.

We have already learnt that the recharge of high level felsic magma chambers by mafic magma is likely to be an important process in the formation of zoned plutons. Such recharge is not restricted to high level, centred complexes but is a ubiquitous feature of plutonism, though manifestly more characteristic of certain global tectonic environments than of others – for example, much more so in plate-edge batholiths than in those related to continental collision. Furthermore, differences in the relative volumes of the two magmas result in a range of phenomena extending from the mixing of two fluids of similar viscosity, through stages of mingling where there is a large viscosity difference, to one where fluid mafic magma is intruded early into fractures in a cooling, quasi-solid, granitic host.

It seems natural to review the evidence for these magma interactions in these terms, starting with the contact relationships between substantial bodies of gabbro and granite, then moving on to those situations where a granitic pluton hosts synplutonic basaltic dykes, or where isolated mafic enclaves are scattered throughout a pluton, and finally to the bulk mixing of magmas at the crust–mantle interface.

GABBRO AGAINST GRANITE

Almost universally the gabbros are the precursors of the granites in any one magmatic cycle. Nevertheless, there is every degree of overlap in time and, as a result, every degree of interaction between the two components. Indeed, the relationships can vary from sharp to semi-gradational even along a single contact, and although this may sometimes be due to the reactants being locally flushed away, it is more often, I believe, the result of the critical nature of the thermal and rheological constraints.

There are some very clear examples of arrested hybridism. One is the compositionally zoned Lamark Granodiorite in the Sierra Nevada, California, and its contact relations with a diverse suite of sizeable mafic intrusions. Frost and Mahood provided a fascinating account linking the resultant interactions with the generation of the small mafic enclaves that are scattered throughout the granodiorite, though they are most abundant at its periphery.

Good evidence of the synplutonic nature of the relationship is seen in the way the mafic intrusions cut across both the compositional zoning and the regional foliation of the Lamark pluton, yet commingle variously with the granodiorite when in contact. Where the compositional contrasts are sharp, abundant mafic enclaves flood the granodiorite, many with the fine-grained selvedge and cuspate contacts indicative of quenching of liquid against liquid. Such mafic blobs show little or no reaction with their host, presumably because they were quenched before they could mix. Where, however, the host is itself relatively mafic, or the local scale proportion of the basic component is large, there is extensive mixing leading to hybridization and homogenization, with diffuse schlieren more evident than discrete enclaves.

The uniformity of mineralogy between the enclaves and their immediate host signifies that thermal equilibrium was reached, whilst the fine to medium grain and equigranular texture of the enclaves, together with the skeletal habit of the minerals, particularly the apatites, are consonant with rapid quenching. A ubiquitous feature is the presence of plagioclase megacrysts within the enclaves, which is discussed in detail later.

The variation in the major elements across the whole spectrum of rock compositions in the Lamark Granodiorite shows the near linear trend to be expected of a mixing process. However, Frost and Mahood claimed that there is a subtle change of slope of the trend line at about 63% SiO_2 (Figure 9.1), which corresponds to the maximum silica content of the hybridized host. They interpreted this to mean that only those parts of the Lamark Granodiorite with SiO_2 less than 63% can have resulted from magma mixing, the higher range being attributed to fractional crystallization. I am not wholly convinced that there is a significant change of slope and indeed this is a good example of the deficiency of the Harker diagram in discriminating between genetic processes. Nevertheless, I am fully prepared to accept that both mixing and fractionation are likely to have been involved.

Frost and Mahood modelled and calculated the physical conditions of a range of mingling and mixing situations, demonstrating clearly the combined effects of temperature, composition, and the importance of the crystal and water content of the magmas. They surmised that the latent heat released from a large body of crystallizing gabbro is absorbed by the remelting of crystals in the adjacent granitic magma and so ensures that the latter remains sufficiently fluid to mix completely with the basic magma. At a point so near to the bulk melting of either component fairly small accessions of heat could temporarily halt or even reverse the crystallization trend. Also, it is a characteristic of pseudoplastics, of which these crystal-bearing magmas are surely an example, that small changes in temperature produce great changes in apparent viscosity.

These workers showed that mafic magma probably interacted with the

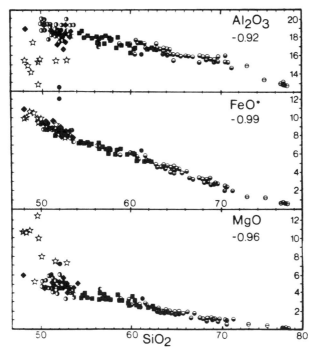

Figure 9.1 Sample of the Harker diagrams provided by Frost and Mahood (1987) for the Lamark Granodiorite and associated mafic rocks. Lamark granodiorite (horizontally half-filled circle); aplite (divided circle); cumulus schlieren (filled circle); hybrid (filled square); hybrid and mafic enclaves (vertically half-filled circles); mafic intrusions (diamonds); hornblende gabbro inclusions (stars). Least-squares correlation coefficients are shown for mafic enclaves, intrusions and hybrids alone. Reproduced with the permission of the authors and the Geological Society of America.

Lamark Granodiorite throughout its entire history of emplacement and cooling, certainly from the stage where globules were being shed into, then mingled and mixed with a crystal-bearing magma, to the stage of the injection of synplutonic dykes. They also concluded that, during the evolution of the granodioritic pluton, increasingly large volumes of basic magma would have had to be injected for any appreciable hybridization to occur.

PRECURSOR APPINITE AND DIORITE AGAINST GRANITE

Despite such clear examples as that just described it is the products of hybridization that are more apparent than the process itself. This can be best illustrated by reference to a characteristic suite of variable dioritic rocks, the appinite suite, associated with, but precursor to, certain granite plutons.

In an example at Ardara, Donegal, the pluton itself contains a scattering of small, mafic enclaves locally concentrated along external and internal contacts. These enclaves show a considerable variation in texture and mode, copying exactly the mixture of rock types within an external cluster of appinitic intrusions. Coupled with the outcrop evidence of direct incorporation of such enclaves from one particular, contiguous, appinitic body, there is little doubt that these igneous enclaves were derived from the appinitic precursor, complete with their pre-incorporation lithologies. Some were even sufficiently rigid at the time of incorporation to break into angular fragments, and there is little evidence of mingling or mixing with their immediate host at the present level of exposure. As a specific model it is easy to envisage the upwelling granitic magma finally catching up with, and deriving material from, a vanguard of more mobile, differentiating, mafic intrusions.

A more general conclusion is that enclaves, though representing comagmatic igneous material, can have a decidedly more complex history than one of direct derivation from a synchronous intrusion. The source material need not be disposed in separate intrusions, as at Ardara, but simply represent the early consolidated parts of an evolving pluton, disrupted and entrained by later pulses. As hinted above, fairly small differences in the time of intrusion relative to the crystallization of the host, coupled with disparity in the relative volumes, will produce a variety of relationships.

SYNPLUTONIC PIPES, NET VEINS AND MAGMA PILLOWS

One such relationship concerns the rise of pipe-like bodies of granite into overlying gabbro which quenches against the pipe material. Elwell described dramatic examples from the Tertiary Centre of Slieve Gullion,

Ireland, and also, with colleagues, from the Cadomian intrusive complex of Guernsey. Similar phenomena from a Caledonian intrusive complex in Donegal are illustrated in Figure 9.2.

Further evidence of the coevality of contrasted magmas is provided by net veining in which the host mafic rock shows a pillowed form with quenched margins and cuspate contacts against clearly intrusive felsitic veins. Here again the granitic material is likely to have been kept mobile by the superheat of the host – so mobile that the felsitic magma back-veins into the mafic pillows. Blake and co-workers provided details of the classic examples of such phenomena in the Tertiary Igneous Province of western Scotland. There, the way in which the pillowing gradationally gives way to angular, straight-veined blocks into the body of the invaded

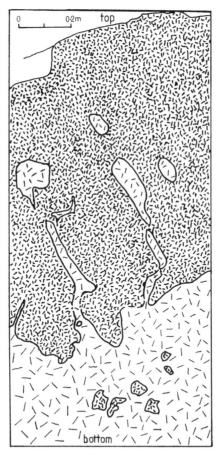

Figure 9.2 Oblique bird's-eye view of a low cliff sloping towards the observer. Pipe-like bodies of granite rise from a sheet of granite underlying a mass of hornblende gabbro. West of Portnoo, Donegal.

basic mass suggests that there was an advancing pulse of brecciation brought about by the explosive release of gas, a process wholly consonant with the subvolcanic environment.

Throughout this discussion we have seen that a common factor determining whether mingling advances to mixing is the provision of extra heat, itself determined by the relative volume of the mafic magma, a finding particularly relevant in the discussion of the phenomenon of synplutonic dykes.

SYNPLUTONIC DYKES

Synplutonic mafic dykes are far more common than might be supposed from published work, though to some extent this omission and misconception has been repaired in a number of recent studies. In the simplest example there is a straightforward cross-cutting relationship where mafic dykes have been injected during the geologically brief interval between two separate pulses of a multipulse pluton. Of particular interest is the common finding that a few members of the swarm cross and penetrate into the second intrusion, only to be disrupted and dispersed as enclaves within the latter.

This phenomenon of the disruption of a mafic dyke within a granitic host is ubiquitous. It was discussed in some detail by Roddick and Armstrong in their 1959 study of synplutonic dykes in the Coast Mountains of British Columbia, and their descriptions can be matched in examples from many Mesozoic circum-Pacific batholiths.

The special features of such dykes are: a recrystallization of the dyke with the production of a microgranular, hornfels-like texture, and with equilibration of the mineralogy of dyke and host; a necking of the dyke along its length, often at the points where it had been displaced by minor faults before the completion of crystallization of the host; a back-veining into the dyke, sometimes of the actual host, sometimes of a leucocratic variant of the latter, at other times of a pegmatitic segregation; and, not least, a dismemberment of the dyke into trains of angular or amoeboid enclaves, the latter often with the familiar cuspate margins convex towards the host. The inference is clear – that basic magma invaded an unconsolidated, yet relatively cooler granitic host.

In the example of the Caulfeild dyke swarm in the Coast Mountains Batholith andesitic dykes with an overall marked regional trend intersect a granodioritic host. Perhaps the most illustrative outcrops occur on the shores of Cortes Island in Queen Charlotte Strait, where narrow, linear trains of angular enclaves represent an advanced stage in the disruption of the regional dyke swarm (Figure 9.3) and where, remarkably, one of these dismembered synplutonic dykes can be seen to be cross-cut by another less disrupted member of the swarm. The variety of interaction is

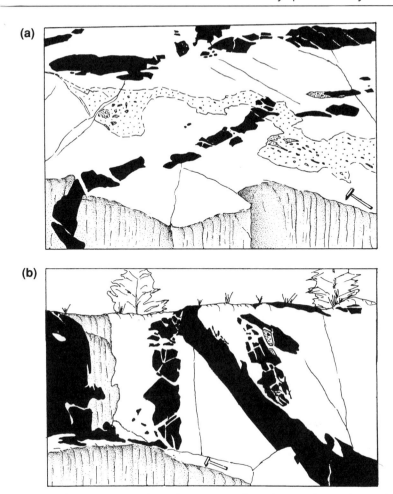

Figure 9.3 Synplutonic mafic dykes exposed on the southern shore of Cortes Island, opposite George Harbour, Queen Charlotte Strait, British Columbia, Canada. (a) Oblique view on edge of cliff of two members of the swarm (black) within a granitic host (white). The one in the foreground cuts across a fluxioned zone of more dioritic composition (stippled), replete with half-digested basic enclaves. (b) View on the cliff-face of a less disrupted member cutting a more disrupted, earlier member of the swarm, apparently along a healed zone of displacement.

also impressive; enclaves with angular and globular shape, either of which is variously hybridized, occur in the same train.

Synplutonic phenomena have also been well documented from the Coastal Batholith of Peru by myself and Andrew Bussell in 1985. A key example is a swarm of basaltic andesite dykes cutting various facies of the Santa Rosa pluton in the Huaura Valley, particularly in the Quebrada El

Carmen. Dykes with chilled margins clearly cross-cut an outer tonalitic facies but then progressively undergo recrystallization, necking, separation, intrusion by pegmatite and then disruption as they enter a central monzogranite facies, when some dykes continue as a string of lobed

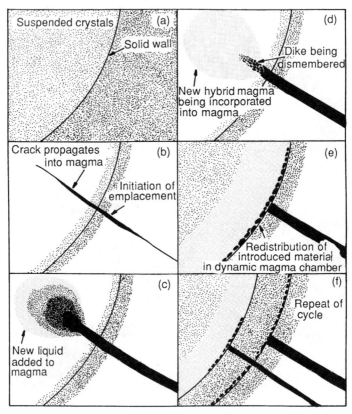

Figure 9.4 Diagram illustrating inferred relationship between mafic dykes and inclusion trains. (a) Wall of crystallizing magma chamber showing solidified material to right and magma containing suspended crystals to left. (b) Propagating crack intersects solid–magma interface. Mafic liquid is then emplaced along the crack. Magma adjacent to the pluton wall behaves in a brittle manner at the time-scale of crack propagation and dyke emplacement. (c) Flow of liquid through conduit, with inferred addition of liquid to the volume of liquid already resident in the chamber. (d) Mixing of new liquid with old liquid, probably accompanied by considerable precipitation of plagioclase and hornblende because of the temperature difference between the dyke liquid and the ambient temperature within the magma chamber. Partly solidified dyke is veined by tonalitic liquids from the magma chamber; pillows of mafic liquid may form, depending on the viscosity contrast between the resident and added liquids. (e) Re-establishment of localized convective system within the magma chamber. Hybrid liquids, suspended crystals and disaggregated dyke rock are redistributed parallel to the pluton walls. (f) Repeat of cycle. Reproduced from Hill (1988) in dedication to the author and with the permission of the *Journal of Geophysical Research.*

lenticles, yet always maintaining the regional trend in a similar way to the Cortes Island example. A dyke origin for these trains of mafic enclaves is not in doubt, with differences in the form of the enclaves surely relating to local differences in the viscosity contrast between dyke and host.

That mafic magma can be injected as a regularly oriented dyke swarm into a still mobile and crystallizing granitic host supposes that fractures develop very early in the cooling history of granitic magmas. There is much independent evidence of this, and early fracturing was recorded by myself and Berger and by Hibbard and Watters. That such quasi-magmas deform by fracture or flow must depend on the rate of application of the stress, but in the seismically active regime of magma emplacement we can surely expect that earthquake shocks will be continuous and promote fracturing whenever the rheological state is appropriate.

Synplutonic dykes are but the end phase of the continuous process of mafic magma injection into a pluton throughout its assembly history. An indication that this is so is shown in Figure 9.3a, which depicts an enclave-rich schlieren intersected by a partly dismembered synplutonic dyke, which in its turn is cut across by a much less disrupted member of this dyke swarm. Such synplutonic phenomena are now widely recognized though variously interpreted, largely because of the multifactorial nature of the processes involved. Elsewhere I have cited examples to illustrate this ubiquity (1991), referring to localities in the western USA, Peru, southern Corsica, Sinai, and also the southern Malay Peninsula, where Kumar referred to such phenomena as enclave dykes. Always the dyke material is more mafic than the host, usually much more so, and always the material of the dismembered dyke can be closely matched with the isolated enclaves in the corresponding host.

As a summary and model explanation of these rather complex relationships I can do no better than reproduce a diagram provided by Hill from his 1988 study of the San Jacinto complex, southern California, which is highly appropriate and self-explanatory (Figure 9.4).

SOME POSSIBLE ALTERNATIVES: PSEUDO-DYKES AND RELICT DYKES

There are other possible scenarios for such phenomena. It is common in contact relationships that angular fragments spalled off from gabbroic country rocks are progressively rounded by preferred corrosion at the edges as they are engulfed and streamed out in the flowing silicic magma. It is also a common observation, wholly consonant with rheological theory, that all types of enclave can be concentrated in such flow streams, especially along internal contacts. Nevertheless such xenolith trains are rarely to be confused with those derived from synplutonic dykes with their maintenance of dyke-like widths and their quench features.

Another possibility is that such trains of enclaves might sometimes represent resisters from wholesale granitization of their host, but I regard this as an untenable thesis for high level plutons. Nevertheless there is a need to probe more deeply where reaction and deformation are more manifest. Thus the conformative mineralogy and the microgranular texture shown by many such dykes may sometimes represent a primary crystallization under amphibolite facies conditions, as would be appropriate to the near-melt temperatures of the granitic host. Such a situation is well known in high grade metamorphic terranes where the accompanying deformation leaves the dyke as boudins of schistose amphibolite within a gneissic host, an example of the Sederholm effect so well described by Watterson from the Greenland basement. In this same vein Bishop in 1963 cautioned that in some deep-seated plutonic environments both the dyke and the granitic host may have been locally recrystallized and remobilized. Even for high level plutons it could be argued that the reaction between mafic dyke and felsic host is but part of a wholesale 'stewing in its own juices', only to be expected in the hot centres of slow cooling plutons, especially where the heat supply is reinforced by magma replenishment. It is in just such silicic cores of plutons that synplutonic phenomena are so often found. Furthermore, deformation is not infrequently involved in pluton and dyke emplacement, and examples cited from Peru and Sinai demonstrate very well how such dyke trains can focus ductile shearing. Within this zone of overlap between magmatism and metamorphism we must expect synchronous dykes to experience all the effects of the continuing evolution of the host, including the continuous recrystallization resulting from synchronous deformation. Simple models are the products of research reports not outcrops!

CRYSTAL TRANSFER: LES DENTS DE CHEVAL

A ubiquitous feature of synplutonic intrusion is the appearance within the enclaves of crystals common to the host, particularly of megacrysts of the feldspars, plagioclase ovoids and ocelli-like aggregates of both plagioclase and quartz.

This introduces a longstanding problem of whether such crystals grew *in situ* within the enclave or represent xenocrysts derived from the host. For a long time I favoured the metasomatic hypothesis, claiming that the near identity between the two megacryst populations, even extending to the physical and chemical zoning of the feldspars, would probably result where there was a free exchange of intergranular fluids within a system in thermal equilibrium. Feldspar porphyroblasts do grow in pelitic xenoliths, but I am grateful to Bernard Barbarin for pointing out that much more often the feldspar megacrysts occur only in those microdiorite enclaves of probable igneous origin, even where there is a mixed population of xenoliths and enclaves. Furthermore, feldspar crystals lie athwart

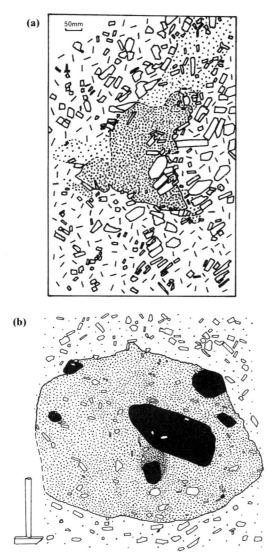

Figure 9.5 (a) Porphyritic biotite granite of the Land's End pluton, Porth Ledden, Cornwall, with mafic enclave grading into dioritic hybrid. Potassium feldspar megacrysts appear as if being inserted into the enclave. Within the hybridized portions the megacrysts can be considered to be xenocrysts. (b) The 'double enclave' phenomenon in the Shap Granite, Cumbria. Microdioritic enclaves lie within the quartz diorite, which is itself enclaved in porphyritic granite. The potassium feldspar megacrysts occur within each facies and can be seen to lie athwart the contacts. The interpretation is that a dioritic hybrid, produced by mixing at depth of mafic and felsic magmas, retained blobs of the mafic magma, with one blob carrying an envelope of the actively forming hybrid. The megacrysts were transferred between the several magmas during transport.

the contacts of enclaves without any sign of reaction within either medium, and such crystals have the appearance of being mechanically inserted (Figure 9.5a). On balance, therefore, I accept the derivation hypothesis as providing an acceptable explanation.

That some kind of reciprocal, mechanical exchange of crystals can take place across magma–magma interfaces was suggested as long ago as 1894 by Grenville Cole, then in 1904 by Harker, and was reinforced by Thomas and Campbell-Smith's study in 1931 of the hybrids and basic igneous xenoliths associated within the Tregastel Granite of the Cotes-du-Nord. To Bailey and McCallien, in 1956, this concept provided an adequate explanation of the occurrence, in the outer basic margins of the composite dykes of the Scottish and Irish Tertiary Province, of potassium-feldspar megacrysts identical to phenocrysts in the central felsite. They envisaged basic magma being squirted through an overlying, less mobile, felsitic magma within the source magma chamber, thereby scavenging the suspended phenocrysts.

Such an attractive thesis can now be shown to be in accord with simple experiments and rheological theory is especially applicable, according to Koyaguchi, when magmas flow together through a conduit, as it the case with composite dykes. However, whereas purely mechanical transfer adequately explains the enclave megacrysts as xenocrysts in subvolcanic environments, the porphyroblastic nature of similar crystals in the enclaves of deeper seated plutons, coupled with the relative infrequency of quenched contacts, suggests that chemical exchange becomes important in these deeper seated environments where the heat balance is surely maintained for long periods. I think it likely that any mechanically transferred crystals will then seed further metasomatic overgrowth within the new medium. Such a multifactorial thesis would explain many of the apparently conflicting textural features which have for so long puzzled petrologists. However, it has to be admitted that all such explanations are sorely strained in attempting to account for the growth history of identical potassium feldspar in the three environments presented by the classical double enclave such as that illustrated in Figure 9.5b; but for confirmation of the xenocryst model I urge the reader to consult a paper presented by Cox and his colleagues in 1996.

THE MAFIC ENCLAVES THEMSELVES

As we have seen, studies of gabbro–granite contacts and the synplutonic dykes provide good evidence of the mafic igneous origin of some of the decimetre-sized, ovoid, mafic enclaves that are so common a feature of metaluminous tonalites and granodiorites. But what of the origin of such enclaves in general?

Although I accept Vernon's thesis (1984) of the ubiquity of igneous origin, at least for the enclaves of metaluminous granites, the evidence of quenching can be equivocal because, as both Bishop in 1963 and Chapman in 1962 caution, we need to be careful of the interpretation of dark margins. Further, although the elongate and streaked-out shapes of the enclaves strongly suggest that they were still molten when enclosed in their host (Figure 9.6), here again we need to be aware of the effects of deformation. The simple fact is that in many plutons the enclaves differ from their host only in their finer grain and more mafic lithology and, furthermore, disaggregation, even to the stage of forming biotite clots, further confounds source identification and the possible confusion with restite. All this is presumably because chemical and mineralogical equilibrium is quickly established by the chemical and physical interchanges.

Such exchanges were generally described by Stephen Nockolds in his studies in the 1930s of the contamination of granitic magmas, from which he deduced that the volatile constituents play a most important part. There are now sufficient modern studies to establish more precisely the pattern of these exchanges, but because they are the result of both chemical and mechanical processes they are not necessarily volume for volume. Indeed, elements such as Cu, Ni, Co, Cr and V may be reduced in the

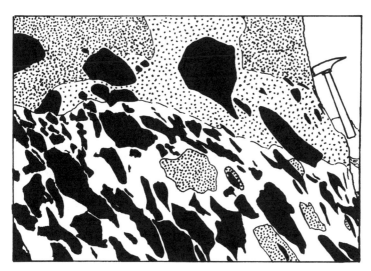

Figure 9.6 Mafic enclaves in granodiorite (unstippled) and heterogeneous hybrid diorites (stippled in accord with colour index). Shows a range of contrasted viscosity phenomena common to mingling and mixing zones: flow elongation of mafic magma blobs; cuspate margins of the latter more evident when in granodiorite host than diorite host; hybridized material becoming xenolithic in the granodiorite; and the formation of 'double enclaves'. Pasillo de enclaves. Playa Canabal, Cangas de Morrazo-Moaña, Pontevedra, Spain. Deseribed by Corretgé *et al.* (1984).

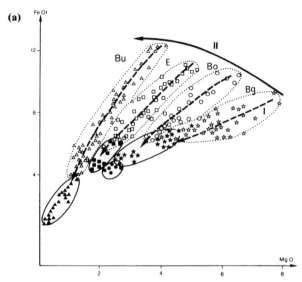

Figure 9.7 (a) FeO$_t$ versus MgO diagram for the Bono and Budduso granitoids and their enclosed magmatic mafic inclusions. Symbols: I (broken arrows), hybridization trend of the enclaves; II, tholeiitic evolutionary trend of the more mafic (less hybridized) hybridized enclaves. The dotted lines delimit the repartition areas of the enclaves; the full lines correspond to the host domains. Enclaves of individual plutons: Bu (△), Buddoso; E (□), Emauro; Bo (○) Bottida; Bg (☆) Burgos. Open symbols, inclusions (enclaves); closed symbols, host rocks. Reproduced from Zorpi *et al.* (1989) with the permission of the authors and Elsevier Science Publishers. (b) Hypothetical model illustrating the phases in the development of normal zoning in calc-alkaline granitic plutons, from the withdrawal from a compositionally stratified acid chamber intruded by a basalt (with a given FeO$_t$/MgO ratio) to the final emplacement of the pluton. (A) Injection of basalt (possibly through a dyke system). Beginning of pillowing and of the ascent of the basalt through the chamber. For the sake of clarity, a simple flat-topped rectangular shape has been assumed for the acid magma chamber and three compositionally different zones have been represented. Thick and thin arrows represent the withdrawal streamlines of the basic and acid magmas, respectively. (B) The basalt influx is continuing but decreasing. Ascent of the basalt which is still pillowing and simultaneous partial removal of the deeper layer of the stratified acid magma. (C) Emplacement of a first magmatic pulse originating from the lower zones of the chamber and withdrawal of the middle zone of the reservoir. (D) Intrusion in the core of the ballooning pluton of a second magmatic pulse originating from the middle zone of the chamber and withdrawal of the upper zone. (E) Third magmatic pulse into the pluton's core (i) and last episode of lateral ballooning (ii). Reproduced from Zorpi *et al.* (1989) with the permission of the authors and Elsevier Science Publishers.

mafic enclave by simple dilution. Nevertheless, certain gross exchanges are obvious on analysis, and there is a concensus among researchers that the elements showing the greatest mobility include K, Na, Cs, Rb and Ba, whereas Fe, Mg, Ni, Cr, Zn and V are the least mobile, with other compo-

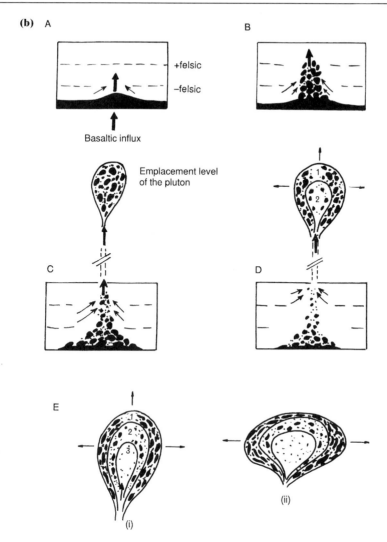

(b)

A

+felsic
−felsic

Basaltic influx

B

Emplacement level
of the pluton

C

D

E

(i)

(ii)

Figure 9.7 *Continued.*

nents showing variable mobility depending on the mineral phases involved: Stimac and his colleagues (1995) discuss this matter in some
detail. Furthermore, it seems that the exchange of isotopes can be quite
extensive, a finding that may well compromise the isotopic tracing of
source material: there are a number of relevant studies (Holden *et al.*, 1987;
Eberz *et al.*, 1990; Pin *et al.*, 1990; Fourcade and Javoy, 1991).

Selective exchanges are motivated by what Johnston and Wyllie referred to as 'up-hill diffusion'. For example, the preferred early nucleation
of hornblende and biotite in the enclave draws in water and potassium.
Visible evidence of such exchanges is the presence of a narrow felsic halo

outlying the biotite-rich rim of some enclaves; the preferred nucleation of biotite has required the withdrawal of iron and magnesium from the adjacent host.

In the main, such studies of the exchange between coexisting magmas have envisaged diffusion as a mechanism of element transfer (e.g. Lesher, 1994). However, Petford and his colleagues, in 1996, explored the possibility of the exchange occurring by advective transfer when melt infiltrates into the partially crystallized microdioritic enclave. The fact that serial sections of selected mafic enclaves from the Ross of Mull Granite, northwest Scotland, revealed pervasive and interconnecting melt channels, coupled with model calculations that indicated that advective transfer in low viscosity granitic liquids would be more effective than diffusion in transporting chemical components, persuaded the authors that this was indeed a viable process.

The basic igneous parentage of many mafic enclaves can be confirmed by reference to the chemical composition because, despite these reciprocal exchanges, the essential signature seems to survive, particularly when identified by certain immobile trace elements. However, the evolutionary history is much more difficult to discern, and sometimes impossible, as in environments such as that of the active plate margin batholiths, where all of the available mafic material, whether extrusive, intrusive, accumulative or residual, has the same heritage – impossible, that is, unless the whole geological context is considered in detail together with the geochemistry.

One such holistic study is provided by Zorpi and her colleagues in an account of the enclave relationships in five zoned granitic plutons from northern Sardinia. There, as is general the world over, the centripetal change from mafic to felsic lithology is paralleled by a decrease in abundance of the enclaves. These workers show conclusively that the composition of the latter relates to that of their respective host and, furthermore, that each population, like the respective host, is typified by a distinctive range in the FeO_t/MgO ratio (Figure 9.7a). It seems that not only was equilibrium reached in each population, but each ratio specifies a single episode of mingling and mixing with basic magma, representing a hybridization accomplished elsewhere and before entering the pluton.

In their modelling Zorpi and her colleagues echo the opinion of many petrologists in envisaging an early stage of compositional stratification in a chamber at a deeper level than the existing plutons – a stratification possibly resulting from crystal fractionation (Figure 9.7b). It is suggested that the injection of basaltic magma into the base of such a chamber introduces sufficient heat not only to remobilize the cooling magma from bottom upwards, but also to promote mixing and hybridization. Upwelling of the mixed magma in the order of remelting provides an explanation of the order of intrusion, the zoning, and also the greater concentration of enclaves and hybrids in the early pluses. Mingling, mixing, hybridization and crystal scavenging processes are likely to be enhanced by the turbulent stirring in the conduits feeding the higher level pluton and here, too, crystal exchange is boosted by shearing in the magma mush. As the pluton distends by central infilling, the early pluses with their suspended enclaves are progressively deformed. Once again the process is conceived to be continuous, though gradually waning, ending only with the intrusion of synplutonic basaltic dykes into a near consolidated host.

This elegant model must depend on the correct identification of the origin of the enclaves, and although I again agree with Vernon's 1984 opinion that most of the fine to medium grained mafic enclaves are probably of igneous origin and represent magma blobs, we have always to be aware of the other possibilities, particularly that they may sometimes represent restitic material derived from the source.

THE PROCESSES OF MINGLING, MIXING AND HYBRIDIZATION

As we have seen, of the several factors that determine the degree of interaction between synchronous granitic and basaltic magma, the most important are the relative volumes of the two magmas and the stage of crystallization of the host. We cannot expect mixing between a crystal mush and the enclaves suspended in it, except when the crystallization process is reversed by the superheat of the mafic intrusion. These points were made years ago by Roobol but have since been well summarized by Barbarin in 1989 in a diagram of the type shown in Figure 9.8.

It is because granitic magmas so often arrive in the higher crust at subliquidus temperatures that mixing and homogenization are far less commonly observed that mingling in outcrop. The exception lies in the volcanic centres where high heat flow and volatile concentration are particularly potent factors. Furthermore, as Sparks and Marshall pointed out, it is in this subvolcanic environment that energetic mixing will most easily result from the collapse of a cauldron block into an underlying magma chamber. Appropriately, what is arguably the type hybrid, Harker's marscoite, is the result of the mixing of basaltic and granitic magmas in a ring dyke within the Skye cauldron. Certainly it is within such subvolcanic magma conduits that we best observe the process of mixing, the magma blobs being streaked out into the flow lines, hybridized to mafic schlieren and eventually intermixed with their host.

As an example from my own experience I comment on Bussell's 1985 account of the relationships revealed in a quench of mixed rocks exposed in a ring dyke of the Huaura Centered Complex of Lima Province, Peru; but the reader must be prepared for the real complexities of Nature.

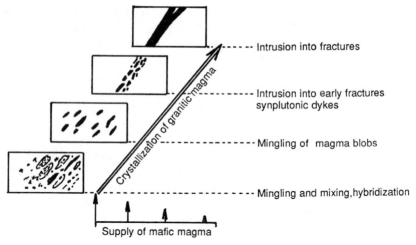

Figure 9.8 Changing interaction between basaltic and granitic magmas at different stages of crystallization of the latter. Modified from Barbarin (1989), Figure 7.

Owing to repeated cauldron subsidence the magmas, which were presumably derived from the top and bottom of a stratified magma chamber, were repeatedly mixed and permeated with metasomatizing fluids – in essence, a dioritic mush carrying phenocrysts of calcic plagioclase and hornblende mixed with a felsic magma carrying crystals of andesine oligoclase, hornblende, biotite and quartz. As a result the hybrid carries a contrasted and incompatible phenocryst assemblage in which the quartz crystals are rimmed with feldspars and are often in the form of ocelli, and where there is every variety of crystal size and form. Heterogeneity is very obvious in the outcrop in a melange of rock types, the diorite occurring as elongate pillows with either gradational or quenched and cuspate contacts against a veritable potage of magma and crystals. Remarkably, that part that Bussell dares to identify as the pristine dioritic component shows chemical trends that support a thesis of fractional crystallization. On the other hand, the obviously mixed rocks show compositional trends with frequent inflections, possibly to be interpreted as resulting from different episodes of mixing.

From this glimpse of the physical and chemical mayhem that is a feature of these complex situations a common pattern emerges wherein fractional crystallization, magma mixing and gas fluxing all play ever changing parts. Perhaps it is unrealistic to expect that we might recognize the separate contributions by the simple plotting of compositional trend lines, yet with careful mapping and sampling this is sometimes possible. Nevertheless, in trying to resolve this complexity we need to be aware of the longstanding controversy on the reality and significance of linear compositional trends – something that, as McBirney points out in his 1980 paper, has never been properly resolved since its initiation by Fenner and Bowen in the 1920s. It really is not possible to confidently interpret straight line variations as due to mixing in calc-alkali, hornblende-bearing rock series. Indeed, I doubt if pure mixing ever occurs uncomplicated by fractional crystallization and metasomatic processes.

To determine the relative importance of the separate effects we can experiment with the behaviour of specific constituents, such as P_2O_5 and TiO_2, or a ratio such as Rb/Sr, or a variation in the light versus the heavy REEs, as in the Ce/Yb–Eu/Yb correlation plot used by Fourcade and Allègre. However, such chemical evidence can only provide constraints, not proof, and is not likely to be sufficiently discriminating unless the sources of the two components are of sufficiently contrasted compositions.

THE SCIENCE OF MIXING

There is a whole branch of science devoted to the kinematics of mixing in shear flows and wakes, in the blending of viscous liquids and the mixing of diffusing fluids, with a nearly universal recognition among its workers that the physical explanations are very difficult. One only has to read the discussion between Davies and Christensen (Davies, 1991) to find how words such as 'stirring' and 'tendril' take on a new complexity in modern dynamic models. However, although I am certainly not able to understand the mathematics of chaotic flows, I can easily appreciate the simple fact that the stretching and folding involved produces exponential growth areas, which explains why interaction is so enhanced by shear flow in a conduit. The experimental work of Blake and Campbell provides some guide to the physical controls and how mixing occurs at interfaces in unstable flow regimes. Further, Koyaguchi has shown that the efficacy of mixing depends on the magnitude of finite strain caused by the circulative motion.

I conclude that despite the complexity of the physical and chemical processes involved, particularly the likelihood of thermal barriers as discussed by O'Hara, magma mixing is an important process in the diversification of granitic magmas. However, we have seen that it requires a continuous input of heat and volatiles to maintain the liquidity of the two components, particularly by reversing the process of crystallization which creates so insuperable a rheological barrier. The transfer of the necessary heat requires the intervention of earth movements that cause the hot volatiles and magma to be pumped and ducted into the active fracture systems underlying volcanic centres, and we have seen how the kinematics of conduit flow enhance the mixing process. Subsidence into the source magma chambers not only acts as the pump, but stirs up the magma.

BULK MIXING AT DEPTH

For all their eye-catching display in outcrop, mingling and mixing in the higher levels of the crust represent but second-order processes in the diversification of the granitic rocks. Although outcrop evidence provides important clues to the nature of these processes, we have to accept, however reluctantly, that the mixing of magmas only takes place in bulk within the deep crust, even more probably at the crust–mantle interface, where true melts are produced. We are rarely privileged to see what is actually happening, so that discussion has largely been based on chemical and isotopic models and their ambiguities. Nevertheless, a revealing study by Hildreth and Moorbath in 1988 has introduced a new perspective by taking advantage of a set of geological constraints unique to a long segment of the Quaternary volcanic arc of Chile. Although this experiment was concerned with volcanic rocks, its findings clearly apply to the generation of their plutonic analogues.

A geochemical analysis of 15 volcanoes equidistant from the Chile Trench shows that there are systematic changes, along 4° of latitude, in the base-level geochemical signatures of the volcanic centres marking the arc. The only clear variables in the geological setting are a progressive northward increase in the thickening and age of the continental crust underlying the arc and a complementary thinning of the mantle wedge. Modifications of the ascending magmas by local assimilation and fractional crystallization – AFC in the modern jargon – are deemed to be relatively small, so that the northward increase in the base levels of K_2O, Ba, Rb, Cs, Th, U and the Rb/Cs ratio, coupled with an increase in the $^{87}Sr/^{86}Sr$ ratio and decrease in $^{143}Nd/^{144}Nd$, are most readily explained by increasing contributions derived from the deep continental crust mixing with the newly evolved mafic magmas.

It is hardly possible to do justice here to either the original database or the persuasive arguments of Hildreth and Moorbath, but the latter convincingly dismiss any important contribution from subducted sediments or any control by along-arc regional changes in the enrichment of the lithospheric mantle. They also argue that these crustal contributions were not derived from the upper crust because the initial $^{87}Sr/^{86}Sr$ and $\delta^{18}O$ values do not agree with derivation from sediments. Furthermore, the HREE values are best explained by assuming the presence of garnet in the source rock, as might be expected at the pressures pertaining at very deep levels in the crust. The full discussion of these matters is complex and not entirely unequivocal, but it does seem likely that the generative process was deep seated and that the source rocks were located near or at the base of the crust, a conclusion that will be echoed time and again throughout these discussions.

Hildreth and Moorbath envisage that deep beneath each large magmatic centre there was a complex zone of melting, assimilation, storage and homogenization (the acronym is MASH). This was located in the lowermost crust or crust–mantle transition, where hot basaltic magmas, ascending from the mantle wedge, induced sufficiently vigorous local melting and mixing to homogenize the isotopic composition of the derivative magmas. The latter will have acquired a base level isotopic and trace element signature characteristic of a particular MASH domain, and constitute a starting base for subsequent assimilation and crystal fractionation.

In my own studies, with colleagues, of the deeper levels of Andean arcs, as represented by the Coastal Batholith of Peru, we have also found the need to ascribe the primary evolution to deep-seated processes in melt cells of considerable extent, though here it was remelting and the subsequent fractional crystallization, and only locally deep crustal assimilation, that established the final rock variations – a finding wholly consonant with the longer crustal residence of plutonic magmas.

In Peru and northern Chile the very old continental crust is revealed in upfaulted blocks as consisting of just the mafic, igneous composition and high metamorphic grade we might expect of a model protolith to Andean magmas, but there it no trace of residua material or of a Phanerozoic melting episode in these particular outcrops. So is there anywhere where we might examine the presumed deep zone of melting, mixing, assimilation, storage and homogenization, the home of MASH?

THE ENIGMATIC MIXING ZONE REVEALED

It is possible that the remarkable Ivrea zone of the Italian Alps provides an almost unique opportunity in exposing an 8.5 km cross-section through the lower crust and uppermost mantle. Not surprisingly the interrelationship between the various components is structurally complex (Quick *et al.*, 1992; Kalakay and Snoke, 1995), but according to work by Voshage and co-workers in 1990, in essence this natural section reveals a sill-like, mafic magma chamber overlain by kinzigites representing the original metasedimentary crust, and underlain by mantle peridotite. Within the 8 km thick section a layered zone of ultramafic cumulates and gabbros passes up into gabbros and then into a type of diorite, the latter being roofed by anatexite (Figure 9.9). In the lowest 200 m of the layered zone, the isotope compositions are mantle-like but highly variable, yet they quickly become homogeneous and crust-like throughout the upper part of the body. This is interpreted as being due to continuous contamination by crustal melts as the magma chamber was repeatedly replenished by hot mantle-derived magmas, possibly having the composition of basaltic andesite. In contrast, the concentrations of the incompatible trace elements increase progressively upwards, presumably in response to the evident fractional crystallization.

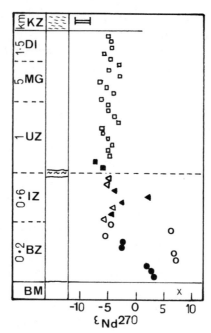

Figure 9.9 ε_{Nd} values at 270 Ma for the Mafic Complex of the Ivrea Zone as a function of stratigraphic position. (ε_{Nd} is the ratio $^{143}Nd/^{144}Nd$ of the rock expressed in relation to that of chondritic meteorites.) A so-called diorite (D1) passes down into gabbro (MG), then a layered series, the upper zone (UZ) of which is composed of norite with anorthositic layers passing down through an intermediate zone (IZ) into a basal zone (BZ) of ultramafic cumulates and gabbro. The Kinzigite Formation (KZ) forms the roof of strongly migmatized amphibolite facies metasediments, the ε_{Nd} value of which is shown (⊨). Underlying the layered series is the Balmuccia Peridotite (BM), representing the mantle. Mafic Complex: (□) UZ-MG-diorite: (■) UZ pyroxenite; (△) IZ gabbro; (▲) IZ pyroxenite; (○) BZ gabbro; (●) BZ pyroxenite; (×) Balmuccia Peridotite. Simplified from Voshage *et al.* (1990), figure 2, with permission from Professor A.W. Hofmann and Macmillan Magazines Ltd.

A model involving the development of a continuously replenished magma chamber, in which the mixing of magmas is a major process in combination with fractional crystallization, is so much in accord with the evidence available in outcrop that it is tempting to attempt a comprehensive explanation. Indeed Voshage and co-workers envisaged that the I-type Hercynian granites of northern Italy could have been derived from lower crustal parent magma chambers similar to those of the Ivrea zone. Thus it is of great interest that in 1988 Boriani and his colleagues considered that the evolution of these particular granites, and the associated appinites, involved a complex history requiring the partial melting of different sources followed by assimilation and crystal fractionation. As a marvellous embellishment of this model, I read that Pin and Sills found that these granites have the same age and Nd/Sr isotopic compositions as the Ivrea ultrabasics and, moreover, have complementary REE patterns!

At this juncture I cannot resist returning to my central theme that magmatism is driven by tectonics, for that is indeed the message of this example from the Ivrea zone – that great volumes of hot mafic magma were squeezed or pumped by dilatancy into the crust along major dislocations during the Hercynian orogenesis. Such a thesis is wholly consonant with Huppert and Sparks' claim that the bulk melting of crustal rocks will only be important if the magmas are in the constant agitation and convection necessary to transfer heat to the interfaces (p. 94), surely the condition of a developing magma chamber in the Ivrea zone.

However, I am not yet ready to rush to this general conclusion, to confuse myself if not my reader, with mysterious acronyms like AFC and MASH in exchange for an actualistic solution to the problems of the generation of granitic magma. Instead I turn to the actualism of the outcrop to examine the special example of mixing represented by the appinitic rocks.

SELECTED REFERENCES

Didier, J. and Barbarin, B. (eds) (1991) *Enclaves and Granite Petrology.* Elsevier, Amsterdam, ch. 2, 3, 11 and 28.

Vernon, R.H., Etheridge, M.A. and Wall, V.J. (1988) Shape and microstructure of microgranitoid enclaves: indicators of magma mingling and flow. *Lithos* **22**, 1–11.

Zorpi, M.J., Coulon, C., Orsini, J.B. and Cocirta, C. (1989) Magma mingling, zoning and emplacement in calc-alkaline granitoid plutons. *Tectonophysics* **157**, 315–329.

Appinites, diatremes and granodiorites: the interaction of 'wet' basalt with granite

10

When the whole [appinite] suite is uniformly known, especially geologically, then there is no doubt that fundamental conclusions on magmatism and orogeny will result. Meanwhile, I can but indicate a few of their most pleasurable aspects.

H.H. Read (1961) Liverpool and Manchester
Geological Journal, *2, 670.*

INTRODUCTION

Of all the examples of the association of basic rocks with granites, that of the appinite suite is one of the most revealing in respect to the multifactorial processes at work in igneous rock petrogenesis. The essential coherence of this close-knit rock suite was first recognized in 1916 by Bailey and Maufe in the district of Appin, Argyll, where, as elsewhere in the Scottish and Irish Caledonides, such rocks are intimately associated in time and place with the Late Caledonian plutons.

Perhaps the most outstanding characteristics of the suite are the dominance of hornblende and the extreme modal and textural diversity. Thus there is every gradation between ultrabasic lithologies, represented by kentallenite, cortlandtite and hornblendite, through a range of hornblende-pyroxene-biotite diorites, to quartz diorite, granodiorite and leucotonalite, the last two often as late stage sheets and veins. Further, certain of the calc-alkaline lamprophyres also belong with the suite. Remarkably, despite this diversity, the order of magmatic evolution and intrusion conforms to the classical sequence mafic to felsic in the manner of a truly consanguineous series, though following two closely related pathways (Figure 10.1).

We are fortunate in having available comprehensive accounts of these rocks in the British Isles; by Bowes and co-workers in Argyll (1967, 1976), by French, Hall, Yarr, and myself and Berger in Donegal (1966, 1967, 1991,

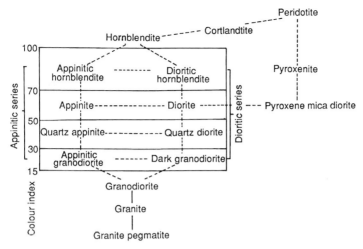

Figure 10.1 Nomenclature of the appinitic suite of Donegal according to French (1966).

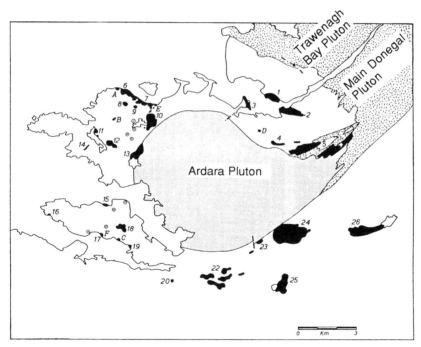

Figure 10.2 Ardara cluster of appinitic intrusions. 1, Cor; 2, Mulnamin More; 3, Roskin; 4, Crockard Hill; 5, Meenalargan; 6–9, Portnoo Group; 10, Naran Hill; 11, 12, Kiltooris Lough; 13, Summy Lough; 14, Rosbeg; 15, Liskeeraghan; 16, Loughros Point: 17–19, Southern Loughros Group; 20, Maghera Road; 21, Glen Gesh; 22, Monargan; 23, 24, Kilrean; 25, Owenea Lough; 26, Lough Anna. Intrusive breccias: A, Dunmore; B, Birroge; C, Crannogeboy; D, Kilkenny; E, Portnoo; and F, Cloghboy. Reproduced from Pitcher and Berger (1972) with the permission of John Wiley and Sons.

1972), and by Brindley, Gupta and Kennan in Leinster (1976). We learn that, for the most part, appinites occur as relatively small, pipe-like bodies, sills and dykes, characteristically assembled in clusters, the location and form of which are obviously controlled by fault lines, joint systems and fold axes. Such a cluster is commonly satellitic to an individual granodiorite pluton (Figure 10.2), though there are exceptions, and whenever relative ages can be determined that of the main part of the cluster brackets the early stages of emplacement of the appropriate pluton, with only a few lamprophyric dykes of later date. According to Rogers and Dunning these appinitic intrusions of Caledonia have only a narrow spread of ages of around 425 Ma, which was a time of major transcurrent fault movement during the post-collisional phase of the Caledonian.

CRYSTALLIZATION FROM A WATER-SATURATED MAGMA

Where the whole range of rock types occurs within the one intrusion, a crude compositional zoning is sometimes discernible, though just as often the types are so chaotically distributed in outcrop, and so transitionally interbanded and mutually enclosed, that their depiction in map form is necessarily subjective (Figure 10.3).

This variation in mode extends in full measure to the mineralogy and texture, the patent inequigranularity being coupled with mantling, complex zoning, sieve and replacement textures. In the dioritic members, the appinites proper, the feldspars are never phenocrystic and are only weakly zoned, and along with any quartz are characteristically intersertal and moulded by the mafic minerals, particularly the hornblende. The latter forms well-formed, stumpy or acicular crystals but these may be fractured, split both by wedges of calcite and plagioclase, or hollow cored by plagioclase (Figure 10.4). Following a consensus view, the common occurrence of skeletal and acicular forms in the minerals, especially apatite, is thought to indicate rapid crystallization under high vapour pressure conditions. Accessories such as apatite and sphene are abundant – which must defeat all simple calculations based on REE budgets – whereas sulphides and carbonates form part of the pristine texture, albeit representing a late stage of growth. Noteworthy are the obvious disequilibrium textures between the felsic and mafic minerals overall, but also expressed as reaction relationships between the latter. Many of the diorites are characterized by rhythmically zoned clots of green amphibole cored with pyroxene and fringed with biotite, and it is likely that many hornblende diorites are the end result of a wholesale recrystallization, presumably under the influence of percolating fluids. Marie Palivcova illustrates the variety of textural forms common to these rocks in her wide ranging review of 1981, concluding that the remarkable constancy in the

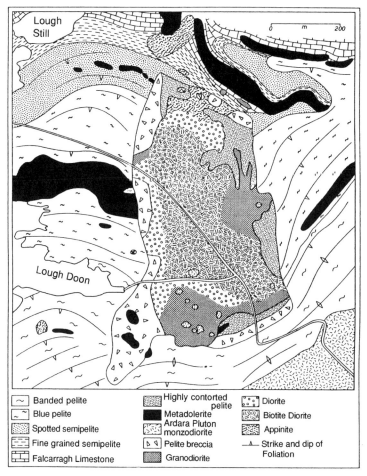

Figure 10.3 Naran Hill appinitic intrusion (No. 10 of Figure 10.2) and its marginal breccia. After French (1966), reproduced from Pitcher and Berger (1972), with the permission of John Wiley and Sons.

textural variability in such rocks the world over suggests a similar crystallization process.

Overall, the evidence points to a continuous history of crystallization occurring under high vapour pressure, and in a volatile-rich environment in which the crystal mush was in vigorous agitation. From their detailed studies Hamidullah and Bowes envisaged olivine, possibly along with orthopyroxene and spinel, crystallizing as an early cumulate phase at depth. This was then intruded and explosively disrupted within gas cored, diatremic pipes, and it is in this energetic environment that the bulk

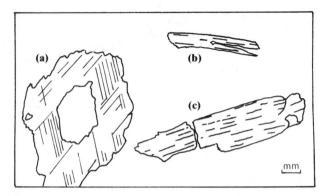

Figure 10.4 Textural features of hornblende in appinitic rocks of the Ardara cluster. (a) Hornblende cored with plagioclase; (b,c) contemporaneous splitting across and wedging along hornblende crystals.

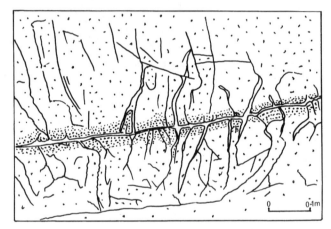

Figure 10.5 Leucocratic veins in hornblende meladiorite host. Feldspathic veinlets emanate from a thin leucocratic vein and ramify into the crystal matrix of the host, becoming continuous with its groundmass. Mulnamin intrusion, Donegal.

of the amphibole and feldspar crystallized in an increasingly high pressure vapour-rich environment. Such a history is wholly consonant with the overall textural relationship – that is, olivine mantled by orthopyroxene and accompanied by spinel; these minerals are then encased by complexly zoned hornblende and by phlogopite, both of which are idiomorphic towards the intersertal felsic components. How intersertal is well demonstrated in outcrops where this plagioclasic infilling forms an anastomosing network of interconnecting channelways linking back to more obvious vein-like fracture fills and discrete sheets of leucotonalite (Figure 10.5). The latest stages of this extended growth history are marked by the appearance of carbonates and sulphides, attesting to post-consolidation gas streaming.

Within the larger intrusions indistinctly bounded ultrabasic masses lie within a chaotic assemblage of variable hornblendic rocks, a distribution fully in accord with this developing model of early formed, olivine-rich cumulates being disrupted and carried upwards in a mafic crystal mush lubricated by a residual fluid of felsic composition.

Estimates of temperatures, by Hamidullah and Bowes, of around 800°C for the appinitic diorites and 1000°C for the ultrabasic rocks seem reasonable for hydrous magmas. The pressure measurements are much more variable, which is not surprising in itself, taking into account the expected fluctuations of P_{H_2O}, but leave me in doubt as to the wisdom of applying amphibole geobarometry in such polyphase systems involving subsolidus re-equilibria. What meaning would the translation of water pressure into depth have in such a fluxed and kinetic system? Certainly the estimates of between 5 and 8 km are at variance with the outcrop evidence unless it is assumed that the hornblende crystals were carried up from a much greater depth.

THE LAMPROPHYRE CONNECTION

I am wholly in agreement with Ayrton, in his 1991 review, that there is a close relationship between appinites and lamprophyres. The appinite clusters are universally associated with dioritic lamprophyre dykes and, furthermore, a few appinitic diorites show chilled margins of these rock types. Although such diorites have long been regarded as the plutonic equivalent of the spessartites and vogesites, I do not think that the composition of such rocks represents that of the primary magma, but rather of an early stage in its magmatic evolution.

A clue as to the origin of this primary magma is provided by Fowler's comprehensive study, in 1988, of the hybrid appinite pipes of Sutherland, Scotland. He showed that in that region the appinites, lamprophyres and syenites form a coherent suite with clear chemical affinities with shoshonitic basalts in general. He suggested that mafic material of this character, with its mantle signature, also probably formed a component of the associated Caledonian granites.

A CHEMICALLY DISTINCTIVE SERIES

The mafic members of the appinite suite have a basaltic affinity, but differ mineralogically in the preponderance of hornblende and biotite, the more diopsidic nature of the pyroxenes, and the relative richness of the mafic minerals in titanium and magnesium. The chemical composition has a distinctly alkalic cast with its relatively high content of potassium, high K_2O/Na_2O ratio, often high titanium and generally higher rubidium, strontium, barium and the light REEs, the latter elements being strongly concentrated into the more felsic members, presumably during hornblende-dominated fractionation in the presence of volatiles. The high values of nickel and chromium suggest derivation from a primitive mantle source, and what little isotope work has been published precludes any substantial crustal component, a finding which we shall see also applies, perhaps surprisingly, to the volatile components.

Crudely linear patterns of chemical variation confirm the evidence of the outcrop that the appinites form a true rock series, at least for all the mafic members (Figure 10.6). Sometimes, however, there is a subtle temporal and compositional hiatus between the latter and the accompanying leucotonalites and granodiorites, with outcrop and textural evidence of mixing, especially so in the Donegal exposures, but whether a particular suite wholly derives from a common parent or represents the mixing of two components is not always easily decided. My own view is that the differentiation process was often interrupted in its late stages by a mixing in of a felsitic liquid either derived from the associated granites or melted out from the crustal envelope at depth, the latter contention strongly supported by Platten's finding, in 1983, of partial melting in a marginal breccia of one of the Scottish examples of these intrusions.

Whatever the petrogenetic model, Hall concluded that the variations within the appinite series are distinct from that of the basic and ultrabasic rocks of both the stratiform and alpine types. Furthermore, in a number of studies, the major elements reveal this high potassium, calc-alkaline parental chemistry of shoshonitic cast, whereas the evidence provided by the REEs and by the Zr/Y versus Zr plot indicates that this parent was of the within-plate-basalt type.

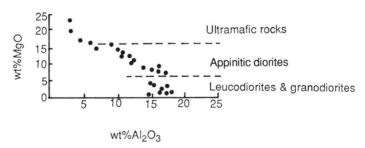

Figure 10.6 MgO—Al_2O_3 plot for appinitic rocks of the Ardara cluster. After Hall (1967) and French (1978).

Thus, taken together, the textural and chemical evidence support an important role for fractional crystallization, whereby olivine was precipitated early, with hornblende then taking over the dominant role and clinopyroxene and a phlogopitic biotite playing minor parts. However, mixing was also involved, with the fractionated products mixing, at various stages in magma evolution, with extraneous granitic melt.

It also seems that these variations within the appinite suite are distinct from those of the associated granodiorite plutons, though, according to Hall, there is one common factor, that both the appinites and the associated granodiorites crystallized under conditions of high water vapour pressures. Of course, this would have favoured the early crystallization of amphibole in the appinites, as has been shown by the experiments of Yoder and Tilley, and also by Cawthorn and O'Hara, who have described the preferred precipitation of hornblende from a water-saturated tholeiitic magma with the consequent suppression of the crystallization of plagioclase. As an added bonus to the emerging general model involving mobile crystal mushes we can expect that such water enrichment will considerably depress the viscosity.

But what of the alkalic signature? It is significant that these same chemical features are common to the slightly younger, yet essentially co-magmatic, Lorne lavas of Argyll. That these, too, are relatively potassium-rich olivine tholeiites, shoshonitic in part, indicates that the latter represent the original magma type, if not the actual magma, with the implication that the origin of the alkalic composition, as already hinted, is to be sought in the mantle and not the crust. But from where did such a magma obtain the volatile content required for the evolution of its appinitic analogue?

EXPLOSION BRECCIAS AND DIATREMES

The highly active environment of these water-saturated basalts is further revealed by the almost ubiquitous association of the appinite suite with marginal intrusion breccias, breccia pipes, dykes and sills, many of the features of which can be best explained in terms of fluidized systems in which the gaseous component is derived from the crystallization of the water-rich magmas themselves.

These intrusion breccias show great variety. The clasts range from man-sized to sand size, from being angular, poorly sorted and local in origin to being rounded, well sorted, exotic and far travelled, as in the distinctive pebble dykes (Figure 10.7a). When the source rock can be identified, the travel distance can sometimes be estimated as, for example, 300 and 600 m for the contents of two pipes in Donegal. The degree of rounding by attrition is often remarkable and is associated with both surface reactions and precipitation, leading to quartzite pebbles with hornblendic rims, and fragments of calc-silicate hornfels with actinolitic fringes.

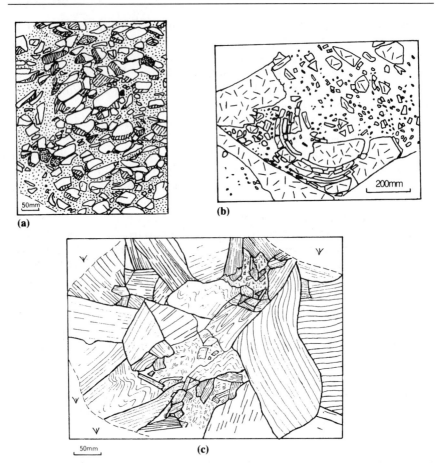

Figure 10.7 Intrusive breccias. (a) Pebble pipe of Kilkenny, Donegal, showing quartzite (white) and calc-silicate fragments (close stipple) lying in a granophyric groundmass. Light shining from right-hand side. (b) Intrusive breccia at Wheal Remfrey, Cornwall, showing spalled granite fragments in tourmalinized micro-breccia. (c) Marginal breccia of the Naran Hill intrusion, Donegal (see Figure 10.3), showing the close fitting of the schist fragments with infill of comminuted schist fragments.

Haloes of fracturing around some of the breccia pipes, as at Polakeeran, Donegal, show dramatically how the granitic country rock was fractured and pervasively penetrated by a sand of comminuted granite and horn-felsed metasediments, the blocks then being chaotically entrained in discrete breccia flows in which a high degree of sorting was attained. Evidently, when the pipes became choked, the blocks packed and inter-locked in a highly characteristic manner, the interstitial fines being blown out by the flow of gas (Figure 10.7c). In some other breccias the blocks are spalled in a manner suggestive of thermally induced fracturing (Figure 10.7b).

The physical relationship between the clasts and true magma is almost universally one in which the brecciation and transport is followed immediately by the infiltration of magma. However, mobile magma can itself form amoeboid enclaves, as was described and discussed in 1984 by Platten, who also provided exquisitely detailed accounts of all these diatremic phenomena. Both he and Bowes commented on the structural control of intrusion and the likelihood that some structures, such as antiformal crests, may act as traps for gas–magma accumulation. The larger appinitic bodies are often bordered by such breccias, albeit always locally derived (Figures 10.3 and 10.7c), and these are best preserved in re-entrants along contacts, the scalloped outlines of which suggest the near superposition of several contiguous, gas-cored diatremes. Platten provided beautiful descriptions of these phenomena from the appinite cluster of Argyll.

Thus the general picture is, as suggested in 1954 by Reynolds, one of gas so rapidly released from a crystallizing magma that it explosively brecciated its country rock conduit, leading to the formation of a fluidized system – that is, a boiling bed of gas and comminuted rock. This tuffisitic mix becomes entrained as the compressed gaseous mixture blasts and cores its way upwards along fractures which are opened by both dilation and gas coring.

I doubt whether these diatremic intrusions often vent to the surface, the gas pressures usually being dissipated in the fractured uppermost crust. Nevertheless, certain outcrops of breccias, such as those of Kilmelford, in Argyll, are so vent-like and so near the basal unconformity below the Devonian red beds with their basaltic lavas that extrusion seems likely. Indeed, so close is the original surface, perhaps even the volcano–plutonic interface itself, that there has even been some confusion between the sedimentary or gas attrition origin of certain clasts in such diatremes. Thus one pipe, exposed on the island of Kerrara, contains rounded cobbles which might just as easily have been derived from a nearby conglomerate as by attrition in the diatreme down which they fell. Such an example illustrates the clear possibility that clasts can rise and fall in gas streams just as Cloos showed in 1941 in his classic study of the Swabian tuff pipes. Perhaps the reader will enjoy with me the concept of diatremic Neptunian dykes!

MAGMA–ROCK INTERACTION: A MULTIPLE SOURCE FOR THE VOLATILES

The key factor in the physical and chemical evolution of the appinite series is the relatively high concentration of water, carbon dioxide and sulphur within a magmatic system fairly high in the crust. We might reasonably expect such volatiles to have been derived from the country rocks, especially as the latter, in both Argyll and Donegal, include such highly fertile lithologies as pyritous black phyllites and marbles. Indeed, there is an almost uncanny relationship between calc-silicate hornfelses and the diatremes which suggests that the carbon dioxide is likely to have been of local origin. The diatremic bodies were certainly hot enough to act as the necessary 'heat engine', for the evidence in the narrow thermal aureoles of a degree of partial melting, indicated to Platten in 1982 that temperatures of at least 780°C were reached. Furthermore, there are haloes of sericitization, chloritization, carbonation and sulphide mineralization in and around some of these breccia appinite bodies which, coupled with the presence of trace amounts of gold, identifies this diatremic environment as belonging to that of the gold porphyries.

That the appinite breccia intrusions can set up a reciprocal exchange of water and carbonates with their country rocks is well demonstrated by a geochemical study of a 150 × 50 m breccia pipe cutting through a marble at Portnoo, Donegal (Figure 10.8). On the basis of $\delta^{18}O$ and $\delta D(^2H/^1H)$ isotope variations, Yarr detected a 40 m wide halo throughout which a significant circulation and exchange of water and carbonates had occurred between the country rock and breccia. However, contrary to my expectations, the system most closely approximates to a primary expulsion of a fluid from the magma in the form of a geochemical front, though this magmatic water then intermixes with locally derived meteoric water. So, again, there is no simple solution and we need to turn once more to a multifactorial model.

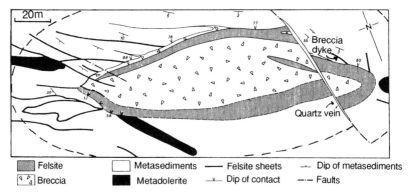

Felsite Breccia Metasediments Metadolerite Felsite sheets Dip of contact Dip of metasediments Faults

Figure 10.8 Dunmore breccia pipe (A of Figure 10.2). Adapted from French and Pitcher (1959), with hydrothermal aureole inserted schematically (as broken line) on the basis of oxygen isotope studies by Yarr (1991).

A MODEL FOR APPINITE PETROGENESIS

From this wide ranging array of data we can surmise that an original potassium-rich, basaltic magma derived, perhaps, from the remelting of phlogopite-bearing, hornblende peridotite in the mantle, early precipitated olivine, orthopyroxene and spinel. This cumulate assemblage, lubricated by a residual magma in which magmatic water was being concentrated, was channelled upwards along faults and fractures, surely absorbing some meteoric water and other volatile substances on the way. Water saturation led to high vapour pressures and the preferred precipitation of hornblende. The consequent release of gas in the shallower levels of the crust was in such volume as to cause explosive fracturing of the conduits, especially where the gaseous magma mush was temporarily ponded in structural or lithological traps as, for example, a bed of close jointed quartzite in a fold core. The water-rich felsitic residuum was literally blown through the fractured aggregate of hornblende crystals, with which it reacted, before finally crystallizing within the interstices of the crystal framework, a process representing a kind of autohybridization.

Satisfying though it may be, this petrogenetic model is incomplete without involving the association of the appinites with granodiorite plutons and taking account of the overall relationship with faulting and uplift so evident in the Late Caledonian of the British Isles. That both granitic and basaltic magmas might be generated in response to faulting is but a return to the central theme of this book, but I believe the connection to be even more intimate. It may be that the two magmas, granitic and basaltic, sometimes mix in the pipes, but I think that much more often the genetic process involves the release of magmatic and meteoric volatiles during the generation and intrusion of granite, and it is these that flux the more mobile, faster upwelling, basaltic magmas. That so often a pluton has risen into or underlies an immediately precursor cluster of satellite diatremes filled with water-saturated basaltic magmas suggests that its attendant cortège of volatiles was channelled into the same conduits as carried the coeval basalts, the gases catching up and mixing with the latter during their journey to the surface (Figure 10.9). Sometimes the granitic magmas themselves finally overtook degassed appinites, when a certain degree of mingling would have occurred as at Ardara, Donegal. However, if this elaborate model involving a mixed origin of the volatiles has any credence it should surely be of general application and capable of being tested by further oxygen isotope studies.

A ROLE FOR IMMISCIBILITY?

The rather special features of appinitic suites, that is to say the wide range of composition within a small compass, hydrous mafic mineralogy and

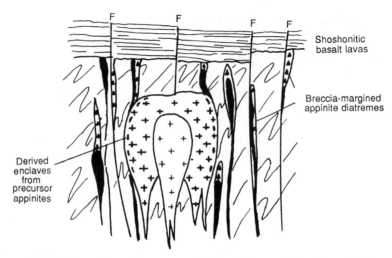

Figure 10.9 Highly schematic representation of the spatial relationship between a multipulse granitic pluton, a precursor cluster of appinite-breccia pipes and overlying comagmatic lavas.

presence of calcite and sulphides, coupled with a certain individuality of the felsic end-member, might suggest that the latter separated out as an immiscible fluid at a late stage in the differentiation process.

Such a thesis had largely lain dormant for a nearly a century until resurrected by Holgate in 1954, and then by Roedder and Weiblen in 1970 and 1971, following their discovery of 'lunar granites' (Roedder, 1979). It was later applied by Bender, Hanson and Bence in their modelling of the evolution of the Cortlandt Complex, New York, where the amphibolitic rocks involved are indeed comparable with the appinites, so that it is of immediate interest that Bender and his colleagues envisaged the unmixing of an immiscible felsic residual liquid now represented by the latest member of that particular suite, a granodiorite. Their thesis was largely based on a certain trace element incongruity between the granodiorite and the mafic members of the suite, which they interpreted as resulting from the partitioning between two coexisting liquids. Clearly, in such a continuously recrystallizing medium as revealed by the textural studies there is little possibility of any evidence of menisci being preserved in such rocks, so that I have to return an open verdict, though with the caveat that the geochemical data might equally fit with a hypothesis involving two components, geologically interrelated, but of different provenance as suggested above.

APPINITES GALORE

Until fairly recently any survey of published work would have indicated that the appinites and their problems were of a provincial character, and it might have seemed that those few appinite suites that had been reported outside Caledonia were of British invention. It is true that Joplin had identified appinites in southeastern Australia in the late 1950s, but only after seeing appinites during an earlier excursion with myself in Donegal.

Now, however, many examples of this rock association have been reported, especially from the European Hercynides, for example, by Palivcova in 1981 from localities in the Bohemian and the Italian Adamello Massifs; by Boriani and co-workers in 1974 from the Italian Alps, and by Enrique in 1983 from the Pyrenees. I have no doubt that we must also include the vaugnerites of the Massif Central of France, reviewed by Sabatier, and also the Nadeldiorites of the Austrians. Thus it becomes obvious that granite-associated dioritic intrusions, with their characteristic acicular hornblendes, are ubiquitous in the Caledonides and the Hercynides of Europe – in fact, whenever and wherever in Europe the tectonic context is that of post-closure block faulting of thick crust. I only avoid including the similar diorites of the earlier Cadomian of the Channel Islands because of the doubts expressed by Wells and Bishop that these are really of the appinite genre as just defined. Certain of these European studies are notable for their concern with texture as a proper guide to the understanding of the complex petrogenetic processes involved, and here the works of Marie Palivcova in 1982 are particularly informative, especially in the discussion of the nature and origin of ocelli in appinitic and vaugneritic rocks.

Outside Europe and with the exception of reports in Australia by Joplin, in Tasmania by Jennings and Sutherland and in Tanzania by Rock there have been few reports of such lithologies. Nevertheless, I am wholly persuaded that this is largely because of a lack of recognition of their petrogenetic significance, particularly in the USA; I have personally recognized such appinitic minor intrusions in association with the Boulder Batholith of Montana. This omission is not entirely without reason because the ultrabasic and basic rocks forming an integral part of the Mesozoic–Cenozoic batholiths of the western Americas show a different relationship with their granitic associates. Although it is true that these are also precursors and often hornblendic in mode, even accompanied by breccias and pebble dykes, the scale, chemistry and tectonic environment are significantly different.

Surveying these world-wide occurrences, I conclude that a thick continental crust is a key factor in the petrogenesis of the appinites proper, with the sedimentary horizons in the upper crust possibly providing a source for some of the volatile constituents. More importantly, it is the thick crust that provides for sufficient residence time for the magmas to undergo the polyphasal crystallization required for the contrasted differentiation of a remarkable gamut of rock types.

Up to now, we have been largely concerned with the processes involved in the diversification of granitic magmas, but it is time to discuss how these different magmas rise in the crust.

SELECTED REFERENCES

Ayrton, S. (1991) Appinites, lamprophyres and mafic magmatic enclaves: Three related products of the interaction between acid and mafic magmas. In: Didier, J. and Barbarin, B. (eds), *Enclaves and Granite Petrology*. Elsevier, Amsterdam, pp. 465–476.

Bowes, D.R. and McArthur, A.C. (1976) Nature and genesis of the Appinite Suite. *Krystalinikum*, **12**, 31–46.

Hamidullah, S. and Bowes, D.R. (1987) Petrogenesis of the Appinite Suite, Appin District, Western Scotland. *Acta Universitatis Carolinae-Geologica*, **4**, 295–396.

Pitcher, W.S. and Berger, A.R. (1972) *The Geology of Donegal: A Study of Granite Emplacement and Unroofing*. Wiley-Interscience, New York, ch. 7, pp. 143–168.

Sabatier, H. (1980) Vaugnérites et granites: une association particulière de roches grenues acides et basiques. *Bulletin de Mineralogie*, **103**, 507–522.

Controls of upwelling and emplacement: the response of the envelope: balloons, pistons and reality

11

The high inclination, therefore, and the quaquaversal dip of the beds around the borders of the granite boss . . . are facts which all accord with the hypothesis of a great amount of movement at that point where the granite is supposed to have been thrust up bodily, and where we may conceive it to have been distended laterally by the repeated injection of fresh supplies of melted materials.

Charles Lyell (1865) Elements of Geology, *p. 723.*

PREAMBLE

It is most unlikely that magmatism is ever a tectonically passive event; melting is initiated, the site of intrusion selected, upwelling motivated and the arrival in the upper crust controlled, all in response to crustal movements. In the broadest sense magmatism is always syntectonic, especially so if account is taken of the whole process time involved in the melting, assembly, intrusion and cooling. Unfortunately, even with the best of outcrop maps, geophysically controlled for the third dimension, we can only observe the arrival stage of this emplacement process, though structural studies of both the pluton and envelope can provide the clues for unravelling the ascent history.

The common categorization of intrusion as pre-, syn-, ser- and post-tectonic is based on relationships observed at just one level in the crust, and representing one brief moment in geological time. For a proper understanding of the correspondence between intrusion and earth movement we need to know how long is the gestation period and how quick the birth. However, even without the exact measurements we can guess that these categories will be artificial. And no less so because the structural relationships that we see are strongly dependent on the rheological state of both magma and country rocks.

In his papers and 1981 book Hans Ramberg has emphasized that the way magmas move, how they rise, the paths they follow and the shapes they assume while in motion are all directly related to contrast in viscosity, a contrast which is likely to change in magnitude during an intrusion history. However, as we have seen, the viscosity of magmas is a complex function that is particularly responsive to composition and crystal content. In addition, the viscosity of the ambient country rocks is influenced by lithology, metamorphic state and previous structure, and a leading role has often been given to crustal depth in this ductility equation. Doubtless the crust behaves very differently at depth from how it behaves at shallow levels, but the direct correlation between crustal depth and the structural ambience of a pluton cannot be as important as envisaged by both Buddington and Read. This is because the ductility contrast relates more to the localization of deformation, and the intensity of heat flux through a particular zone in the crust, than to simple depth. Indeed, the most cogent explanation for the observed variety in emplacement mechanisms lies in the changing interplay between pluton-centred forces, including buoyancy, and the ambient tectonic forces. Thus in the natural experiment we must search for connections between mechanism and tectonic environment, particularly in those revealing situations where plutons of similar composition were intruded close together, at the same crustal level, and at virtually the same time, yet by different mechanisms.

In discussing the changes to be expected in the physical properties of granite magmas as they rise in the crust, I concluded that for much of their life history in coherent plutons these viscid, crystal-carrying polymeric fluids must often behave as stress-supporting Bingham bodies, that is before they lock to crystal frameworks. I turn now to discuss how these changes will be recorded in the resultant rock.

FABRIC AS A GUIDE TO FLOW OR TO DEFORMATION?

If granitic magmas have a strength they will respond to the application of stress by pure shear, the resultant strain being recorded by changes in shape and orientation within the crystal fabric. From the Arzi-type diagram (Figure 11.1) we can expect a continuous progression of fabrics in response to the changing rheology with time as magmatism gives way to autometamorphism, a matter discussed early on by Berger and Pitcher in 1970 and by Barrière in 1977. However, in my opinion the fabrics of what I take to represent free flowage are rarely identifiable in granitic plutons. The well-organized linear fabric that we map, and often label as a *magmatic foliation*, is normally an indication of strain, and induced by pure shear of a progressively strengthening, viscid, crystal suspension – probably not long before the crystal framework finally linked up. Continuing deformation takes the form of plastic strain, which is commonly represented by the development of a strong foliation in the solidified

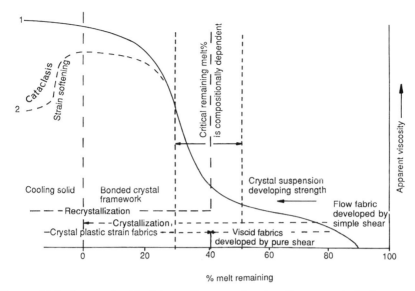

Figure 11.1 Composite 'Arzi'-type diagram showing the expected changes in rheology during the crystallization of a granitic magma. After Tullis and Yund (1977), Arzi (1978), van der Molen and Paterson (1979), Marsh (1981) and Capais and Barbarin (1986).

peripheries of plutons. With this in mind we can expect a different response to deformation before and after the formation of the feldspar framework as expressed in terms of pre-full crystallization and crystal plastic strain fabrics, respectively. But the progression is not easily observed in its entirety because the later stages are often so pervasive in their overprint.

Its recognition requires a detailed examination of the fabric of each mineral component and is complicated by mimetic growth. An example is Tribe and D'Lemos' study, in 1996, of the fabric development of the quartz diorites of the Channel Islands, UK. In these rocks a magmatic state deformation fabric is only locally preserved because of overprinting by solid state deformation fabrics, both stages in a synplutonic process. The latter fabrics were formed at the moderate temperature of 400–500°C, and are marked by microcracking of plagioclase and amphibole, the local ductile bending and recrystallization of the latter minerals, marginal recrystallization and myrmekite growth in the margins of potassium feldspar, the ductile deformation and recrystallization of quartz, and also biotite recrystallization within microshears. From the absence of high temperature deformation fabrics in their example the authors identify a discontinuity in the progression from the magmatic to the solid state. Their explanation of this involves a synplutonic deformation developing progressively inwards and following the advance of the crystallization front from the periphery to the core of a pluton. Any continuing deformation is concentrated at the outer contact, this being the zone of maximum anisotropy, and results in a hiatus in the fabric history. In reviewing such phenomena Tribe and D'Lemos drew attention to the possible effect of high strain rate and slow cooling, both of which might lead to a greater continuity in the fabric progression in appropriate situations.

In general terms there are some well-established criteria for attempting to distinguish mineral foliations formed by fluid flow from those formed by deformation and again I refer the reader to the most useful reviews by Paterson, Vernon and Tobisch in 1989 and Paterson and Vernon in 1995. As for myself, I am uncertain how to define flow in view of the rheological complexity already commented on and I find the distinction especially difficult when many of these criteria are distinctly equivocal. Among these criteria perhaps the presence of a magnetic anisotropy in an otherwise randomly oriented fabric may well suggest an early phase of wholesale flowage. Schlieren orientations may be similarly interpreted, but in granites with strong mineral fabrics such schlieren orientations are likely to have been transposed. The orientation of well-formed crystals is often taken to indicate laminar flow, but we have to be aware not only of mimetic growth within an established foliation, but also that euhedralism can result from porphyroblastic growth. The lack of orientation in the quartz interstitial to aligned feldspars is said to prove that flowage occurred before the attainment of the critical melt percentage, yet we know that the equigranular recrystallization of quartz is a common feature of granites. Further, the growth of late stage phases within pressure shadows is thought to suggest the presence of a residual interstitial magma, but this I regard as highly subjective in view of essentially similar textures in metamorphic rocks. In the field of crystal plastic strain we are on safer ground with a whole range of familiar metamorphic textures for identifying high- and then low-temperature solid-state deformation.

The simple fact is that we ought to expect polyphase fabrics. As an example I quote the work of Marre and Pons at Queriqut, Pyrenees, where early formed biotites exhibiting a linear planar fabric are interpreted as representing an early stage of viscous flow. A planar orientation of what are, texturally, late potassium feldspar megacrysts represents a later flattening within a stress supporting crystal fabric. Such interpretations of mixed crystal assemblages are open to criticism, yet the general message is for a return to the microscope, especially now that we might use serial sections to map out any change of shape during growth.

Opinions clearly differ about the origin of mineral foliations in granites and certainly each example requires separate assessment. I repeat my contention that such foliations are more often formed in response to synplutonic deformation and are not often the result of wholesale flow such as might occur during intrusion through a conduit or by convective overturn in a magma chamber.

THE FOLIATION IN OUTCROP

That this is so is particularly obvious in outcrops where foliations traverse external and internal intrusive contacts and lithological and facies boundaries (Figure 11.2a and c), cross mineral schlieren, lie axially to folded schlieren, lie common to both granite and aureole (Figure 11.2b), or are shared by a microdiorite enclave and host. Furthermore, these orientations can easily be described in strain terms of planar and linear fabrics – that is, as S, $S > L$, $S = L$, $L > S$, L. The degree of strain is most easily measured from the ellipsoidal shape of deformed microdiorite enclaves (Figure 11.3a and b), which, as they often represent blobs of synplutonic magma, are the most likely of the enclosed material to have had a ductility near to that of the host. With this in mind the measurement should only be made on those enclaves sharing the foliation of the host. The total strain is recorded from the time of inclusion and there are some complicating factors in that the viscosity ratio, enclave to matrix, will vary with cooling time and blob size, and, furthermore, the degree of distortion is related to the packing density of the blobs – which is yet another indication of the pseudoplastic condition.

When the elongation of the enclaves is seen to be slightly shifted from that of the preferred mineral orientation this does not necessarily signify a phase of early flow but is more often the result of a different competency response to deformation (Figure 11.3c). In fact, this obliquity indicates the sense of shear as well as a measure of strain intensity, providing a good example of the use of quantitative intrusion strain analysis as developed by Fernandez and Tempier in 1971. But its interpretation requires great care, as is demonstrated by the studies of Tikoff and Teyssier. These workers made a distinction between the rotation of enclaves in a viscous medium (the so-called 'tiling' of Fernandez et al., 1983) and their imbrication in the solid state, and by their computer-generated modelling of the latter process they were able to make inferences on the rheology of the Sierra Nevada granodiorites when the rotations occurred.

Yet more information of this type is provided by the differential spacing of crystals as measured by the nearest neighbour method developed by Hanna and Fry. Further, there is the form of the foliation itself, particularly where the dominant S-plane is intersected by a crenulation cleavage, the 'C' or cisaillement plane, in which case the intersection angle again measures an increment of the strain and also records the sense of shear (Figure 11.3d).

For an introduction to published work on such structural methods I refer to reviews by Hutton (1988a), Fernandez and Barbarin (1991) and Paterson and Vernon (1995). I would simply emphasize that the strain data, incremental or overall, must be set in the time-frame of pluton emplacement. In the case of multipulse plutons, as noted above, consolidation inwards of the first pulse necessarily provides a ductility progres-

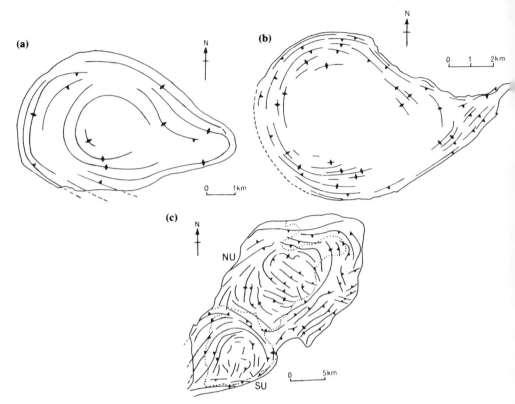

Figure 11.2 Foliation trajectories (expressed in standard form) in a range of plutons. (a) Flamanville pluton, Manche, France (Martin, 1953). Overall structure is concordant with the walls, but locally crosses contacts. According to Brun *et al.* (1990) the foliation cuts across the mafic enclaves without refraction, thus attesting to bulk solid-state deformation. (b) Ardara pluton, Donegal, Ireland (Pitcher and Berger, 1972). Overall structure concordant with walls and foliation in aureole but somewhat discordant to facies-change boundary. In detail, however, the foliation crosses re-entrants and mafic enclaves and is an essential component of gneissic periphery. (c) Plouaret multiple pluton, France, according to Guillet *et al.* (1985). Overall folial structure generally concordant to the main outer contact in the northern unit (NU), but crosses this in the southern unit (SU); everywhere it crosses pulse boundaries (stippled).

sion and this responds to the synplutonic deformation – often caused by the continuing magma infill – by an inward decrease in strain intensity (Figure 11.3b). Marking this fall-off in strain the foliation weakens and the degree of flattening of the dark enclaves lessens, features common to many plutons. In some ideal examples patterns of incremental strain can also be established from the differing response of dykes of different relative ages of intrusion (Figure 11.4).

That the mineral foliation and lineation is not obviously deflected

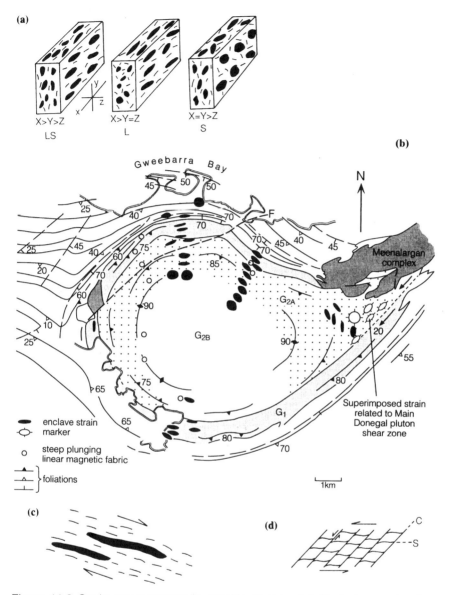

Figure 11.3 Strain measurement in granitic plutons. (a) Strain types showing, qualitatively, the variations in shape change and orientation in both mafic enclaves and phenocryst alignment. (b) Ardara pluton, Donegal. Shows mineral foliation within pluton, with representation of the strain measured from the microdiorite enclaves by Holder (1979). Main regional foliation shown in country rock pelites with strain in the aureole measured in deformed andalusite by Sanderson and Meneilly (1981). Dashed lines represent emplacement-related D_4 structures. (c) Obliquity between aligned enclaves and host foliation, the shift effect providing the sense of shear (but see text p.187). (d) C–S fabric with 'S' the cisaillement plane providing the sense of shear. Magnetic lineations after King (1966).

across internal pulse contacts denotes near rheological identity, again implying a hiatus of sorts between solidification and deformation. But in my expectation of a contact between a liquid-filled framework and a crystal mush I would not expect any recordable difference in rheology. Furthermore, internal contacts are so often welded and subtle, and show such an acute angle between contact and foliation (Figure 11.2b, but compare 2c), that deflections are not easily measured. Such are the complications! But we are steadily becoming aware of the messages retained in the outcrop.

Of course, the real challenge is in identifying the cause of the deformation, whether it be regionally or pluton generated, and it is the wall rocks that provide the most complete history of deformation, especially as the magma itself may not have sufficiently crystallized to record the earliest stages. Indeed, there are good examples where a very strong fabric in the envelope contrasts with the lack of a discernible deformation fabric in the

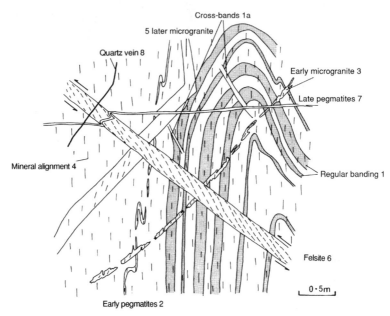

Figure 11.4 Schematic diagram showing chronological relationships of banding, mineral alignment and the various dyke suites in the Main Donegal pluton. In order of decreasing age: (1) regular bands; (la) cross bands; (2) early pegmatites; (3) early microgranite; (4) mineral alignment; (5) later microgranite; (6) felsite; (7) late pegmatites; and (8) late quartz veins. Illustrates well the time relationship of the principal mineral foliation of the granite: post-banding, post-early pegmatites and early microgranite, quite separate from the foliation in the felsite and wholly previous to the later pegmatites and quartz veins. Reproduced from Pitcher and Berger (1972) with the permission of John Wiley and Sons.

pluton, and the Arran diapir of Scotland is a case in point. On the other hand, highly competent lithologies within the wall rocks may resist the imprint, and the presence of pre-emplacement fabrics certainly complicates its analysis, but the record is there to be read by the enthusiast.

That the deformation is truly synplutonic always needs to be tested by reference to mineral growth histories within both the igneous and aureole rocks. Here the relation of the aluminosilicate porphyroblasts to new foliations is particularly relevant. By such comparative studies it is possible to unravel the separate effects of the external forces and the squeezing, pumping and buoyancy forces associated directly with the pluton. However, as always, there are complications, not least in the effect of lithology.

EFFECT OF LITHOLOGY

That the lithology of the envelope rocks is an important factor in determining the ductility contrast is well demonstrated where a single pluton shows a contrasting response to different types of wall rocks. At Ploumanac'h, Brittany, greywacke and older granite produce concordant and discordant relations, respectively, whereas at Qernertog, southern Greenland, semipelitic schist and a gneiss induce this same differential effect. At Papoose Flat, California, a thin bedded shaly horizon locates the locus of forceful distension by magma intrusion.

Such a competency control is reinforced by chemical effects such as the release of water, which causes hydraulic weakening, whereas the release of carbon dioxide from calcareous rocks may well lead to preferential freezing of the magma in contact. In general the cortège of hot volatiles released from the aureole must have a profound effect on the ductility of the country rocks above and around a rising magma. It is likely that granitic magmas would quickly freeze solid if they were not following an early established hot conduit.

This loss of material from the country rocks, particularly of silica by solution in the water, is greatly increased by tectonic strain, and here again a strong lithological control is evident, argillaceous rocks having much greater volume losses than volcanic lithologies. I suspect that in many strongly deformed argillaceous aureoles over half the bulk is lost, which represents a significant contribution to the space problem. Furthermore, the destination of this vast amount of water and silica is rarely taken into account.

Such a lithological control is particularly evident on the regional scale. Thus the array of high level plutons making up the Coastal Batholith of Peru is sharply cut out of a massive roof carapace of volcanic rocks. However, where plutons intrude less competent lithologies they shoulder aside the enveloping sediments. This is the general pattern of the emplacement of the plutons of the deeper-seated Sierra Nevada Batholith, where they intrude the metasediments of the Palaeozoic clastic apron of the western USA.

At deeper crustal levels the ductility contrast might well be reduced as a consequence of the higher ambient crustal temperatures. However, this depth control is far too simple a concept to use as a guide to the relative level of granite plutons in the crust, particularly to label them as epizonal, mesozonal or catazonal on the basis of a perceived ductility contrast. There are so many examples where contrasted mechanisms of emplacement operate at the same crustal level and at the same time that the granite process is better discussed in terms of heat flux and rate of deformation.

A MERE FOOTNOTE ON THE EFFECTS OF CONTACT METAMORPHISM

This line of reasoning leads directly to a consideration of the thermal state of the enveloping rocks during emplacement as recorded by contact metamorphism. Notwithstanding its importance such a discussion would require a major digression into the complex field of metamorphic studies. Only by an overall understanding of the mechanisms of grain coarsening, new mineral growth, metasomatism, anatexis and deformation, with insights into the roles of fluids, of permeability and porosity, and of deformation, is it possible to model the thermal regime associated with a pluton. Fortunately such a review is available in the form of 14 essays assembled by Derrill Kerrick in 1991, which together provide a fascinating insight into the complexities of the metamorphic process as acted out in the natural laboratory of contact aureoles.

Here I simply emphasize one particular point touching on a central theme of this book, that synplutonic deformation is also vital in determining the nature of contact aureoles, controlling not only the grain coarsening but the time and nature of mineral growth. This may not be a popular view, but in my early work with Read (1963) on the aureoles of Donegal, I was impressed by the way the contact effects reflected the contrasted mechanisms of emplacement, even to the degree that crystallization aureoles coincided almost exactly with deformation aureoles. This is presumably because of the enhancement, by the provision of foliations and fractures, of the mobilization and permeation of hot fluids. This leads to the production of the lubricated, ductile skin to an intrusion that, as I have just suggested, greatly facilitates emplacement.

DUCTILE SHEAR BELTS, HEAT FLUX AND GRANITE INTRUSIONS

I consider it generally established that the emplacement of granite is related to crustal fracturing on all scales, with a particularly strong correlation with major strike-slip fault systems; examples will appear in the sequel. Both Leake, in 1990, and Watson, in 1984, have argued that the connection is genetic, the former envisaging that the uplift and depression of the fault-bounded crustal blocks, consequent on compression, must lead to partial melting of both crust and mantle.

There is little doubt that some major fractures can extend right through the crust and into the mantle, as indeed is strongly suggested by the ubiquitous appearance of mantle-derived mafic rocks early in the history of nearly all granite complexes. Other major fractures, such as the listric thrust faults of thin skin tectonics, may slide deep enough within the crust to collect and channel crustal magmas. Furthermore, deep reaching faults not only provide the conduits of the precursor hot basaltic magmas and volatiles, but are likely to initiate melting in the upper mantle or lower crust.

This thesis holds generally for all the global tectonic environments, but it is in the regions of continent–continent collision where the interplay of deformation and high heat flux is most dramatically displayed. Particularly relevant is the finding that the resultant plutono-metamorphic zones are often steeply inclined and comparatively narrow, indicating that the thermal gradients are steep and so cannot represent a simple response to depth. Rather they are related to major ductile shear zones produced in response to oblique slip during transpressive continental collision.

The wide ranging review by Hutton referred to earlier mentions megashears in Brittany, northern Portugal, Galicia, central Spain and Newfoundland as examples of where such hot shear zones form the loci of synkinematic granites. These zones are represented variously as broad belts of high grade schists into which the magmas arrived just after the peak of metamorphism, as in northern Portugal, or as narrow contact-schist aureoles flanking syntectonic granites such as those of Donegal and Leinster, Ireland.

Studies of ductile shear belts in general reveal a marked contrast between the intrusion mechanisms operating within the ductile core of a belt and outside it in the brittle flanking blocks. Examples are provided by the studies of Pitcher and Berger in 1972, and Hutton in 1982, both from Donegal, Ireland, and by Davies in 1982 from Saudi Arabia, whereas Castro in 1986 described how such differing emplacement mechanisms can also relate in time to the evolution of a major shear zone, as in Extramadura, Spain. But how is it that granitic magmas can intrude such compressive fracture zones? We turn first to consider concordant plutons in general before attempting to answer this question.

CONCORDANT PLUTONS: DIAPIRISM OR BALLOONING?

By analogy with both simple experiments and natural salt intrusions a model was generated by Grout, in 1945, of a rising magma blob or diapir, motivated by buoyancy, piercing the crust and shouldering aside the country rocks and finally mushrooming out at high crustal levels (Figure 11.5a). Among modern studies Bateman, in 1989, summarized the field criteria for the recognition of such classical piercement diapirs, Dixon and Cruden, in 1975 and 1990, respectively, both provided the experimental analogues, while Marsh, in 1982, formulated a Hot Stokes' model for the ascent motivated by buoyancy. Marsh found that the ascent velocity will largely depend on the viscosity within a softened contact zone, revealed in nature by the narrow, highly deformed innermost aureoles of many diapirs. However, it seems that the drag which develops during ascent must seriously impede movement, and only the previous warming of the conduit will allow such a hot blob to rise. Marsh concluded that although such blobs are to be expected in the mantle, they are unlikely to exist in the crust (but see p. 198). Furthermore, it is likely that any such diapir isolated from its source will quickly freeze solid, and Clemens and Mawer in 1992 confirmed by calculation that the long distance transport of a plutonic blob of granitic magma is not viable on thermal grounds. But the discussions continue with the calculations becoming more and more sophisticated as new parameters are included. Thus Weinberg and Podladchikov, in 1994, concluded from their modelling that a granite diapir rising through a 'thermally graded power law crust' could rise to shallow crustal levels while remaining molten!

I find it difficult to judge the validity of this mathematical modelling but on balance I think that this Groutian hot blob should be abandoned in

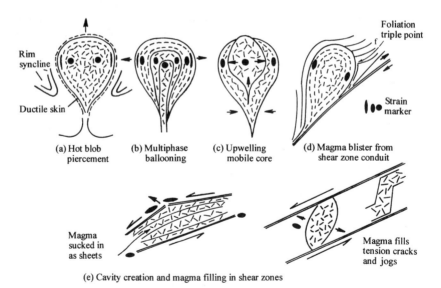

(a) Hot blob piercement   (b) Multiphase ballooning   (c) Upwelling mobile core   (d) Magma blister from shear zone conduit

(e) Cavity creation and magma filling in shear zones

Figure 11.5 Schematic illustrations of the concepts of piercement diapirs; multipulse, ballooning plutons; upward welling mobile core; 'fault-blister' plutons; and cavity-filling plutons in shear zones.

favour of more realistic models based on observable structural data. A moment's reflection on the relative thickness of the average crust and the expected dimension of a pluton will show that there is rarely room for the tailed Montgolfier balloon so beloved of geological cartoonists.

It was appreciated by Hans Cloos, in 1925, that the strain markers within the contact zones of so-called diapirs more often indicate horizontal distension than vertical piercement. As a consequence Cloos advocated a mechanism of ballooning by the continuing infill of magma (Figure 11.5b), a thesis enthusiastically adopted by both Akaad and Martin in their post-war studies, in 1956 and 1953, of the Ardara and Flamanville plutons, respectively (Figure 11.2a and b). In another similar study, in 1955, of the Bald Rock pluton, California, Compton envisaged the internal foliation developing incrementally in response to successive infills by new magma in the core of the pluton, the border phases remaining sufficiently mobile to deform and recrystallize under the stress created by dilation. Such models of incremental distension, coupled with intense deformation of the peripheral rocks, have now been firmly established by the use of strain measurement as suggested by Ramsay in studies of the Chindamora Batholith, Zimbabwe, in 1975 and 1989, and used to good effect by Holder in a follow-up study of the Ardara pluton, Donegal, in 1979 (Figure 11.3b), and by Bateman in a structural analysis of the Cannibal Creek pluton, Queensland, in 1985. From the measurement of progressive strain it is possible to calculate the incremental distension of a ballooning pluton, modelled as due to the repeated influx of magma (Figure 11.5b). It is perhaps unfortunate that in 1988 Cruden showed that, depending on the rate of ascent, such simple flattening strains could also be generated by a conventional piercement diapir! If so, perhaps the only certain evidence of piercement may be the upward displacement of country rocks at the equatorial level of a pear-shaped pluton – an almost impossible section to identify in practice. Despite Cruden's valid criticism, the ubiquity of uniaxial flattening within plutons coupled with the simple tightening of pre-existing folds, particularly as so well exemplified by Bateman's 1989 study of the aureole of the Cannibal Creek pluton, Queensland, and Holder's 1979 study of the aureole of the Ardara pluton, Donegal, satisfy me that ballooning is a central process in the arrival stages of pluton emplacement.

A MODERN DEBATE ON BURSTING THE BUBBLE

According to Paterson and Vernon in their 1995 paper, provocatively entitled 'Bursting the bubble of ballooning plutons', the strain measurements as presently used apparently fail to solve the space problem. New structural analyses of a number of well-known examples of granite diapirs by these authors, and also by Fowler and Paterson in 1993, showed that only about one third of the space can be accounted for by ductile shortening or by the return flow of the wall rocks. This is a problem that I have lived with for a long time (Pitcher and Berger, 1972, p. 182).

Commenting briefly on Paterson and Vernon's critique I agree that most nested 'diapirs' are multipulsal and, furthermore, it is a central tenet of this book that they are also syntectonic, though it would seem that I differ in opining that forceful emplacement itself contributes in good measure to the tectonism. Concerning the magmas themselves I do not think that they are often as crystal poor on arrival as Paterson and Vernon believe. There is so often a lack of textural change across internal contacts, and even against the country rock contact, that I contend that many are crystal rich, though each situation must be judged on its own evidence. The criticism that the microdioritic enclaves are not necessarily good strain indicators of lateral expansion has to be taken seriously, since blobs of mafic magma could deform in the feed magma. But if only those enclaves are measured that share the foliation of the host, I think we can accept the measurement as a record of the post-arrival strain.

The crux of Paterson and Vernon's argument is that the magma chamber does not grow in any large measure by distensional infilling of the centre. They consider that the lack of evidence for substantial subsolidus deformation *throughout* the solidifying or solidified peripheral pulses denies major extension as a provider of the necessary space. Subsolidus deformation is concentrated at the strongly anisotropic country rock contacts, where it appears in the form of a superposed fabric. These are a very valid points.

I have only seen but briefly many of the examples cited by the authors, so that I must rely on a more than passing acquaintance with the multipulse pluton of Ardara to focus my contrary opinion. In this pluton, as in many other examples, the pulse contacts are transitional, albeit rapidly so in the case of the peripheral member, so that the time and rheological differences between the pulses must be small. Thus I doubt that it is possible to determine the separate response of strain on these adjacent materials.

Concerning the central problem of the reliability and significance of the strain measurements themselves, a very detailed study by Molyneux, in 1995, while confirming Paterson and Vernon's view that the deformation of the internal, main mass of the Ardara pluton was confined to the magmatic state, yielded no evidence that the resulting pervasive fabric was formed by any superimposed regional strain. Indeed, Molyneux's measurements, based on the shape of mafic enclaves (which were derived from the satellitic appinite intrusions) and also on the Fry crystal-spacing method, showed that the strains represent pure flattening in accord with the ballooning model first proposed by Akaad in 1956 and elaborated by Holder in 1979. Moreover, de-straining of the enclaves showed that some

80% of the area of the pluton could well have been formed by a ballooning mechanism. It seems that only the steeply inclined magnetic fabric is left to record the initial upwelling of the magma, for this would not be transposed by lateral extension. Thus I remain content with the ballooning hypothesis, especially as there is no evidence at all for any of the alternative methods of material transfer such as stoping, doming, return flow of wall rock or earlier emplaced magma, or assimilation on a massive scale.

Nevertheless it is perhaps puzzling that there should be so often a mismatch between strain estimates in pluton and aureole, those within the pluton being so much the greater. Thus, in the example of the Cannibal Creek pluton, Bateman calculated, in 1985, the extension to be 70% within the periphery of the pluton as against 30% in the aureole. The mismatch is probably even greater on the premise that the highest strains in the aureole should be recorded in a fairly narrow inner zone, as predicted in 1982 by Marsh, who reasoned that the temperature dependence of the viscosity of the wall rocks will concentrate deformation in this position. However, the strain measurements in the immediate aureole probably fail to record the regionally distributed stresses, envisaged by Bateman to amount to a 1% shortening for over 45 km away form the pluton. Similarly the emplacement of the Ardara distensional diapir was associated with a widespread system of gentle folds which also represented a regional dissipation of the stress. Clearly the accommodation of magma in the crust has to be viewed in regional and kinematic terms and cannot be exactly determined.

A RETURN TO MULTIPLE PREJUDICES?

Of course, diapiric intrusion may operate very deep in the crust. If so, it carries the convenient corollary that such hot blobs might arise by the growth of irregularities at a viscosity contrast interface within or at the base of the crust, when the expected wave-like form of these irregularities offers one explanation of certain regular pluton and volcano spacings. Nevertheless, continuing the theme of this essay, I follow Vogt (1974) in attributing such spacings to location at the intersections of major crustal fracture systems, controlled by lithosphere thickness. That thickness is a key factor is confirmed by Rickard and Ward's finding that such spacing follows a bimodal pattern, with long arrays of plutons perhaps reflecting melting at great depth and short groupings or nests of plutons the ponding of magma beneath the upper brittle layer of the crust (see Figure 11.13, 3).

All kinds of combinations of such processes are possible (Figure 11.5); piercement and ballooning are not exclusive. One attractive elaboration mentioned to me by Gastil is the possible mushrooming of a piercement diapir by the rise into its consolidating head of mobile magma residing in the tail (Figure 11.5c). However, the clear fact is that mushroom geometries and quaquaversal inclined floors are not easy to demonstrate. Thus it is intriguing that a fresh look at Wegmann's key example of diapirism – the Rapakivi plutons of south Greenland – by Hutton and others in 1990 has introduced a realistic alternative. In three-dimensional outcrop at least one mushroom turns out to be one-sided, the roof and floor representing a flat, ramp-like sheet intruding an extensional shear zone. So there was no rising blob here and perhaps none anywhere else!

Another view of shape evolution was expressed in 1986 by Castro, who argued that the shape of certain ballooning 'diapirs' results from squeezing induced by regional shortening (Figure 11.5c). In his model the narrowing of the waist represents the throttled throat of a magma conduit and not the tail of a buoyant blob; the tilted Bergell pluton of southern Switzerland and northern Italy, described by Rosenberg, Berger and Schmid in 1995, may provide an example of just this mechanism. It is certainly easy to envisage compression at depth propelling either the whole of a quasi-solid pluton or its still mobile interior upwards, to balloon out at higher levels. Furthermore, I see no reason to doubt that sometimes plutons may dome and punch into their roofs, emporte-pièce, as envisaged by Barrière for the Ploumanac'h pluton. And Corry has shown that roof uplift may be a viable emplacement mechanism at shallow crustal levels.

THE SPECIAL CASE OF MAGMA BLISTERS

Some other plutons model best as magma blisters emanating from a feeder shear zone (Figure 11.5d). That of Ardara, Donegal, has already been mentioned and the Papoose Flat pluton, California, as discussed by Paterson and co-workers in 1991, also by Law and colleagues in 1992, is yet another possible example, though here the timing of the associated deformation is in contention. The structural studies by the Rennes school of the 'mise en place' of the diapiric plutons fringing the ductile shear belts of Brittany provide a wealth of field, laboratory and modelling data supporting a general thesis of the synkinematic intrusion of magma along shear zones, often to be squeezed out as blisters (Figures 11.5d and 11.6).

From these examples we can easily understand that whether the stresses are pluton generated or merely represent a natural concentration of external forces around a rising hot-spot is often unanswerable. I doubt, in fact, whether such alternatives exist because they are patently consequential and reciprocal, especially when viewed from the point of view of my general thesis – that magmatism is driven by tectonics.

And if magma is so driven and not floated up as hot blobs – diapirs now being dead – how does it enter the crust?

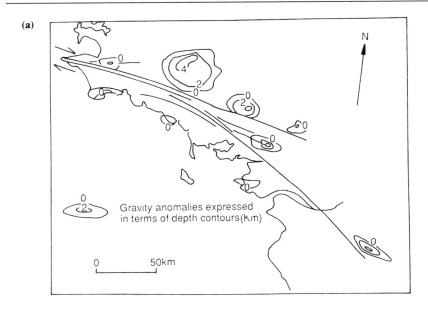

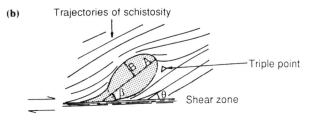

Figure 11.6 (a) Leucogranite plutons of Brittany (as depicted by their gravity anomalies) in relation to the South Brittany Shear Zone (depicted as linears). The plutons are interpreted as blisters emanating from the shear zone during movement. (b) Structural pattern of a model plutonic blister with some of the parameters used in its description. Adapted from Vigneresse and Brun (1983).

THE SUCKING-IN OF MAGMA: CAVITY-FILL EMPLACEMENT INTO EXTENSIONAL SHEAR ZONES

The shear zone model can be developed on the basis that tensional opening within such a zone will provide potential space for the accumulation of magma (Figure 11.5). Thus, according to the review by Hutton, the principal reason for the contrasted mechanisms of emplacement of the eight plutons that make up the multiple batholith of Donegal is the differential response to a sinistral shear movement which is demonstrably synchronous with intrusion. The essence of this model is that magnitudes of strain and displacement vary along the acceptor shear zone in such a way as to cause either a bowing out of one side of the zone, so leading to an axial split, or a pull-apart (Figures 11.5e and 11.11). In the former instance one wall bends aside to provide a potential cavity, thus allowing magma to sheet up into the core of the movement zone. Of particular interest is the extension of this thesis to include the associated stresses operating in the flanking crustal blocks and, by doing so, to provide an overall connected explanation of the various modes of emplacement apparent in outcrop.

Extensional cavity filling has also been advocated for what might be regarded as a more conventional pluton, that of Strontian, Scotland (Figure 11.7). This was modelled by Hutton (1988b) as a sheet intruding a listric extensional fault, the space being created by the relative movement between hanging wall and foot wall. Such mechanisms might well apply on the grander scale of batholiths, as I shall suggest in the sequel. For example, a study by Hutton and Ingram, in 1992, of the great sheet of tonalite, over 1000 km long, that forms a major element of the batholithic complex of Alaska, indicated that this too was intruded into a ramped fault zone. I repeat my belief that sheeting or dyking is the principal

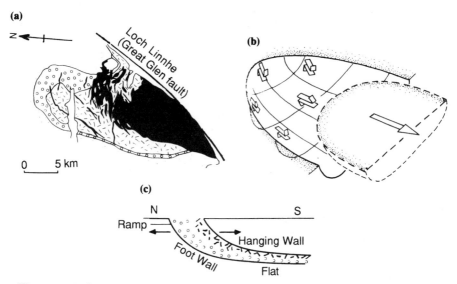

Figure 11.7 Strontian Granite. (a) General geology: country rock (dotted ornament); tonalite (circles); granodiorite (dashes); biotite granite (black). (b) Block diagram showing nature of foot wall to intrusion with shear sense observations indicating the horizontal southward movement of an overlying hanging-wall block of country rock. (c) Generalized north–south section. Adapted from Hutton (1988a), with the permission of the author.

mechanism by which magma enters the upper crust and I envisage plutons being fed by dykes, often repeatedly, from sources far below.

So attractive are these new models (Hutton, 1992) that we need to recall that what appear to be true granitic diapirs have often been regarded as characteristic of high grade and presumably deep-seated, metamorphic terrains. Yet it is within such environments that synchronous deformation is universal, implying that the concordances and tadpole-like shapes are much more likely to have resulted from regional ductile shear mechanisms than from piercement. I suspect that many dome in dome, fold interference structures in basement terranes have provided a structural locus for synkinematic intrusion. Brun in his 1980 restudy of the gneiss domes of Finland, unravelled the complex patterns of interference structures consequent on the progressive inflation of neighbouring 'diapirs'. Once again the latter cannot satisfactorily be modelled as rising hot blobs.

ZONE-MELTING AND GRANITIZATION: A DECLARED PREJUDICE

At this point I rid myself, if not the reader, of two other models for the emplacement of granitic magmas. Firstly, I know of no field evidence for the melting of roof rocks in bulk, though melting on a trivial scale is recorded from some high level subplutonic cauldrons. Furthermore, the energy requirements render the model untenable except, perhaps, in relation to the very lowest crust (cf. Fountain *et al.*, 1989). Certainly there is little evidence for the bulk assimilation of the envelope at high crustal level, so I reject zone melting outright, at least at levels of the crust presently exposed. Secondly, I also reject the thesis of wholesale metasomatic replacement for the emplacement of discrete bodies of granite, but here there is a proper case to answer, which I will address later.

DISCORDANT PLUTONS: FRACTURE PROPAGATION AND CONTROL OF INTRUSION

There remain many examples of plutons apparently devoid of recorded evidence of deformation both internally and externally. These, 'les massifs circonscrits', are the downward enlarging stocks which are particularly characteristic of brittle crustal environments, best exemplified by the multiple centred complexes representing the roots of volcanoes.

Despite the apparent passiveness a fracture control is generally obvious on all scales and examples are legion (Figure 11.8). From my own experience I quote the plate-edge, dominantly extensional environment of the Mesozoic Andes, where both the marginal, volcanogenic basins and the great linear arrays of granitic plutons that intersect them are precisely located by major rift faults, the latter having a dextral component of shear, sometimes also taking on the character of hot, ductile shear belts even at this high subvolcanic level in the crust. (Figure 11.9). A feature of Mesozoic Andean tectonics is the episodic rhythm of short periods of compression and long periods of relaxation (Figure 11.9), which might well be related to changing rates in sea-floor spreading.

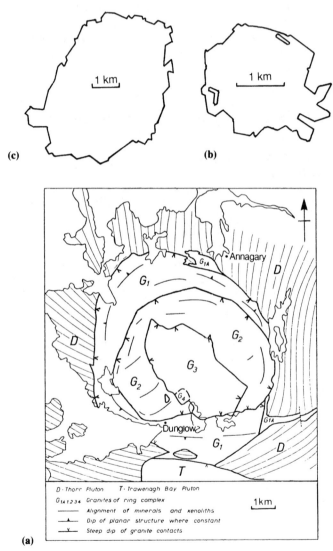

(c)   (b)   (a)

Figure 11.8 Fracture-controlled contacts. (a) G_3–G_4, Rosses multipulse pluton, Co. Donegal, Ireland. After Pitcher (1953, 1992). (b) Sitarah, Saudi Arabia. After Agar (1986). (c) Cerro Condormarca, Ancash Province, Peru. After Bussell (personal communication).

In detail, plutons cluster at fault intersections, contacts are universally angular and their magmas were emplaced by cauldron subsidence initiated by pull-apart opening. Furthermore, the linear dyke swarms clearly represent intrusion into a highly orientated, tensional fracture system. Fracture propagation and intrusion were often near synchronous, with movement occurring between successive magma pulses (Figure 11.9d), or synchronous, with fractures propagating ahead of dykes, some even generated during the consolidation history.

Examples multiply of this structural control. I refer to just one more because of its relevance to a major intra-continental thrust belt; the Boul-

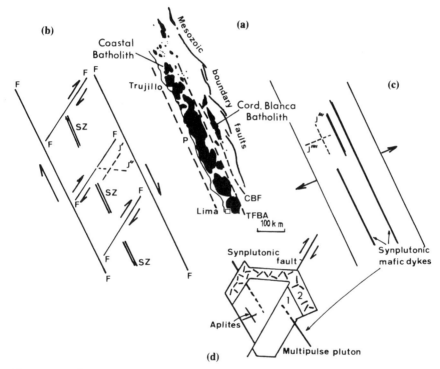

Figure 11.9 Plate margin example of the structural control of emplacement involving the Coastal Batholith of Peru. (a) Outline map of the northern part of the batholith to show its containment within strong dextral linears: the axis of the Tapacocha Fold Belt (TFBA) and the Paracas Terrane Boundary (P). Also depicted is the Cordillera Blanca Fault (CBF) which controlled the location of the batholith of that name. (b) The fracture system during a transpressive phase: faults (F), joints (J) and shear zones (SZ). (c) The fracture system during a relaxation phase, mainly axial faults and joints. The combination of fractures controlled the location of plutons (at fault intersections), decided the form of the contacts (joints and faults), outlined the stoped blocks, and provided access for the basic dyke swarms. (d) Schematic representation of the structural control of a two-pulse pluton intruded during fault movement.

der Batholith of Montana, described by Schmidt and co-workers. Three long established, deeply penetrating, regional fault sets not only control the shapes of the constituent plutons but generate the space by pull-aparts between shear zones.

On a smaller scale a similar fracture control of the Rosses pluton, Donegal, is particularly revealing in that each of four successive magma pulses was guided by synplutonic fractures, each linear representing a tangent to the centred complex (Figure 11.8a). Here, the fact that the same fracture system persists at all the scales of observation, maintaining a self-similar fractile formulation, strongly suggests that the fracturing was a response to the interplay between regionally and pluton-induced stress systems (Pitcher, 1953, 1992).

Much work has been published about these guiding fractures of intrusions, beginning with Anderson in 1942. The aim has always been to calculate and map the stress trajectories consequent on the updoming caused by a combination of thermal expansion and magma pressure. Anderson's model diagram has been repeatedly reproduced but, in outcrop, the reality is this interplay between regional and pluton-induced stress systems, with the varying interference of pre-existing structural grains. Thus outcrop shape, whether as lenses, tablets, boxes or cylinders, depends on the degree to which the plutonic stress field can override the regional. From the theoretical point of view this interplay was adequately modelled by Odé in 1957, followed by Muller and Pollard in 1977, in an attempt to explain the various orientations of pluton-associated dyke swarms, whether tangential, radial or linear, and their analyses fit well with synplutonic fracture systems in general. Other theoretical studies confirm the outcrop finding that the boundary fractures of plutons and ring dykes should be steep and outwardly inclined.

Along these bounding fractures sheet-like apophyses of granite can often be seen in the process of prising away slabs of the wall or roof rocks (Figure 11.10), but here I think the splitting is as much due to thermal spalling as to simple hydraulic fracturing. However, the separation of water vapour from the crystallizing melt must generate tremendous mechanical energy, a matter fully discussed by Shaw in 1980. The importance of such fluid pressure in fracture formation is dramatically demonstrated by the way the cracks are so often penetrated by vapour-rich magmas that brecciate, comminute and then entrain the wall rocks in the form of intrusion breccias and tuffisites. This is especially so in the subvolcanic regime of centred complexes, where fractures are initiated and old fractures regenerated, even passing across contacts into still consolidating magma – all in response, I believe, to the seismic activity accompanying such high level intrusion.

On the larger scale, fracture systems resulting from this interplay of forces isolate crustal blocks. Density differences, magma and gas loss into

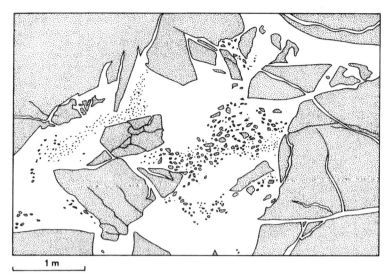

Figure 11.10 Example of piecemeal stoping. Northern contact of Maria Cristina pluton, south of Rio Huarmey, Lima Province, Peru. Tonalite (stippled); monzogranite (white). Field sketch by John Myers.

the peripheral rocks, and the seismic pumping of magma up and around the blocks ensures that they will subside; this is the proper environment of major stoping, where rock pistons and flat-topped, bell-jar-shaped plutons really do exist, as in the Coastal Batholith of Peru (Cobbing and Pitcher, 1972) and the ring complexes of northern Nigeria (Jacobson *et al.*, 1958), to quote but two examples. Such blocks must often be broken up, but the larger the sinking fragment the less the cooling effect it has on the magma, which is why the piston mechanism works at all, certainly why we rarely see a jumble of small blocks in deeply eroded plutons. Dramatic though examples of piecemeal stoping can be (Figure 11.10), I doubt that this process often provides much of the space for emplacement. On the other hand, major stoping is obviously an important mechanism, and fragmentation on this scale, when combined with multiple intrusion, can provide fairly complex histories of emplacement. Overall, I envisage nested arrays of multipulse plutons extending downwards as a stack of foundered blocks and interconnected cauldrons that reach deep into the crust. None of the individual cauldrons needs to be very thick; none would be bottomless. The whole process represents but a transfer of mass within the lithosphere.

This tendency for collapse can continue during consolidation, especially if heavier, more mafic rocks lie at depth. Excessive crustal loading, coupled with contraction, results in a central sag within the congealing roof of a pluton. The potential void sucks in new magma or intercrystalline melt into either flat-lying lenticular, 'blind' sheets or the curved flange-type of intrusion suggested by Walker. The Tumaray pluton, Lima Province, Peru, provides a spectacular example of such a sheeted roof zone involving the segregation of more highly differentiated magma, whereas the flange form is well displayed in the Mourne Mountains Complex, Northern Ireland. According to Roberts such a sag mechanism best explains the many examples where internal arcuate contacts enclose intrusions of progressively decreasing area, so providing a realistic interpretation of nested plutons.

Thus we see that even the discordant plutons are essentially syntectonic. The tectonically active fracture systems control intrusion and it is the resulting gravity collapse coupled with seismic pumping that propels magma upwards, certainly in the upper crust.

THE IMPORTANCE OF GIANT DYKES AND SHEETS

As we have seen there is an increasingly strong appeal for space creation within active, developing fault systems with the consequent intrusion taking the form of dykes and sheets. Some large plutons are obviously constructed from a confluence of such sheets; examples are the Main Donegal Granite (described by myself and Read in 1959, myself and Berger in 1972, and Hutton in 1982), and the Ox Mountains Granite (as reported by McCaffrey in 1992), both from Ireland, and both representative of steeply inclined sheet complexes. There are other plutons that are fringed by an extensive network of dykes, such as the immensely thick, gently inclined sheet of the Manaslu Leucogranite of the Himalayas, described by LeFort in 1981. Gentle inclination, as in an example from Greenland reported by Hutton and co-workers in 1990, can lead to a single sheet outcropping over hundreds of square kilometres.

I agree with Hutton and co-workers that the virtual coalescence of multiple sheets may, as in the Main Donegal Granite, produce a relatively homogeneous body, only recognizable as composite when country rock septa intervene (Figure 11.11a). Thus it is possible that sheeting is far more common a process in the assembly of plutons than is generally thought. But, yet again in these discussions, enthusiasm for a particular model has to be tempered by the reality of the outcrop, in this case by the simple and undeniable fact that by far the majority of plutons in the world are unaccompanied by vast sheet complexes; neither do they show evidence of *multitudinous* pulse supply. However, even if the contact zone is unsheeted and the main body not multiply constructed, it is my view that it was most likely to have been fed, at depth, by a plexus of thick sheets. They may not have to extend far to the source and, furthermore, their magmas would have the mobility associated with juvenility. On this interpretation the arrival shape is simply a final structural accident and

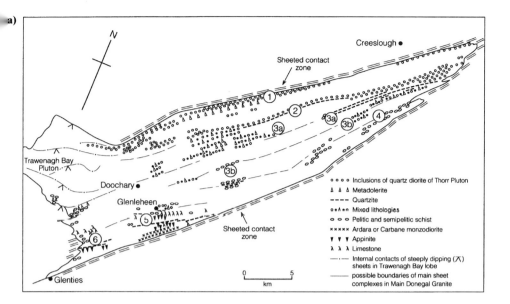

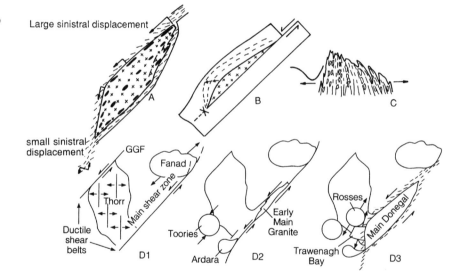

Figure 11.11 (a) The ghost stratigraphy of the Main Donegal Granite, Ireland. The trains of xenoliths divide into zones of different country rock lithologies reflecting the external stratigraphic relations. They represent septa or disrupted septa effectively dividing the pluton into separate granite tongues that appear to be contiguous with the major sheets making up the Trawenagh Bay lobe. Modified from Pitcher and Read, 1959. (b) Donegal granites. A–C, Main Donegal Granite. (A) General strain patterns. Black ellipses represent relative strain relations in pluton. (B) Crack opening model. (C) Northwest–southeast generalized cross-section showing coalescing sheeted nature of intrusion giving rise to raft trains, septa and roof pendants. D1–3 represent the evolution of the Donegal nested grouping of plutons as detailed in the text. GGF, Great Glen Fault. Adapted from Hutton (1988a) and Pitcher and Berger (1972).

may bear little relationship to the method of magma transport – which is often under the control of an active shear zone.

This new emphasis on the role of dykes in the supply of magma has led to a re-examination of fracture propagation and channelling as a control of magma ascent, the thermal history of dyke ascent, and the physical nature of the magma involved. It is obvious that the flow of magma through feeder dykes requires that it have a relatively low viscosity, much lower than in the case of bulk transport in migrating diapirs: to avoid freezing, the magma would have to travel fast, and possibly the dyke progressively increase in width (Johannes and Holtz, p. 126). Of course granitic dykes exist in outcrop but do they ever remain open conduits with sufficient throughput to supply, even as a swarm, the vast volumes required to fill a sizeable pluton?

There are, indeed, granitic dykes with a sufficient compositional difference between margin and core as to suggest that magma continued to flow through the hot centre as in a true conduit, and some of these even exhibit a chilled margin which might have acted as an insulation. But it is very rare to see any evidence of the contact thermal effects that I would have expected from long replenishment – quite the reverse, and in my experience most granitic dykes represent a short lived, one-off injection of magma. But then I have never wittingly seen dykes beneath a pluton!

Treating the problem from a theoretical standpoint and laudably taking account of a range of factors, Clemens and Mawer argued from their calculations that it was possible for granitic magmas to ascend as dykes via self-generated fractures, such magmas being sufficiently inviscid to travel through these fractures without suffering thermal death by freezing; and remarkably rapidly. This hypothesis was echoed in 1994 by Petford, Lister and Kerr, who, besides emphasizing the importance of fault systems in creating conduits and repeating the criticisms of the diapir model, attempted to calculate the critical dyke width required for the transport of granitic melt through the crust, settling on an estimate of 2–20 m.

One can discern a certain scepticism in Rubin's 1995 contribution to these discussions. Being of the opinion that previous calculations did not tell the full story, he critically assessed the extensive range of controlling factors involved such as the birth of a dyke, its growth, the effect of non-Newtonian flow, the roles of viscoelasticity, viscous heating, latent heat release, tip cavity suction, and volatile exsolution. In his conclusion Rabin did not rule out the possibility that most granitic magmas could be transported through the crust in dykes, but considers it premature to rule out other forms of ascent.

If only we were able to see beneath plutons and so explore the role of possible feeder dykes! Occasionally, where there has been very deep canyon erosion, as in the western Cordilleras of Peru, the feeder ring dyke of a centred pluton is exhumed, but otherwise this is rarely possible, though I am intrigued by Barbara John's 1988 reconstruction of the tilted multiple pluton of the Chemehuevi Mountains, California, as a plate-like body ponded below a thrust and fed by dykes.

INTERRELATIONS AND CONNECTIONS

We should not be surprised at the frequency of the natural interplay of the several mechanisms – buoyant rise, tectonic pumping and cavity opening – because there are plenty of examples to demonstrate that these apparently diverse mechanisms can together be involved in the emplacement evolution of a single pluton. Thus Brun and his colleagues, in 1990, proposed changing ascent mechanisims for the Flamanville granite: lower crust diapirism, middle crust fracture exploitation, and upper crust ballooning. According to Bergman, at Ålva in Finland, ballooning diapirism was followed by cauldron collapse at the same crustal level. Then in the felsic–mafic complexes of the Scottish Hebrides and N. Ireland shouldering aside and subsidence are but consecutive phases of an emplacement evolution. As a general model for the latter Walker (1975) envisaged a felsic diapir venting magma, degassing and cooling as it approached the surface. In this waning stage, loaded down by the infill of basaltic magma below, the drag produces cauldron subsidence at the higher level.

More generally Gastil (1983, p. 273) and others explain circular outcrops and the similar diameters of associated diapirs and cauldrons, as exemplified in the batholith of Baja California, on the basis that it is the doming above the diapir, of whatever type, that generates the arcuate fractures that locate the overlying cauldron. Whether this connection is general is difficult to prove in the absence of a denudation series, but my own view is that cauldrons and distension diapirs are more often alternatives depending on the competency of the ambient crust.

What is important is that stresses and relative competency can vary widely within relatively short distances and at the same level in the crust, as is so well demonstrated by the Donegal studies of myself and Berger in 1972. There, Hutton's all-embracing model (1988a) envisages the Thorr and Fanad plutons filling dilation zones developed between two regional shears (Figure 11.11b). The Toories and Ardara plutons apparently ballooned into a compressional tip area of the shears, the latter pluton developing as a distending blister sprouting from one of the shear zones. As the main shear zone developed, the Main Donegal Granite was sucked into an extensional sector in the form of confluent sheets, and continued to be deformed as it cooled. Meanwhile, outside this shear zone and within the flanking block, the associated extensional stresses located the fractures controlling the collapse cauldron of Rosses and fracture filling of Trawenagh Bay. A complete and compelling thesis indeed!

In the light of these evolutionary structural studies made by both Hutton and Castro (see especially 1988a and 1987, respectively) we can largely dispense with hot blobs, at least within the middle and upper crust, and substitute, where appropriate, model mechanisms of sheeting, ballooning, sagging and subsidence to explain how granitic magmas are sucked, squeezed, pumped or floated into the upper crust. But what shapes do they take up on arrival?

PLUTONS IN THREE DIMENSIONS: DIFFERENT CRUSTAL ENVIRONMENTS

To model the shapes of plutons we need to combine structural mapping with gravimetric, seismic reflection and magnetic surveys. We can no longer be content with the 'gravimetric' cylinder standing proud in an amorphous crust!

Vigneresse in 1990 discussed the considerable difficulties in surveying granite plutons geophysically, opting for three-dimensional gravimetric methods. Even so, because the resolution rapidly weakens with depth, the form of the floor of a pluton can rarely be modelled without ambiguity. In his holistic approach to shape determination Vigneresse found that the alkali granites of the cratons, disposed in the form of subsidence cauldrons, such as those of southern Norway, northern Nigeria and Corsica, yield the least clear cut data, presumably because of the density complexity resulting from the mingling of both granitic and basaltic magmas as they welled up around the foundered blocks. Luckily there are sufficient three-dimensional exposures of such cauldrons to provide a clear picture of short cylinders, either circular or polygonal, standing vertical in the upper crust.

The deep seismic sounding of active caldera volcanoes provides an analogue model for the understorey of such subsidence cauldrons – that of a flat, cake-shaped magma chamber, large or small, within 5 km of the surface. It is perhaps not surprising that, according to Iyer, the largest of these subvolcanic chambers are confined to regions of silicic magmatism, implying that it is only siliceous magmas that commonly pond in the crust. Of even greater interest, however, is Iyer's further conclusion that some of these magma ponds have mantle floors and are not pockets of melt isolated within the crust.

Within the roof zones of the plate margin batholiths of the circum-Pacific zone such subvolcanic cauldrons can be seen either to be cut out of strong lithologies or to balloon and shoulder aside weak lithologies. Again they map out as relatively thin plates fed by ring dykes. At deeper levels the plutons of such batholiths nearly all behave as balloons, some would say as ballooning diapirs. But what of their roots?

For the Coastal Batholith of Peru it is reasonable to model the whole granitoid complex as passing directly down into a dense underplate which possibly consists of an up-arched complex of mafic dykes, gabbros and altered volcanics. Beneath the western part of the Sierra Nevada Batholith this is also likely, or so I interpret the geological findings of Paul Bateman (1981) and the geophysical surveys of Oliver, though in this latter example we are faced with the prejudice of some American workers that granitic magmas must rise to the level of neutral buoyancy, to pond as thin batholiths in the upper crust. Perhaps the hypothetical flat bottoms are represented by the flat reflectors discovered by Lynn and co-workers.

It is within the mobile zones of the continental plates that sheet forms are most likely. Thus within the regional shear zones of Brittany the granite bodies geophysically map out as ellipsoidal, flat cones with simple dimensions of 18–25 km diameter and 6–10 km thickness, with a possible stalk representing the point of infill. Brun and Pons showed that the attitude of these and other examples of flattened mushrooms is primarily controlled by that of the shear zone into which they were injected, though their present shape results from the interference between pluton distension and the synchronous shear movement.

In the USA, in Maine and New Hampshire, the granites of supposed deep-seated emplacement were modelled by Hodge and co-workers as thin sheets, and I suppose that these represent magma penetrating along flat thrusts. Within continental collision zones, the environment of thin-skinned tectonics, it is probable that batholiths commonly take the form of thick sheets bottomed by décollement planes. Thus, along the edge of the Variscan front in southwest England, the Cornubian Batholith can be geophysically mapped as a flat cake, about 10 km in thickness, lying in a crust nearly three times as thick. Shackleton and his colleagues proposed that these granites were generated in the orogenic core and injected outwards above a major décollement as a sheet-like body, from which protru-

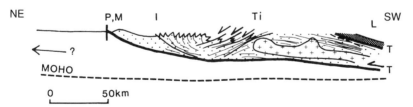

Figure 11.12 Section showing a model for the stages of Variscan evolution of southwest England. Lower Permian. P, Pembroke; M, Mendips; I, Ilfracombe; Ti, Tintagel; L, Land's End. No vertical exaggeration. Dotted, Upper Palaeozoic sediments; cross-hatched, Lizard Complex; crosses, granite; T, major thrust or décollement; thin lines, cleavage. After Shackleton *et al.* (1982), with the permission of the authors and the Geological Society of London.

sions, probably representing separate pulses, moved upwards to form the presently exposed cupolas (Figure 11.12). Similarly, in Montana, according to Schmidt and co-workers the Boulder Batholith is thought to be floored, at 17 km depth, by the basal décollement of the Montana thrust belt.

A Himalayan example is the synkinetic Manaslu Leucogranite, which can be mapped as a lenticular slab up to 8 km thick lying within the top of a 12 km thick package of crystalline rocks thrust over the Indian plate. According to France-Lanord and Le Fort the source of the magmas lies within these underlying metamorphic rocks, a matter I will later take up in some detail.

From these and other studies it is easy to envisage the melting of granite in thrust-thickened zones followed by its injection outwards into the thrust fronts. It is always the synchronous earth movements which control the whole process from genesis to final intrusion, and it is always the faults and thrusts which determine the arrival shape. With this prejudice I now review structural control on the global scale.

THE GLOBAL TECTONIC FRAME AND THE MECHANISMS OF EMPLACEMENT

Vast volumes of tonalite and granodiorite magma appear during ensialic spreading at active continent–ocean plate boundaries. Here, above a mantle asthenolith, and in a zone of high heat flux, deep reaching fault systems tap hot, relatively dry, mantle-derived magmas which rise high into the crust in the form of a stack of interconnected magma chambers created by various combinations of cauldron subsidence and ballooning distension, their relative importance depending on the ambient crustal lithology and the degree of pre-heating of the fault conduits. Overall, upwelling largely results from the foundering of the volcanic carapace. This is one environment in which there is a reasonably clear connection between the cycles of magmatism and the alternations of compression and extension (Figure 11.13(2)).

Within the cratons, compressive movements are concentrated into deep-seated ductile shear belts and wrench faults which suck in magma into jags, pull-aparts and great tension fractures (Figure 11.13(4), (5)). I doubt whether the lower crust is often sufficiently ductile to accept hot blob diapirs, and instead I envisage magma being channelled upwards through a complex of intersecting thrusts and faults.

In the extensional regime of the intracontinental rifts and swells the highly fluxed magmas characteristic of these special environments rise rapidly along the deep-seated fractures, often ponding out along them, but also reaching up into the very roots of the volcanic centres (Figure 11.13(3), (4)).

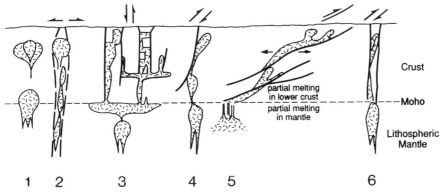

Figure 11.13 Highly generalized cartoons showing various models for the ascent and emplacement of granitoids. All begin with initial diapiric detachment and uprise of melts: (1) continued diapiric uprise in the (?) absence of tectonics, leading to final arrest due to density equilibration followed by late ballooning; (2) uprise into major vertical tectonic extensional fault system, magmas rise to high levels, with uppermost crustal ponding and cauldron/caldera behaviour; (3) diapiric uprise arrested by viscosity strength changes at Moho – this leads to lateral spreading with possible late spawning of crustal plutons; (4) magma (? as a diapir) may sometimes rise into middle crust, intercepting an intracrustal strike-slip fault zone, leading to elongate plutons with late ballooning; (5) magma arising along listric fault/shear zone from zone of melting in thrust-thickened crust, leading to listric granite sheets and possible generation of asymmetric cauldrons and calderas; (6) uprising melts intercept transcrustal vertical transcurrent fault/shear zone – jogs, pull-apart and large tension-gash features create space for magma ponding; note that in all these scenarios the 'source region' is arbitrarily located in the lithospheric mantle. Modified from Hutton (1988a), with the permission of the author.

In great contrast is the continent–continent collision regime, where thrust-thickening leads to the slow build-up of heat and to regional plutono-metamorphism. At depth the lower crust is likely to represent a veritable structural mélange of thrust-separated wedges. Partial melts forming within this mélange will be located and channelled by these structural discontinuities so that the crystal mushes intrude in all directions, as much laterally as horizontally. This is the environment of the granite series of Read, which supposes a direct genetic connection between ultrametamorphism and granite production, and it is here that we might expect hot magma blobs sometimes to segregate and ascend diapirically into the lower crust, but only to be intercepted by synchronous fractures, such as the great décollements basal to the thrust wedges, or transpressive shear zones, along which the magma is finally channelled into the upper crust (Figure 11.13(5), (6)). Of course, this is a story book picture, but it moves us away from the simplistic view of a ductile and brittle, structurally layered crust through which hot blobs rise buoyantly until they get stuck in the cold canopy!

SELECTED REFERENCES

Hutton, D.H.W. (1988) Granite emplacement mechanisms and tectonic controls. *Transactions of the Royal Society of Edinburgh: Earth Sciences*, **79**, 245–255.

Paterson, S.R. and Vernon, R.H. (1995) Bursting the bubble of ballooning plutons: A return to nested diapirs emplaced by multiple processes. *Bulletin of the Geological Society of America*, **107**, 1356–1380.

Vigneresse, J.-L. (1990) Use and misuse of geophysical data to determine the shape at depth of granitic intrusions. *Geological Journal*, **25**, 249–260.

On the rates of emplacement, crystallization and cooling

12

The speed of plutonic processes . . . is one of their fundamental qualities and has to be integrated with the speed of tectonic and other mechanical operations going on in the crust.

Herbert Harold Read (1949) Quarterly Journal of the Geological Society of London, **105**, 102.

INTRODUCTION

We know very little with certainty about the times required to melt, segregate, crystallize and cool granitic magmas. Calculation is bedevilled by imponderables such as the effect of crustal depth, tectonic environment and mode of emplacement, and the extent to which fresh draughts of magma thermally recharge the system. It is extremely unlikely that a pluton ever represents a closed system insulated from the enveloping heat plume within which it rises, and from its source of magma replenishment. Nevertheless, it seems worth while to attempt to identify constraints and assign minimum time spans to the several parts of the evolutionary history – that is, the initial partial melting and the melt assembly, its ascent and intrusion, and its crystallization and final cooling to the ambient temperature of the country rock. However, I do so in the reverse of this natural order in accord with the reliability of the data.

MEASURING THE COOLING TIMES

We might calculate the cooling stages in a simple static system by balancing the heat lost by conduction with the heat gained by crystallization, but this would represent a gross oversimplification. A more sophisticated model was provided in 1982 by Spera, who took into account not only the expected decrease in heat loss with increasing crystallinity, but other relevant factors such as crustal depth, pluton size, bulk melt composition and, most crucially, the availability of water, the circulation of which is so effective in conducting heat. This is not the place to rehearse Spera's calculations, but it is salutary to note that his estimates of the time taken for solidification of a typical pluton from the liquidus to the solidus temperature varied greatly with the assumed water content, decreasing ten-fold between 0.5 and 4 wt.% water.

The time of cooling from near consolidation to the ambient country rock temperature can be more realistically determined directly by using the different closure temperatures of mineral isotope systems. Values have now been determined for much of the cooling range. For example, zircon and titanite have relatively high closure temperatures of >600°C for U/Pb systems, whereas biotite has a low temperature for both Rb/Sr and K/Ar systems, 300 and 280°C, respectively. Lower still is the potassium feldspar value of 150°C for the Ar system. The difficulties in interpretation are legion, especially as closure times vary with the rate of cooling, but the potential is there. Particularly so if, as mooted by Lovera, Richter and Harrison, we may accept that the aliquots of argon locked up in the separate domains of a single potassium feldspar crystal offer a means of determining the full sweep of a cooling history between 400 and 150°C.

As an example of such methods Cliff established a cooling pattern for the Glen Dessary Syenite, Scotland. From this it can be deduced that the rate of cooling decreased from about 30°C/Ma in an early stage to about 10°C/Ma in the interval over which micas became closed for argon, then waned much more slowly to the ambient temperature of the country rock. The overall slow cooling seems to represent a special case of a pluton intruded deep in the crust and at the peak of regional metamorphism (Figure 12.1). In contrast, Graeme Rogers told me that the shallower seated, post-regional metamorphism Ratagain Complex had a maximum cooling rate of about 80°C/Ma. This calculation was based on a crystallization age, determined on U/Pb in zircon, of 425 ± 3 Ma, representing a closing temperature of about 600°C, whereas a mean age of 419 ± 3 Ma provided by biotite–whole rock determinations represents the time that the complex cooled to less than 300°C. For comparison I report Paterson and Tobisch's 1992 determination of cooling rates of at least 75°C to as much as 250°/Ma for certain high level plutons of the central Sierra Nevada. In general we should not be surprised that rates are so much faster in the upper crust; especially so if uplift was rapid. Thus for what may be the youngest exposed pluton on Earth – the 2.2 Ma Takidani Granodiorite in the Japan Alps – Harayama calculated average cooling rates of between 360 and 550°C/Ma.

Slow cooling deep in the crust is dramatically accelerated during uplift and exhumation. As an example Harrison and Clarke took this into account by first deriving a cooling curve by calculation and then confirming its validity from data obtained from the isotopic systems. On this basis

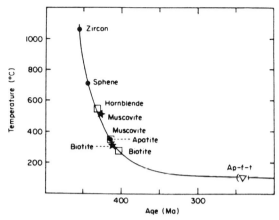

Figure 12.1 Cooling pattern for the Glen Dessary Syenite. Geochronological data are from van Breeman *et al.* (1979). Closure temperature values as follows. U/Pb ages (filled circles): zircon, closure temperature unknown but some U/Pb systems survive magmatic temperatures; sphene, 550°C (Cliff and Cohen, 1980); apatite, same as K/Ar muscovite, about 350°C (Cliff and Cohen, 1980). Rb/Sr ages (filled stars): muscovite, about 500°C (Jäger *et al.*, 1967); biotite, about 300°C (Jäger *et al.*, 1967). K/Ar ages (squares): hornblende, 500+°C (Harrison, 1981); muscovite, about 350°C (Purdy and Jäger, 1976); biotite, about 300°C, 40°C less in relation to Rb/Sr in biotite (Harrison and McDougall, 1980). The open triangle represents the mean apatite fission track age N of the Great Glen Fault (from Hurford, 1977); although this age is much less precise than the others it does indicate a rapidly waning cooling after biotite closure. Reproduced from Cliff (1985) (in which all references are quoted in full) with the permission of the author and the Geological Society of London.

they modelled one particular pluton as only finally cooling, and then relatively quickly, after regional uplift and erosion had brought it up to a shallow crustal level about 7 Ma after its original deep emplacement. The cooling history of the Karakorum Batholith in the Eurasian margin of the Himalayas may well conform to this pattern. According to Schärer and his colleagues there were two distinct pulses of magmatism at 25.5 and 21.4 Ma, as determined from U/Pb in zircon and monazite, findings which, coupled with the radiometric ages determined on the micas and K-feldspar, showed that the leucogranites cooled from 750–650°C at 25–21 Ma to 250–350°C at 6 Ma, a long cooling period only terminated by rapid uplift and exhumation of the cover of the batholith. I suspect that many plutons remain for long periods at relatively high temperatures, providing ample opportunity for subsolidus reactions to occur.

Concerning that early stage of the cooling process, when melt is often still arriving, it would seem to me that the precision of dating techniques is not yet sufficient to record separately the time spans of the different magma pulses that feed a typical multipulse pluton. However, Kistler and Fleck, in 1994, attempted to date the three pulses of the Tuolumne Pluton of the Sierra Nevada, finding them to be 2 million years apart in a 7 million year intrusion history; this despite the fact that the contacts are quite subtle and lack any textural expression!

Turning to non-radiometric ways of placing constraints on the timing of consolidation, and considering first the rates of crystal growth, I note that Cashman, in 1990, cited values for potassium feldspar of 1 mm per 100–1000 years, which translates into a 20 mm long megacryst in a granite growing in 2000–20 000 years. In contrast growth periods several orders of magnitude longer were reported by Christensen and DePaolo in a radiometric study of the Bishop Tuff magma reported below. It is hard to draw any firm conclusions, but Paterson and Tobisch, in a broad ranging discussion in 1992, argued that the time required for the change over from magmatic to solid-state behaviour is but a relatively short part of the entire crystallization history of a pluton, a conclusion I intuitively accept.

It seems, therefore, that the progress of crystallization may not take overlong in geological terms. Setting aside the obvious difficulties in interpreting isotopic results, the narrow spreads of ages recorded by Mukasa, P.A. Wilson, and Beckinsale and co-workers (1985) from within individual rock suites of the Coastal Batholith of Peru, allow only a few million years for the entire crystallization–differentiation process. This is illustrated by the zoned pluton of Santa Rosa, Peru, in which there is near agreement of a U/Pb zircon age with a Rb/Sr age based on a whole rock isochron derived from the entire rock suite.

A more realistic check on this time span and also the overall life of a high level magma chamber can be obtained directly from dating the single cycles of volcanism within the overlying caldera. For the North American caldera complexes these cycles range from 250 000 to 700 000 years, representing large parts of single differentiation events. More specifically, in a radiometric and isotope study of the rhyolites of Glass Mountain, Long Valley, California, Halliday, Mahood and their colleagues determined that three separately differentiated magmas resided and crystallized in the crust under Long Valley for minimum durations of 0.7, 0.5 and 0.3 Ma, respectively. Still in California, Christensen and DePaolo utilized Sr isotopic systematics of phenocrysts and glass from the Bishop Tuff to measure the time span of the evolution of the supply chamber, arriving at a figure of between 300 000 and 500 000 years. Then Van den Bogaard and Schirnick, by employing laser probe ^{40}Ar/^{39}Ar dating of quartz phenocrysts, determined that the Bishop rhyolite magma had existed ~1.1 Ma before its principal eruption. Several of these authors remarked on the long times involved in relation to calculated time spans based on conductive heat loss and opine that heat must have been continually added to the magmatic system by the influx of basaltic melt.

The total life spans of these North American calderas ranges from 1.5 to 4 Ma, so that, on the assumption that the volcanic cycles equate with magmatic pulsing in the feeder pluton, we also have a measure for the multiple assembly of a nested pluton in the upper crust. Taking into account difference of crustal depth, this estimate is in concert with that of the 7 Ma assembly history of the deeper-seated multipulse pluton reported earlier.

RATES OF ASCENT OF MAGMA

It has long been assumed that buoyancy forces motivate the upwelling of silicic magmas, a concept which has spawned a gamut of mathematical modelling. Furthermore, it is generally understood that magma channelled through, or guided by, continuously propagating cracks will rise much faster than that rising as a magma blob through a structurally homogeneous medium. Of course, we have to bear in mind that the rheology of the crust must change with depth, with obvious consequences on the mechanics and rates of emplacement processes.

Marsh's 1982 calculations not only confirm the effect of crack propagation on increasing the rate, but also underline the need for any conduits to be pre-heated, and so pre-weakened, before any considerable volume of felsic magma will rise at all. Taking into account a whole range of determinants, Marsh provided estimates of the ascent velocities of sizeable diapirs, finding that these increase a hundredfold after local warming.

The difficulties in identifying and then quantifying the various factors controlling the rise of hot blobs are formidable, particularly as these focus critically on the width, rheology and temperature of a narrow contact boundary layer. Nevertheless, Mahon, Harrison and Drew made the attempt, recognizing the need to treat viscosity as both temperature and strain rate dependent, when they were able to derive a mathematical expression for a varying velocity of ascent through crust of changing rheological character. More recently Weinberg and Podladchikov, by treating the wall rocks as a power law fluid, estimated that a granite diapir could ascend into the upper crust with the rapidity of 10 to 100 m/ year!

Except for this latter calculation, the several available estimates are of the same order, averaging about 2 m/year, so that the timescale of an initially hot, but then continually cooling, diapir rising over crustal distances could be as short as 10 000 years, surely remarkably fast in geological terms. I cannot judge the validity of these figures, though they appear to me to represent maximum velocities that may only apply to vigorously convecting systems, a condition that I would dispute on the basis that this would obliterate all source-inherited heterogeneities.

More importantly, such sophisticated studies indicate that hot blobs do not easily rise through the crust, unless within a pre-existing thermal plume, and, as Clemens and Mawer expressed it so graphically in 1992, must suffer thermal death and so lock up solid in the lower crust. As I do, Clemens and Mawer hold that the idea of the long distance diapir transport of granitic magmas in the crust is not viable on thermal and mechanical grounds. It is only in the hot ductile zones of the deep crust and mantle that the migration of magma blobs becomes a real possibility.

I have already concluded that granitic plutons grow and are more often supplied via dykes intruding along cracks, both major and minor. But at what rate? The calculations of Clemens and Mawer, referred to at length earlier in this book, show that the times needed for a thick dyke to fill a pluton are 'impressively brief', and since a sizeable pluton might be fed by a number of such dykes this would be an efficient method of rapidly transporting magma and building batholiths; one of Clemens and Mawer's estimates is just 900 years. This is really speedy and I choose to remain sceptical of such rapid rates of ascent of silicic magma.

Calculations aside, it is obvious from field-based studies that the emplacement of granite as dykes, or even as diapirs, is far from being a static process, which argues strongly for the view that it is the rate of the associated deformation, particularly as expressed in fault movement and cleavage formation, which is the controlling factor. In this matter I echo the opinion of Hanson and Glazner, in 1995, that this is the really important control of the rate at which magma rises in the crust, more so than the intrinsic physical properties that are expressed as buoyancy.

That magmas are allowed into the crust as a result of and at the rate of opening of the crust was clearly appreciated by Billings as long ago as 1942, when he reported in his seminal textbook, *Structural Geology* (p. 295), that potential cavities, resulting from horizontally directed tectonic forces, are believed to have filled by magmas as rapidly as they formed. It is also one of the central tenets of this book. In an example familiar to me, the Main Donegal Granite of Ireland, Paterson and Tobisch, in 1992, calculated that, on the basis of a 5 cm/year homogeneous slip in a fault zone as wide as the presently exposed granite, enough space could be made for the magmas in 900 000 years. Such a calculation requires magma to be intruded over the entire period, which is wholly in accord with the clear evidence of continuous sheet intrusion into an active shear zone. Hanson and Glazner extended these ideas by invoking major crustal extension in the emplacement of batholiths in response to strike-slip on major faults at transpressive plate margins. Although I later conclude on structural grounds that extension normal to the tectono–plutonic trend can account for only a fraction of the space requirement (p. 242), it is surely significant that a possible rate of ~10 mm/year normal-to-fault movement (Hanson and Glazner, p. 214) would permit a 10 km dyke-like pluton to fill in just 10 000 years, that is if intruded continously, or at least in lengthy draughts revealed as magma pulses in outcrop. However, I believe that, other than in obviously sheeted complexes, the assembly of multipulse plutons is often a much more discontinous process.

In summary, it seems obvious that the rate of ascent of granitic magmas is a function of the mode of intrusion, and where plutons within the middle and upper crust are fed by dykes, the infilling of each of their component pulses could take as little as a few thousand years. Diapiric rise would take much longer at this level, perhaps failing to make any progress at all, but in the lowermost crust and mantle the rate might only be an order of magnitude different from that of dykes. Depending again on crustal level, particularly in respect of the thermal state of the immediately enveloping rocks, complete solidification of a sizeable pluton is likely to take one or two million years, and cooling to the ambient temperature of the crust much longer. It seems that siliceous magmas can ascend surprisingly quickly but consolidate and cool very slowly to ambient crustal temperature, that is unless their cover is removed by rapid uplift.

TIMES OF GENERATION AND THE ACCUMULATION OF MAGMAS

I have only opinion to offer on how much these periods are extended by including the initial phase of deep-seated melting and magma accumulation. I envisage such generative processes as being relatively rapid in the high thermal fluxes of the arcs and rifts, but much slower in the metamorphic regimes of the collision zones, especially where we are dealing with the possible evolution of anatexitic magmas as the ultimate consequence of ultrametamorphism.

As, in the latter instance, such processes usually involve the separation of granitic melt during regional deformation, and also because the rate of formation of a migmatite could hardly exceed that of the foliation into which the melt segregates, we might again seek some guidance from estimates of the rate of this deformation, as represented by the rate of cleavage development. Using data calculated by Pfiffner and Ramsay, Paterson and Tobisch suggested, in 1992, that cleavages may form in 2–4 Ma at a strain rate of 10^{-14}/s, or as quickly as 0.2–0.4 Ma at a rate of 10^{-13}/s. However, in the context of the hot core of an orogen these rates might be greatly exceeded.

The time spans of the development of the mineral assemblages of regional metamorphism must vary greatly, depending in large measure on the thermal input of any associated magmatism. It is likely that the tens of millions of years often recorded for the duration of an orogenic belt represent the integration of many shorter thermal events, sometimes superposed, at other times migratory and diachronous. To accept any generalized figure is to grasp at straws, but to carry this discussion forward I accept as reasonable the well-argued conclusion of Cliff, Yardley and

Bussy, in their radiometric age study of the possibly diachronous metamorphim of Connemara, western Ireland, that *in any one locality* the observed assemblages developed and partially cooled over a time span of the order of 1–2 Ma.

At this point I am tempted to fall back on inspired guesses and to recall Fyfe's estimate, in 1973, based on buoyancy considerations, that it may take more than 6 Ma for sufficient magma to segregate and assemble into a magma blob that would begin to rise, diapirically, in the lower crust. However, as I have already opined, I do not envisage buoyancy being the main driving force, at least in the middle and upper crust, where, for the most part, magmas are opportunistically sucked or pumped into opening fractures. My guess is that a granitic magma pulse generated in a collisional orogen may, in a complicated way involving changing rheologies of both melt and crust, take 5–10 Ma to generate, arrive, crystallize and cool to the ambient crustal temperature.

EPISODICITY AND PERIODICITY OF EMPLACEMENT

Almost any examination of relevant age data indicates that the intrusion of granitic magma is a discontinuous process, essentially episodic, though many workers seek to establish a cyclic regularity – that is, periodicity. Of course, such episodicity is merely a reflection of the nature of geological processes in general, though especially true of the tectonism that inevitably controls magmatic activity.

Concerning the dating of the intrusion of regional arrays of plutons it is obvious that, until recently, the vicissitudes of the radiometric method have tended to produce a smear of ages over any particular magmatic belt. Thus, as presently reported, it is difficult to discern any episodicity in the arrival times within the scattered arrays of the Late Caledonian, fault-uplift related, plutons of the British Isles, even though within a single array, such as that of Donegal, the overall span of 13 Ma is resolvable into two close knit groups of intrusions dating at 418 and 405 Ma, respectively. Similarly, in a fault block context in the Appalachians, the South Mountain Batholith divides into three plutonic events spanning 9 Ma.

In the different tectonic environment of the marginal, continental arc represented by the Sierra Nevada of California, the reality and significance of magmatic episodicity has been vigorously debated over several decades, initially with an unwarranted reliance on the K/Ar method of dating. Kistler originally envisaged five major intrusion events at approximately 30 Ma intervals, each event taking 10–20 Ma to complete. Further work then filled the gaps, and despite Kistler's cogent reminder that it was necessary to weigh the radiometric data in relation to magma volume, it seemed that only when relatively small areas were considered, of the order of 50 000 km^2, did plutonism appear to be episodic.

Since that time careful surveying coupled with extensive U/Pb zircon dating seems to show that only one locus of magmatism existed at any one period of time and that this shifted episodically eastwards throughout the Mesozoic. Such shifts may well have been caused by events at the plate boundary and, appropriately, Glazner correlated the plutonic episodes in California with periods of oblique subduction and trench-parallel transport of terranes along intra-batholithic faults.

In the analogous example of the Peruvian Coastal Batholith my colleagues and I have shown that there was much less of an eastward migration of the locus of plutonic activity than in California and elsewhere, so that it is easy to see that, within the single plutonic lineament of the Coastal Batholith, the magmatism is distinctly episodic with quiescent periods longer than 15 Ma. Nevertheless, along the 1500 km of the axis of the batholith the several distinct segments show different age sequences, with only a few episodes in common.

The reality and causes of the episodicity of magmatic events at this edge of the South American plate has again been attributed to plate movements, with various workers seeking a correlation between the spreading rate and tectono-magmatic activity, even to the extent of proposing fairly detailed itineraries. A problem in making such a direct correlation is the unlikelihood that the response of magmatism to tectonism is so immediate. Nevertheless, studies by Reagan, Herrstrom and Murrell, using U-series nucleide abundances in conjunction with ^{10}Be/Be ratios to estimate the timescale of the entire process of slab dewatering, melting, magma transport and the differentiation involved in the evolution of certain basaltic suites in Nicaragua, suggested that this takes a mere 50 000 years. We may have to think again!

On a more global scale there is certainly a correspondence between the onset of the most voluminous upwelling of magmas and the high spreading rates and plate reorientation of the mid-Cretaceous (Figure 12.2). Furthermore, there is no doubt of the connection between intrusion events and the episodic compression–relaxation rhythms, themselves reflected in the resurgent strike-slip movement of faults, but it is difficult to discern any more than a gross connection with the global tectonic process; the between-segment variation would seem to be too great. Of course, it could be argued that mere differences in the inclination of a Cretaceous subduction zone, similar to that existing today, would be sufficient to introduce a second order variation into the overall global tectonic control.

If we take a broader and holistic view of the evolution of the Andes during the Mesozoic it is easy to identify several large scale cycles involving a precession of events starting with extension involving basin formation, submarine volcanicity, burial metamorphism and gabbro intrusion, the latter overlapping with the compression that initiates basin inversion.

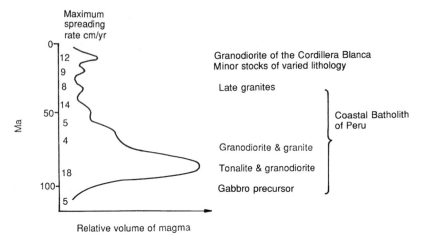

Figure 12.2 Schematic representation of the relationship between volume of plutonic magma intruded and seafloor spreading rate in the Western Cordillera of Peru.

This is followed by alternating phases of compression and relaxation associated with the emplacement of granitic plutons and basic dykes, these minor cycles progressively waning and terminating in uplift. It is as if rifting allows basic magmas to surface rapidly, although in compression such magmas are held back long enough to initiate a cycle of crustal melting and fractionation. The intrusion of the differentiated magmas only follows the relaxation of the compression. All this, however, assumes that the upper and lower crust are responding in unison to global tectonic events.

According to Aguirre and co-workers in 1974 such grand cycles have a periodicity of the order of 40 Ma in northern Chile, but clearly many more determinations of the initial and terminal events are required. Rogers reported spot ages of 187 and 155–158 Ma for Jurassic volcanism and plutonism, respectively, in the Cordillera de la Costa of Chile. The assembly times of segments of the great batholiths of the Andes may be of this order; it is 30–40 Ma for the bulk of the Coastal Batholith of Peru, with intrusion continuing for a further 30 Ma into the following cycle. On the other hand, as hinted earlier, the lesser cycles, as represented by the intrusion and evolution of the superunits (supersuites), are comparatively short, of the order of 1–2 Ma, and rarely exceed 4 Ma.

DIFFERING RATES IN RESPONSE TO TECTONIC ENVIRONMENT

Summarizing, we may expect the rates of intrusion of plutons and the assembly histories of batholiths to reflect the type and duration of the coeval tectonic events. Thus I envisage the intrusion of the hot, mobile magmas of the A-type granites of rift volcanism, involving the dyke-controlled filling of cauldrons, being rapidly accomplished with the over-all magmatic episode lasting only as long as rifting continues. I-type magmas (and some S-types) emplaced in response to post-tectonic uplift and block faulting may take longer to assemble and intrude, yet the whole episode may be quickly terminated as uplift and erosion redress crustal instability. A similar, relatively short emplacement time for the individual M- and I-type plutonic suites of the arcs contrasts with the long, sub-duction controlled assembly time of the entire batholithic ensemble. In great contrast the relatively cool S-type crystal mushes of the granite series, generated in response to ultrametamorphism associated with plate collision, may well take the longest of all to generate, migrate and intrude. What we lack are hard data, without which such concepts are mere entertainments. But we can surely conclude that Nature allows more than sufficient time for most physico-chemical processes to run to completion. Armed with ideas on the processes involved in the generation and intru-sion of granitic magmas, and with some appreciation of the time spans involved, we can now turn to examine the different granite types in their separate tectonic niches.

SELECTED REFERENCE

Paterson, S.R. and Tobisch, O.T. (1992) Rates of processes in magmatic arcs: implications for timing and nature of pluton emplacement and wall rock deformation. *Journal of Structural Geology*, **14**, 291–300.

Plagiogranite and ferrogranophyre: extreme differentiation in contrasted situations

13

(a) OCEANIC PLAGIOGRANITE: GRANITE IN THE OCEAN FLOOR

One of the most intriguing rocks to be afforded the title of granite is the micrographic intergrowth of quartz and plagioclase that occurs as bands, pods and veins within the upper parts of cumulate gabbros and basaltic dykes of ophiolite complexes, and which has also been recovered by sea-floor dredging and drilling.

The plagiogranites of the Aves Ridge in the Caribbean, as described by Walker and others, exemplify some of the textural oddities of such rocks, for there they consist of close packed aggregates of highly zoned plagioclase mantled by pure albite, set in a felsitic, sometimes grano-phyric, groundmass of quartz, potassium feldspar, albite, amphibole and chlorite. The late stage crystallization of albite and the presence of veins lined with albite and epidote attest to the circulation of a sodium-rich, hydrous fluid phase.

According to Coleman and Donato in their 1979 review of the plagiogranites in general, the latter differ from most other granitic rocks in having a very low potassium content, and heavy REE distributions that resemble those of mid-ocean ridge basalts (MORBs). However, it has to be admitted that such rocks are not easy to study geochemically, owing to the effects of extensive hydrothermal leaching, as noted in the Aves Ridge example, and which possibly accounts for the low potassium. In this environment many elements are labile and interpretations depend on the retention of the immobile trace elements. Nevertheless the MORB-like character indicated by the latter is wholly in keeping with their close association with mid-oceanic dolerites and basalts, as was well demon-strated in a study by Kanaris-Sotiriou and Gibb of cores through the oceanic crust of the Faeroe-Shetland basin. This study showed conclusively that bands of plagiogranite within certain dolerite sills resulted from the progressive and contrasted differentiation of a MORB-type magma, representing the separation of the small percentage of silicic residuum held in the dolerite. This accords well with the experimental findings of Dixon and Rutherford, which indicated that plagiogranite may well represent the separation of a late stage immiscible liquid, a suggestion of the greatest interest in view of the general rejection of immiscibility as a central process in petrogenesis.

Ancient plagiogranites are well known from uplifted fragments of oceanic crust, a prime example being the Cretaceous ophiolite complexes of the Oman and Troodos, where the bands, dykes, sills and pods are so intimately associated with, and often transitional into, the gabbro-dolerite cover of the mantle pyroxenite and dunite, that a local origin by contrasted differentiation is undeniable.

The Troodos exemplifies well a high level, multiple magma chamber developing at a spreading axis (Robinson and Malpas, 1990; also Figure 13.1a, b). In the upper levels of its several magma cells plagiogranite occurs in the form of en echelon pegmatitic veins and pods, sometimes culminating in larger sheets some 100 m in thickness and of kilometric extension – sheets which grade laterally into variably textured gabbro, the transition possibly representing a phase of incomplete segregation. We are seeing here an extreme and contrasted form of differentiation operating in a hydrothermal system open to seepage of sea water, and one where potassium is lost by leaching. Most investigators invoke some form of contrasted crystal fractionation (Barbieri et al., 1994), and although this is how the process may well have started, the specialized composition and rapid transitions into the parent rock, without any dioritic intermediary, strongly suggest that a hydrous residual melt was eventually segregated and secreted by some form of filter pressing mechanism, which may well have involved some degree of immiscibility.

There is another possible origin. Flagler and Spray, in a study of a tectonic fragment of ancient oceanic crust in the Canadian Appalachians, modelled the plagiogranite bands as resulting from a multistage process involving, first, the amphibolitization of the suboceanic crust on low angle shear zones, followed by partial melting, which led to the segregation of felsitic bands bordered by amphibole-rich restite. Although metamorphism and deformation certainly occur in the oceanic crustal environment, I cannot avoid the suspicion that the deformation in this ancient tectonic fragment was later superimposed during the long subsequent history.

Whatever the complications it is certain that the plagiogranites derive from MORBs which have themselves derived from mantle sources. Further, I like to think that some keratophyres represent the extrusion of such

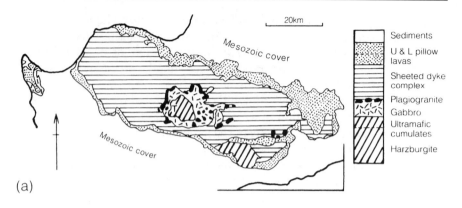

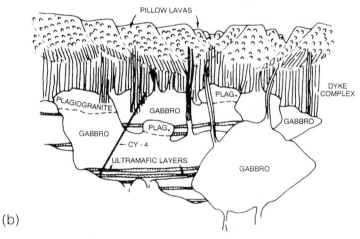

Figure 13.1 (a) Troodos Complex, Cyprus, showing location of the plagiogranite outcrops (black) within the gabbro–dyke complex. T. Troodos village. Adapted from Gass and Smewing (1973), Robinson and Malpas (1990). (b) Multiple magma chamber model for the Troodos Complex (from Robertson and Xenophontos, 1993, with permission of the authors and the Geological Society of London). CY-4: position of a drill hole.

plagiogranite magmas into the overlying oceanic sediments. Finally, trivial though these rocks may be in volume, their origin provides a blueprint for the process of highly contrasted, low pressure fractionation. Moreover, if thin, near-continuous films of plagiogranite are subducted as an integral part of the oceanic crust, what may their contribution be to plate-edge granitic magmatism?

(b) FERROGRANOPHYRES: GRANITE IN THE TOPS OF CONTINENTAL GIANT SILLS AND ULTRAMAFIC COMPLEXES

AN EXTREME FORM OF DIFFERENTIATION

Within the stable continents there are many giant sills of dolerite and layered complexes of ultramafic rocks in the tops of which there are lenses of ferrodiorite and ferrogranophyre. Of no great volume they, like the plagiogranites, nevertheless present an extreme bimodality that challenges explanation. There are particularly good examples of such granophyre-topped dolerite sills in Tasmania, described by McDougall in 1964, and in the Gettysberg region of Pennsylvania, described by Grossenbacher and Marsh in 1991, while the nature of the hedenbergite granophyres in the upper part of the Skaergaard intrusion of East Greenland is well known from the classic studies of Wager and Deer in 1939, Wager in 1960 and Wager and Brown in 1968.

The origin of these contrasted silicic rocks has been explained in terms of either extreme crystal fractionation or liquid immiscibility (Natland, 1991), and, just as in the case of the plagiogranites, it is not easy to decide the issue. An origin by extreme differentiation in a system involving iron enrichment, as originally advocated by Wager, Deer and Brown, can be justified, even to the extent of explaining the highly contrasted composition on the basis that a late stage precipitation of iron oxides would cause an abrupt enrichment of silica (for a full explanation see Marsh, 1996, p.17, and Natland, 1991, p.84). On the other hand the experimental work of Dixon and Rutherford, in 1979, showed that immiscibility can occur in iron-rich systems, thus supporting McBirney and Nakamura's earlier suggestion, in 1974, that the granophyres represent the separation of a immiscible silicic fluid within the Skaergaard intrusion.

There is also the likelihood that some granophyres in these situations have arisen as a result of the partial melting of the country rock, a possibility I will return to in discussing the origin of the epigranites (p.270). However, it is not a thesis easily applicable to those within-pluton ferrogranophyres in which the general and isotopic composition is in balance with the host, thus demanding an intratelluric origin. The problem is then how to explain the segregation of this highly contrasted siliceous material, especially as it seems unlikely that residual fluids would have seeped through a great thickness of crystal mush to the top of the intrusion (p.74). Thus Marsh argues, in his 1996 Hallimond lecture, for a dynamic and continuing process under the title of the 'granophyre lifecycle'.

THE GRANOPHYRE LIFE-CYCLE

The cycle begins in the unstable solidification front within the mafic magma body, where, ideally in side-wall situations, the differentiated fluid produced in the frontal layer of this advancing front, and in pockets within it, is envisaged as detaching in the form of blobs that rise by buoyancy to the head of an advancing column of magma, to eventually collect against the chilled margin at the roof, where the blobs coalesce into sill-like lenses that may even flow and intrude the carapace.

I am attracted to such a dynamic model involving continuous extraction and collection of relatively tiny amounts of late stage differentiate, and one which neither denies the possible involvement of immiscibility or assimilation in producing the siliceous fluid, but I question whether such iron-rich liquids would have a sufficient buoyancy, and be sufficiently different in viscosity not to mix with their host (Marsh, 1996, p. 17). Furthermore, although I have sometimes seen discrete pockets of granophyre 'rising' in mafic bodies, I would expect mixing to be a more usual.

Whatever their origin, the occurrence of these highly contrasted differentiates in such small volumes, in a rather special compositional niche, strongly supports the widely held view that granite in any great volume is rarely a direct differentiate of mafic magma.

SELECTED REFERENCES

Coleman, R.G. and Donato, M.M. (1979) Oceanic plagiogranite revisited. In: Barker, F. (ed.), *Trondhjemites, Dacites, and Related Rocks*, Elsevier, Amsterdam, pp. 49–168.

Robertson, A. and Xenophontos, C. (1993) Developments of concepts concerning the Troodos ophiolite and adjacent units in Cyprus. In: Prichard, H.M. *et al.* (eds), *Magmatic Processes and Plate Tectonics, Geological Society of London Special Publication* No. 76, 85–119.

Marsh, B.D. (1996) Solidification fronts and magmatic evolution. The 1995 Hallimond Lecture. *Mineralogical Magazine*, **60**, esp. pp. 17–19.

The reader is also directed to:

McBirney, A.R. (1996) Mechanisms of differentiation in the Skaergaard Intrusion. *Journal of the Geological Society of London*, **152**, pp. 421–435.

Cordilleran-type batholiths: magmatism and crust formation at a plate edge

14

There is general agreement that batholiths are to be found only in, or on the immediate borders of, mountain-built regions. This rule is so general that it may be called a law.

Reginald Daly (1912) Canadian Department of
Mines Memoir, **38**, p. 725.

INTRODUCTION

When, in 1906, Reginald Daly completed his monumental task of mapping across the North American Cordillera at the 49th parallel, he had recorded the first detailed account of a cordilleran batholith. It was this experience that led him to propose his law about the association of batholiths with mountain belts, the explanation of which, as we now know, lies in the generation of magmas and new crust at plate margins.

Such great volumes of granitic magma are produced in this apparently simple tectonic context that the study of the cordilleran batholiths directly addresses the problem of the origin of calc-alkaline tonalites and granodiorites in general. To this end my colleagues and I set out to examine the Coastal Batholith of Peru in all its aspects: its tectonic setting, anatomy, composition and the origin of its magmas. This study provides the basis for a general discussion on what I regard as a relatively uncomplicated example of a cordilleran batholith.

Firstly, however, it is necessary to cross the usual hurdle of nomenclature. For the description of granitic bodies I consider that Cloos' 'pluton' is a more internationally acceptable term than 'stock and boss', and I use it for largish bodies of the order of 1–200 km² in outcrop, built from one or more pulses of magma yet clearly circumscribed by its country rock envelope. A batholith is an array of such plutons, a description particularly appropriate for the large multiple intrusions found in orogenic belts, just as Suess intended and Daly discovered. This is not simply an exercise in semantics; magmas clearly take repeated advantage of crustal flaws – those of a multipulse pluton, for example, of a fault intersection, those of a linear array of plutons, of a more fundamental crustal lineament. Disposition, shape and size are structurally decided. Thus, although there are many common features, each pluton and batholith requires a separate description and interpretation.

There is no better example of this recommended usage than that of the linear arrays of plutons making up the Mesozoic Cordilleran batholiths of the western Americas. In the explanations for their close association in time and place with volcaniclastic basins, their location marginal to a plate edge, and the initiation of granitic magmatism immediately after the cessation of sedimentation and volcanism, we can discover a great deal about the nature and origin of the granitic rocks.

At this point I should perhaps remind the reader that the French equivalents for pluton and batholith were, until fairly recently, batholith and massif, rather the reverse of the English and American usage and echoing the Central European convention whereby batholith ranks below pluton. If the reader wishes to delve further into usage and nomenclature, I recommend reference to Paul Bateman's views, in 1992, on the North American consensus.

BATHOLITHS AND MARGINAL BASINS

The Pacific margin of South America, at least that large part of it south of the Gulf of Guayaquil, is one of the most clearly demarcated and continuously active plate margins in the world. There, on the flank of a great downbowing of the oceanic Moho, but within the continental lip, a string of immense granodioritic batholiths and their precursor gabbros intrude the root of a coeval volcanic arc (Figure 14.1; Pitcher, 1988).

Despite the obvious appeal of a subduction-related model, the determinable tectonic regime for much of the Mesozoic was one of extension associated with major block faulting, and only on its far eastern flank does thin-skinned, thrust tectonics play any significant part. Such an extensional regime was early initiated in the Permo-Triassic with the development of a synsedimentary graben in the Eastern Cordillera of Peru and Bolivia. It then continued in the Western Cordillera, where it resulted in the thinning of the continental crust and the development of fault-bounded marginal basins all along the Pacific margin from mid-Ecuador to Patagonia (Figure 14.1). Indeed, Patagonia, with its well-known Rocas Verdes ophiolite sequence, provided Dalziel with particularly compelling evidence of actual splitting apart. This extensional phase largely ended with a late Albian compression contemporaneous with a world-wide increase in sea-floor spreading rates which, Larson (1991) told us, was coupled with an extraordinary upwelling of heat and deep mantle

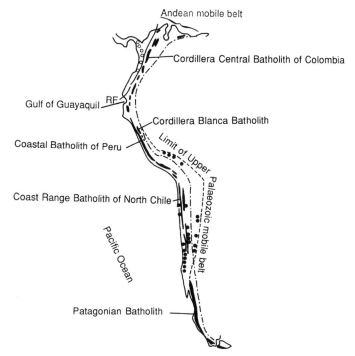

Figure 14.1 Mesozoic–Cenozoic I-type batholiths of the Andean mobile belt. A line of possible M-type plutons is shown to the west of the Romeral suture (RF) in Colombia. The outer limit of the Upper Carboniferous mobile belt is shown to emphasize the fact that the Andean is superposed on the former. The important belts of Upper Palaeozoic granitoids are indicated. Batholiths and plutons: (○) Cretaceous–Cenozoic M-type; (◣) Jurassic–Cenozoic I-type; (●) Upper Palaeozoic S-type.

material in the Cretaceous Pacific basin. It coincided not only with batholithic events in Peru but also with the emplacement of the Sierra Nevada and Peninsular Ranges Batholiths of the Californias. After this the continental margin became increasingly stabilized and cratonized, albeit subject to relatively minor cyclic periods of compression, relaxation and uplift well into the Cenozoic.

These long, relatively narrow, marginal basins show a marked polarity of their sedimentary infill, with clastics derived from the eastern continental interior contrasting with the submarine lavas and detritus of a western volcanic arc. I hesitate to label such basins as 'back-arc' because they do not resemble truly back-arc basins such as that of the Bransfield Strait. Neither do I think that the Gulf of California provides a convincing analogue. Rather, the Andean basins are due to pull-aparts at the margins of a continental plate, and McKenzie, in 1985, envisaged that the thinning of the lithosphere and the complementary upwarping of the mantle leads inevitably to adiabatic melting of the latter. Hence the evident basaltic volcanism.

THE EXAMPLE OF THE HUARMEY-CAÑETE BASIN

The Lower Cretaceous Huarmey-Cañete Basin of Peru provides a prime example (Figure 14.2). Pillow and sheet lavas, hyaloclastic breccias and flysch-like volcaniclastics interbedded with exhalative mineral deposits are intimately associated with dykes, sills and late gabbro plutons. The latter betoken a continuous accession of basaltic magma into a new, growing crust, with early dykes acting as feeders to the submarine lavas.

A feature of the Huarmey Basin, together with its analogues in Chile, Ecuador and Colombia, is the non-deformative burial metamorphism producing the zeolite–prehnite–pumpellyite-bearing greenschist facies rocks described by Aguirre and Offler. The telescoped nature of the depth zones, coupled with the presence of high temperature zeolites such as

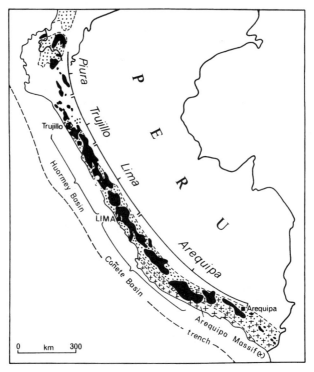

Figure 14.2 Coastal margin of Peru showing the location of the Mesozoic basins (stippled) and the Coastal Batholith (black). Also shown is the ancient Arequipa Massif with its thin cover of Mesozoic volcanoclastics and flows (+.) and the petrological segmentation of the batholith.

wairakite, attests to a contemporary high heat flux of the order of 300°C/ km, exactly as we might expect in such a volcanic belt. Furthermore, slicing through these volcanic piles are narrow, steeply dipping, ductile shear belts involving strongly deformed amphibolite facies rocks, and these I consider represent deep-reaching 'hot faults', which acted, along with the dyke swarms, as heat conduits.

In Peru, we are fortunate in being able to see the ancient Precambrian massif that was ruptured and pulled apart during the formation of the marginal basin (Figure 14.2) and also to trace out the ductile shear zones within it. Furthermore, Jones provided good geophysical evidence for a deep crustal arch of high density material rising, diapirically perhaps, into this old basement – an arch that can be convincingly modelled as representing an underplate of new basaltic crust extracted from the mantle during the extensional phase (Figure 14.3). The intrusive gabbro plutons and their associated dykes give some clue as to its nature and, indeed, in southern Chile and South Georgia similar gabbro–dyke complexes have been identified by Dalziel as forming the deep floors of analogous basins.

Concerning the composition of the volcanic infill of the Huarmey Basin, Atherton and Webb described how the predominant calc-alkaline basalts and basaltic andesites become more potassium-rich towards the continental margin of the basin, an expression of an almost universal trend in cordilleran magmatism. Of even greater petrogenetic significance is a pronounced upward depletion in the light REEs, coupled with an increase in the K/Rb ratio and a decrease in the Rb/Sr ratio. A possible explanation of these systematic variations is that the initial splitting of the continental

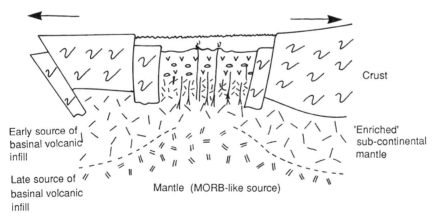

Figure 14.3 Cartoon depicting the origin of the Huarmey Basin (with its inland sea) involving extension and subsidence with the splitting of the crust and subcontinental mantle to give the vertical chemical variation shown in the basaltic infill of the marginal basin (**Vo**) and its gabbroic underplate (close stipple). After the hypothesis of Atherton (1990).

edge first plumbed a source with marked calc-alkali chemistry of within-plate character. Then, as opening progressed, access was provided to a deeper level, representing a more depleted MORB-like source, a finding consonant with the tapping of the different levels of a layered, subcontinental mantle (Figure 14.3).

In 1990 Atherton drew a most instructive comparison between this type of Andean basin and that associated with the spreading rift zone of Iceland, the evolution of which had been quantitatively modelled by Palmason. A notable aspect of this modelling is the calculation of the crustal heat flux as mantle magma intrudes between the parting lithospheric plates (Figure 14.3), with the finding that temperatures compatible with the melting of hydrated basalt would be reached at depths as shallow as 5–7 km. This comparison and its calculations provide intriguing support for a general model for the generation of cordilleran magmas.

BASIN AND BATHOLITH COINCIDENCE

We have seen how rifting controlled sedimentation and vulcanicity whereas continuing subsidence produced depth-related burial metamorphism. The story continues with the intrusion of great dyke-like bodies of gabbro towards the end of this extensional phase, an event overlapping with resurgent faulting, folding and uplift. Then followed a remarkable switch to silicic magmatism with the upwelling of voluminous granitic magmas along the axes of the marginal basins (Figure 14.2). Nevertheless, injection of basaltic andesite continued in the form of dense swarms of synplutonic dykes.

Charrier, also Aguirre, and Levi and Aguirre (1973, 1983 and 1981, respectively) have shown that the Mesozoic Andean represents a repeat of such magma-tectonic rhythms of overlapping events, each motivated by the interplay of extension and compression, themselves a response to global plate tectonics. It is easy to imagine that extension will promote the upwelling and extrusion of magma, whereas compression will inhibit it and so provide a sufficient residence time for the voluminous ponding of magmas and their subsequent fractionation.

THE PRECURSOR GABBROS

The upwelling of mantle-derived magmas at around 105 Ma in the form of the precursor gabbros links the terminal phase of basin formation and the initial stage of the assembly of the granitic batholith at 102 Ma (Figure 14.5).

The rock assemblage, troctolitic gabbro–olivine gabbro–diorite, together with accompanying basaltic dykes, provides evidence for a complex pattern of differentiation and syncrystallization deformation operating at the same relatively shallow levels and along the same axial lineament as the succeeding granites. Nevertheless, despite this obvious precursor relationship, there is a clear contact hiatus between the early gabbros and the granites, a contrast strengthened by the absence of any mineralogical or geochemical continuum. The mafic intrusives were evidently generated independently of the granites and certainly do not represent the cumulate fractions from tonalitic melts. Even though dissected by later intrusions, considerable remnants of sills, dykes, plug-like bodies and great, vertically walled gabbroic plutons can be identified in which repeated injection and deformation have contributed to an almost chaotic complexity.

As an example, Regan unravelled the history of the Huaural Gabbro pluton. This involved a staged evolution, with an initial static accumula-

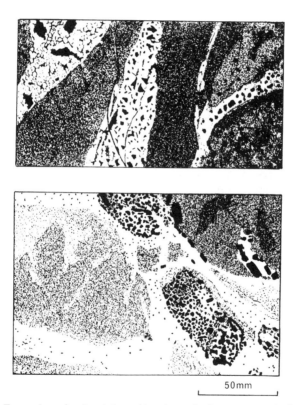

50mm

Figure 14.4 Examples of net veining of basic rocks in response to the invasion of a mobile leucosome. The host rock has, in many instances, reacted in an essentially competent fashion but has experienced variable amounts of hybridization and the growth of poikiloblastic amphibole (black). From the Rio Seco section of the Huaural Gabbro pluton, Lima Segment, Coastal Batholith of Peru. Drawings by Peter Regan.

tion of crystals, registered by peridotitic lenses and accumulative banding, disrupted by both the reintrusion of the crystal mush and the intrusion of fresh magma (Figure 14.4). Despite these complexities a differentiated sequence of primary gabbros can be recognized involving olivine, plagioclase, augite and pigeonite (which later inverted to orthopyroxene), with primary amphibole appearing at a late stage, presumably as P_{H_2O} increased on crystallization. Both cumulate and non-cumulate textures exist but it is common for the plagioclase to provide an open network of cumulus grains within which the mafic phases take an interstitial role. In strong contrast are zones in which the gabbro has undergone a subsolidus recrystallization due to a synchronous, hot deformation, locally intense enough to produce strong direction fabrics and a segregational banding.

A post-consolidation amphibolitization and hybridization, acting independently or in concert, locally caused a wholesale reconstitution of the pre-existing lithologies. This whole process clearly predated the intrusion of the granitoids, at least at the level of present exposure.

As for the source of the bulk of the fluids responsible for this auto-hybridization, we probably need to consider a general model whereby the basic magmas rose into the marginal volcanogenic basin, thereby contributing to the high geothermal gradient, promoting hydration reactions and hot fluid circulation in the country rocks. Such fluids locally mixed with the gabbro magma, leading to the crystallization of hornblende and the separation of a highly mobile felsic differentiate which reacted with and even reintruded its host during the final stages of consolidation.

The complexities of this tectonically influenced history of emplacement and crystallization do not obscure the fact that the chemical evolution is essentially one of the crystal fractionation of a high aluminium, olivine tholeiite, the scatter of compositional values reflecting the cumulate nature of the process. Furthermore, all the geochemical data confirm the primitive nature of the tholeiite and also its mantle derivation. Thus from the compositional point of view these basic rocks belong to the volcanogenic basin, yet structurally they link with the granitic batholith. It is the switch over to silicic magmatism that is so dramatic, and it was this that convinced Atherton (1985) that basic magmas were involved in the remelting of this new crust and the production of tonalitic magmas.

ANATOMY OF A BATHOLITH: A MULTIPLE INTRUSION

The Coastal Batholith of Peru illustrates well the general nature of silicic magmatism of Cordilleran type, particularly because it forms a single plutonic lineament with only a little eastward migration. During its life of 65 My (~100–35 Ma) the master lineament continued to be exploited, possibly because once traversed by magmas further pulses followed the same hot, permeated and weakened pathway; the early channelway had become self-perpetuating (Figures 14.5 and 14.6).

Overall a 1600 km long linear array of hundreds of intersecting plutons was stoped out of the string of marginal basins of which that of Huarmey forms an integral part (Figure 14.2). First came the bulk of the tonalites and granodiorites, followed by lesser volumes of granodiorite and granite in the form of short-lived pulses of increasing felsitic character. Even so, mafic magma was always available in the form of swarms of dykes which often show synplutonic features. The magmatic activity was locally discontinuous and its detailed chronology differs among the several distinct segments of the batholith (Figure 14.2). This questions any appeal to a precise global tectonic control such as short-term changes in sea-floor spreading as advocated by Frutos. Nevertheless, Soler and Bonhomme made a good case for a second-order relationship between magmatism

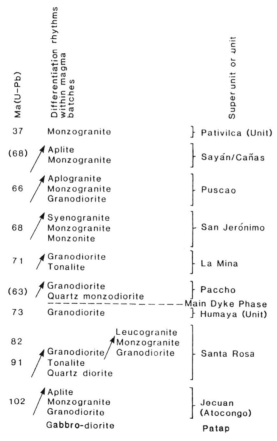

Figure 14.5 Chronology of the superunits of the Lima Segment of the Coastal Batholith of Peru expressed as differentiating magma batches.

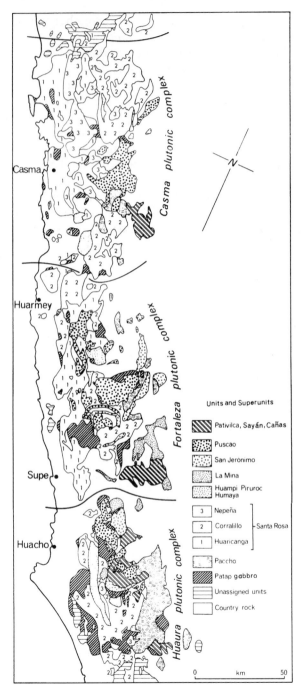

Figure 14.6 Outline map of part of the Lima Segment of the Coastal Batholith of Peru. Shows well the multiple character of this linear batholith.

and the rate of plate convergence, and there is certainly a gross correlation between the onset and the volume of granitic plutonism and the mid-Cretaceous reorganization of the Pacific plates.

Of the rock types, calc-alkaline, magnetite-bearing, I-type tonalites and granodiorites predominate, though the compositional spectrum is locally

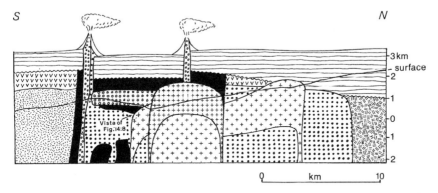

Figure 14.7 Interpretative but controlled section across the centred ring complex of the Huaura, part of the Lima Segment of the Coastal Batholith of Peru. Shows intersecting plutons and ring dykes of granites, granodiorites (various stipples) and gabbro–diorite (black), roofed by pre-batholith volcanic rocks and overlain by post-batholith volcanics: in contention is the degree to which the plutons ever contributed to the latter.

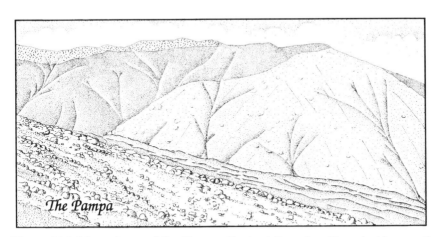

Figure 14.8 View of part of the Huaura Centred Complex, Peru, looking south-west from Quebrada La Toma, southwest of Sayán. The top of the hill exposes the San Jerónimo Granophyre (stippled) as a flat sheet overlying a foundered central block of mafic rocks (dark stipple). The Sayán monzogranite (light stipple with huge boulders) represents a later intrusion stoped out of the mafic block. Height of mountainside up from the pampa is about 500 m (see section of Figure 14.7).

widened to include both potassium-rich quartz diorites and evolved granites.

The basic outcrop pattern of this multiple batholith is of great, lenticular, multipulse plutons of tonalite–granodiorite intersected by smaller, circular, polygonal or rectangular plutons of granodiorite and granite, all with knife-sharp contacts (Figure 14.6). Steep sides and flat tops are the rule, so there can be no explanation for emplacement other than by the subsidence of huge crustal blocks and their fragmented remnants, the outlines of which were clearly controlled by contemporary fractures, both major and minor. Thus the cross-section of Figure 14.7 is not as schematic as might seem, as will be evident from the vista of Figure 14.8. On a smaller scale, blocks were spalled-off from the roofs and foundered into the magma chambers. So, overall, the batholith is the result of crustal collapse in a fault-block tectonic regime. However, it cannot be regarded as a great dyke because there are discontinuities whereby the country rocks 'bridge' the batholith, dividing it into spaced plutonic complexes. Faulting controlled location, enabled subsidence and provided the conduits for the magmas but made little contribution to the space requirement.

SUITES, SUPERSUITES AND PULSES

The granitic rocks group naturally into well-defined, time-separated, consanguineous rock suites, each with its own identity as defined in terms of chronology, modal and chemical composition, textures, enclave populations and dyke-swarm association (Figure 14.5). I was constantly amazed in the field at this specific rock physiognomy, which led to instant recognition of each particular suite in tens of separate plutons distributed over hundreds of kilometres. For a suite with such a regional distribution (p. 25; also Cobbing *et al.*, 1977) Cobbing and I coined the term superunit, which is equivalent to the supersuite of White and Chappell, the sequence of Bateman and Dodge and the intrusive suite of the North American Stratigraphic Code.

Equally noteworthy is the finding that each of the rock units making up a superunit normally has intrusive contacts, so representing a distinct and separate surge or pulse of magma. Thus it seems to me that the zonation of many plutons is more likely to represent a multipulse infilling than differentiation *in situ*, the magmas arriving from depth partly or wholly differentiated. Nevertheless, as William Taylor showed in 1985, some other plutons in the Coastal Batholith are so gradationally zoned, particularly in depth, that their evolution *in situ* seems inescapable. In yet others the central rest magma has clearly come unstuck and extruded into the envelope of the pluton. Clearly all possibilities exist, but with the important generalization that the processes of differentiation continued during the whole time of emplacement, always following the same evolutionary paths and yielding the same products in all modal and textural details. This surely emphasizes that the specific signature of each superunit was early established.

In any one transect of the batholith there are several of these superunits, often in an ordered succession and increasingly more silicic (Figure 14.5). Furthermore, changing assemblages along the 1600 km axis of the batholith mark a distinct compositional segmentation (Figure 14.2). On the basis of these outcrop patterns of superunits I envisage that, at depth, each segment of the batholith consisted of a beaded string of deep-seated melt cells, generally independent in space and time, yet laterally most extensive and most voluminous at the earliest stage of the batholith's history.

EVOLUTION OF THE MAGMAS

The compositional, mafic to felsic variation within each superunit, expressed either modally or chemically, is smoothly curvilinear, with little deviation from the trend line (Figure 14.9a). Although the major elements show little individuality, a specific chemical signature is often very well expressed by the trace elements (Figure 14.9b), particularly by the REEs.

McCourt, also William Taylor, and Atherton and Sanderson have explored in detail this aspect of the geochemistry (1981, 1981 and 1985, respectively), concluding that these systematic variations are best modelled as magmas undergoing crystal fractionation. In such a model, precipitation of plagioclase is of prime importance, with variations in the subordinate parts played by hornblende with or without magnetite, and biotite. Only in the most evolved rocks do potassium feldspar and quartz play an important part.

By reference to various trace element ratios, as manipulated by the Wright and Docherty programme, it is possible to calculate the composition of the crystal crops required to satisfy these trends (Figure 14.10). In the single example illustrated, the strontium versus rubidium relationship demonstrates the overriding importance of plagioclase loss, whereas the inflection in the barium versus rubidium curve indicates the incoming of biotite with the consequent removal of rubidium. The results of such chemical modelling are remarkably consonant with the textural evidence, providing an important clue to the mechanics of the actual crystallization process.

Mason's studies of mineral growth, in 1985, particularly the zonation of the plagioclases, attest to a crystallization history related to successive periods of residence in interconnected magma chambers at ascending levels in the crust. For example, the cracked, calcic cores of many plagioclase crystals are unconformably overgrown by contrasted zones of

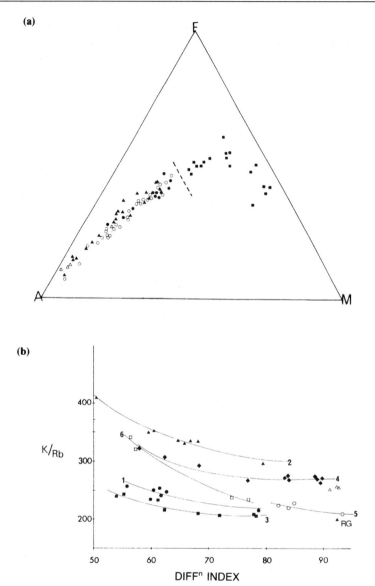

Figure 14.9 (a) AFM diagram for the individual superunits of the Lima Segment. (■) Patap gabbros and diorites; (●) Paccho and (○) Santa Rosa tonalites and granodiorites; (▲) ring complex granodiorites and granites; (△) Pativilca granite. Note the hiatus between the granitic rocks and the gabbro diorites. After McCourt (1981). (b) Comparative plots of K/Rb versus differentiation index for the main superunits of the Lima Segment. (1) Paccho; (2) Santa Rosa (Huaura); (3) Santa Rosa (Nepeña); (4) Puscao–Sayán–Cañas; (5) San Jerónimo; (6) La Mina. The Pativilca superunit is represented on the K/Rb plot by the open triangles and RG is the Red Granite isolated within the Santa Rosa outcrop. After McCourt (1981).

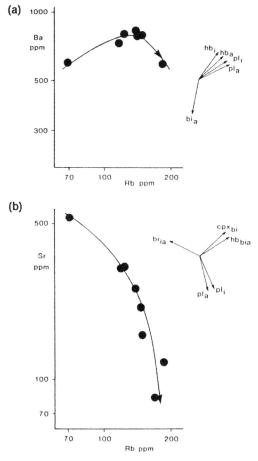

Figure 14.10 Example of the kind of analysis of the chemical data derived from the Señal Blanca superunit. (a) Ba versus Rb and (b) Sr versus Rb, providing a model of the mineral separations required to satisfy the actual trends. Vectors calculated from the K_D values of Pearce and Norry. After Atherton and Sanderson (1985).

more intermediate composition. These then evolve about 10 percentage points towards albite during the precipitation of much of the hornblende (with or without pyroxene) followed by biotite. Plagioclase continues its growth along with quartz and orthoclase. So neat is this picture of cotectic crystal growth that it seems inconceivable that the major process involved can be other than fractional crystallization – but not directly from basaltic magma, as is shown by the evidence of the compositional hiatus between the gabbros and tonalites in outcrop, a gap inadequately bridged by quartz diorites of no great relative volume.

There is also evidence of other processes at work. As already noted, a characteristic feature of this and also analogous batholiths is the presence of swarms of coeval, basaltic dykes, showing that mafic magmas were always available. That these magmas mingled with the tonalite and granodiorite magmas is demonstrated by the frequency of synplutonic features, when the dykes contribute to the populations of microdiorite enclaves. However, evidence of true magma mixing, involving the production of hybrids, can only be seen to have occurred on a significant scale in the subvolcanic centred complexes.

In texture and mineralogy these granitic rocks conform to relatively high temperature, water-undersaturated magmas lacking in late pegmatitic differentiates and vein systems – that is, except in the roofs of certain of the highly evolved granitic plutons. Such findings are in accord with the model of upwelling of still differentiating magmas to a high, subvolcanic level in the crust where they were emplaced in a brittle fracture regime. Appropriately the high temperature thermal aureoles are relatively narrow and static in character, so that we can surmise that the final cooling is likely to have been rapid – a conclusion neatly confirmed by the near identity of the K/Ar and zircon U/Pb ages in those superunits not complicated by resetting.

The generally high level of emplacement is confirmed by the excellent examples of the plutonic–volcanic interface provided by the several centred complexes which, with their granophyres, ring dykes and breccia pipes, are clearly the basal wrecks of caldera volcanoes. Furthermore, the porphyry copper breccia pipes also attest to this shallow-seated environment. Nevertheless, I suspect that few of the plutons of the batholith represent substantial feeder magma chambers; rather it was the synplutonic dykes that provided the conduits for likely coeval extrusives. Although intruded into the root of an arc, the batholith was not a substantial source of arc volcanics. The relationship between volcanism and plutonism in Cordilleran batholiths is that they are alternate rather than strictly coeval.

ORIGIN OF THE MAGMAS: A TWO-STAGE MODEL

The compositional and isotopic data are wholly in accord with a primary mantle source for the magmas. For example, the ε_{Nd} data range from -1.4 to -2.0, which are outside crustal values. Furthermore, Beckinsale and co-workers showed that there is a remarkable uniformity in the initial strontium isotope ratios which cluster around 0.7042, and which equal those of the precursor gabbros. Mukasa and Tilton reported that the dominant lead component in all the superunits appears to have come from an isotopically homogeneous reservoir – that is, undepleted mantle. As for $\delta^{18}O$ values, two specific studies show increases from a typical mantle value of 5.7‰ to 7.0 and 7.6‰, respectively, during the evolution of the rock suites, which is an indication of a limited interaction with meteoric water.

Crustal contamination is more evident where the outcrop of the Coastal Batholith demonstrably lies within the ancient basement itself (Figure 14.2), and indeed there are compositional changes along the plutonic axis as the batholith enters this basement in the region of Arequipa. Even so Mukasa, Beckinsale, and Le Bel, on the basis of lead and strontium isotope studies, show how limited is its scale and how highly selective is its character (all 1985). That it is so limited, even where the batholithic rocks are in direct contact with the old crust, shows conclusively that the bulk of the within-basin magmas never resided for long in contact with that crust. From such data we can surmise that the old crust had been differentially thinned and perhaps even split open along the megalineament, the widest extent of this being under the marginal basins, where there existed an insulated passageway for the magmas.

At this point I digress to mention one particular superunit, that of Linga in the Arequipa segment, because it has a particular message with respect to petrogenetic models in plate margin environments. All its rocks are monzonitic, ranging from monzogabbro to monzogranite, yet, as Agar and Le Bel pointed out, there is little in its age, rock texture and geochemical behaviour to distinguish it from its potassium-poor analogues – that is, except for an increased abundance of fluid inclusions associated with a low grade copper mineralization. Nor do any of the strontium, neodymium or oxygen isotopic ratios provide any evidence for more than a mild crustal contamination of the same order as other superunits in the vicinity. Furthermore, the fact that this monzonitic superunit outcrops on the Pacific flank of this multiple batholith is the reverse of any expectation based on the conventional view of a low dipping subduction zone. Perhaps such a location favours the thesis, advanced by Le Bel and others, that subducted oceanic material was involved, but I reject this on the basis that the Linga superunit is so much an integral part of the multiple Coastal Batholith in time and place. My own guess is that the potassium-rich Linga magmas were derived via a potassium-rich extract from a subcontinental, compositionally layered mantle, perhaps tapped by an especially deep-reaching fault system.

Returning to the central theme of the petrogenesis of the predominantly potassium-poor granitic rocks, it is obvious that the precursor basaltic rocks played the key part, particularly in building new crust during their extensional phase (Figure 14.11). Patently this basaltic underplate must

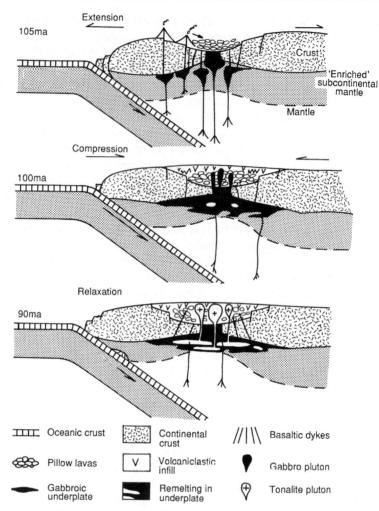

Figure 14.11 Schematic diagram illustrating the formation of a crustal underplate at an Andinotype margin; the underplate then provides the source for the tonalitic magmas.

have been the source of the tonalitic magmas and, indeed, there is ample experimental evidence that partial melting of a tholeiite will provide a tonalitic magma. We have also learned that the upwelling of magma into a rift between lithospheric plates would ensure temperatures sufficient to cause remelting quite high in the crust.

The key to the switchover in the composition of the magmas must lie in the change of tectonic regime – in the ending of the long period of extension and the stabilization of the new crust. Fresh basaltic magma could only then intrude during brief renewals of extension. Compression and buoyancy led to uplift of this new, hot crust along resurgent faults. As a result partial remelting occurred in response to the reduction of pressure, and tonalitic melts separated out in voluminous batches. As they rose they were held in residence at various levels in this thickened edge of the continental plate until tectonically released, certainly for sufficient time to enable sidewall crystallization to promote fractionation.

To what extent this partial melt separated from a refractory residue at the source, or carried much of this residue up with it, is a matter of conjecture. I favour the former possibility because I do not interpret either the mafic minerals or the enclaves in the intrusives of the Coastal Batholith as restite; to me the mineral textures and assemblages are wholly explicable as crystal precipitates, and the enclaves clearly recognizable as contributions from coeval basic magmas.

This primary remelting phase was followed sequentially by others, each providing a new melt which then underwent fractional crystallization as it was pumped up through interconnecting chambers created by block subsidence. Although it is possible that the source melt of each rock suite represents a new but different degree of fractional fusion of the basaltic underplate, the fact that later melts become more silicic overall suggests that a degree of refining may have gone on, and that the early produced tonalites and granodiorites may themselves have been partially remelted during the later phases.

For the driving force of this complex process I place great emphasis on the role of tectonics. I think it possible that the same deep-seated faults that located the precursor basin also located the sites of magmatism and triggered the deep melting, and then acted as channels for the melts intruding between the fault blocks. That they continued to do so during the episodic periods of compression and relaxation that continued throughout the long life of the batholith is well demonstrated by Bussell's 1975 finding that the late centred complexes were not only located at fault intersections but were active during repeated resurgent movements on these same faults.

Such a conclusion is obviously sympathetic to the central theme of this book and the reader will be prepared for a certain overemphasis. It may come as a surprise that so little reliance has been placed on the subduction process in view of the manifest connection of the rifts and faults with the most obvious plate boundary in the world. However, all this geologist saw in the field were steep faults and their effects, and the plate boundary is, itself, a major transcurrent fault bringing a 2000 Ma old ancient crust into direct contact with the oceanic Nazca plate. I will have to suppose that subduction did occur during the Cretaceous if only in deference to convention (Figure 14.11), but the essential structural element is the great transcurrent fault system, albeit driven by the approach of the Nazca plate.

THE CORDILLERA BLANCA BATHOLITH: FROM THE ABYSS WITHOUT CHANGE

Magnificently exposed in the high sierras of Peru, well to the east of the Coastal Batholith, is a very young batholith, the 10–3 Ma age of which shows it to be the end phase of continental growth (Egeler and DeBooy, 1956; Cobbing *et al.*, 1981; Atherton and Sanderson, 1987). It is flanked by a great, Andean-trending fault that obviously controlled its location (Figures 11.9 and 14.1; and Petford and Atherton, 1992), perhaps even tapping the sources. Furthermore, it lies above the >60 km deep keel of the Andean crustal prism, the lower part of which is thought to have been thickened by the magmatic accretion of basaltic rocks.

The light-coloured tonalites and leucogranodiorites of this mountain batholith are starkly cut out of basinal black shales of Jurassic age, in which it has created a relatively simple contact aureole. Of the many interesting features of this multiple and composite body there is one especially germane to the present discussions, that is, although the leucogranodiorites have some S-type characteristics, there is a virtual lack of geochemical evidence for any crustal contamination. These seemingly S-type features include marginal apophyses and aplite veins bearing tourmaline, andalusite, sometimes cordierite and spinel; local strong interaction with the contact sediments involving a degree of metasomatism, albeit localized, marked by tourmalization and greisenization; the inclusion of hornfelsed xenoliths in the intrusive, and the late growth of muscovite in the latter, associated with a mild peraluminosity in the most silica-rich types. Despite these apparently diagnostic features Atherton and Sanderson found that the bulk of the rocks of the Cordillera Blanca Batholith, including the leucogranodiorites, are geochemically similar in many respects to those of the Coastal Batholith, having clear I-type characteristics and compositions and isotopic characters wholly consonent with a mantle derivation *without* crustal contamination. In fact the composition of this high alumina, trondhjemite–tonalite–dacite series accords with an origin by partial melting of a newly underplated, basaltic lower crust.

The authors explain the localized S-type characteristics as being due to late stage, subsolidus, fluid interaction with the graphitic shales of the envelope at high level in the crust, a thesis favoured by the presence of widespread tourmaline and a clear spatial association with fault-induced deformation. It represents but a superficial overprint on the primary I-type character of a magma consolidating at high level.

I am entirely in accord with these findings. They carry the important message that granitic magma can rise through extremely thick crust without reacting with the latter. And these particular magmas, tapped by a developing deep-reaching fault, must have been intruded exceptionally rapidly.

CORDILLERAN BATHOLITHS IN SOUTH AND NORTH AMERICA

Of the string of like batholiths that core the Andes from Colombia to Patagonia, the Patagonian Batholith is nearly identical to that of Peru, both being of long duration (that of Patagonia spans 155–10 Ma) and having like dimensions, nearly identical compositions and a similar history, and with a close temporal and spatial relationship with a transcurrent, fault-controlled, 'back-arc' basin, here with an observable floor of sheeted gabbros.

Between Peru and Patagonia similar batholiths occur (Figure 14.1), but in contrast to other segments of the Mesozoic Andes both the volcanic and plutonic belts of north and central Chile show a marked easterly migration during much of the Mesozoic and Cenozoic. This is accompanied by compositional changes of rather disparate character, but certainly involving a general increase in the initial strontium ratios from 0.7022 to 0.7077. The low values were obtained from intrusives on the Pacific flank and were considered by McNutt and co-workers to imply a subcrustal source for these granitic magmas. In a significant study in 1988, embracing north Chilean plutonic belts from the Late Palaeozoic to the Cretaceous, Pankhurst and co-workers showed that, with decreasing age, Sr_i decreases from >0.7100 to 0.7038, and ε_{Nd} changes from -10 to $+4$, which they interpret as due to a decreasing contribution of crustal melts in their mixing with mantle-derived magmas. Such systematic changes in these ratios are characteristic of Cordilleran-type, marginal batholiths and we shall need to return time and again to consider their significance.

The Chilean sector provides even more highly significant information. According to Tarney and his colleagues, in 1987, deep erosion into the roots of the Andean Belt only reveals older, more potassium-poor and less siliceous tonalites, and not zones rich in cumulates or refractory residues. The generative process must always have been located very deep in the crust, most probably at the crust–mantle boundary, if that could ever be defined in this environment.

THE PENINSULAR RANGES BATHOLITH

Moving northwards into Mexico and California, the Peninsular Ranges Batholith provides a further key to the nature of the evolution of granitic rocks within an arc environment (Figure 14.12). Unlike the situations in

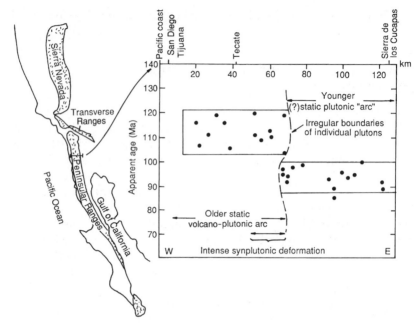

Figure 14.12 Age profile across the Peninsular Ranges Batholith showing the marked time-step across a zone of synplutonic deformation. Modified from Silver *et al.* (1979) and Walawender *et al.* (1990).

Peru and Patagonia, but analogous to that in Chile, there is, according to Silver and co-workers in 1979, a contrasted development between a western zone representing an older static arc, and an eastern zone representing a younger arc, perhaps eastward migrating, the former lying within a fringing volcanic arc and the latter within a clastic apron built on the continental lip.

It is with the intrusions of the western zone – Gastil's 'gabbro belt' (Gastil, 1975) – that the Peruvian Coastal Batholith compares most closely, in sharing a volcaniclastic basinal environment, in its association with precursor gabbros, in exhibiting a diverse spectrum of granitic lithologies conforming to separate intrusive sequences, and also in having geochemical characteristics which equally firmly identify the source as being of high aluminium basalt in bulk composition. The comparison between the Peruvian Coastal Batholith and the western zone of the Peninsular Ranges Batholith is indeed close, yet the much greater proportion of true granite in the former highlights an important distinction between a batholith emplaced inside the lip of a continent and another outboard of it. It seems that key factors are the thickness of the ambient crust and the nature of the mantle source.

There are several findings of singular interest about the Peninsular

Ranges Batholith. One is the gabbros which, according to Walawender and Smith, show, as in Peru, multipulse injection, late dyking, deformation and a form of hybridization. Early cumulate textures predominate in the plagioclase–olivine–orthopyroxene assemblages of the norites, which grade locally into quartz diorites with the incoming of hornblende, biotite and quartz. Again, as in Peru, this change is marked by great textural variation, and the presence of pegmatite lenses, net veining and comb layering suggests near volatile saturation of the magma mush at a late stage in its evolution. Once again the chemical data preclude simple fractionation as a sufficient explanation of this variation and it seems that the gabbroic magmas were locally enriched in elements such as potassium, rubidium, strontium and barium, which seem to have been leached out of the volcanic country rocks.

Another feature common to the two batholiths is the finding by Silver and Chappell that the primitive geochemical signatures of these sequences were imprinted in the source regions, as in Peru.

There is, however, a structural contrast in that the western zone of the Peninsular Ranges Batholith suffered a greater degree of ductile shearing than its Peruvian analogue. Significantly, one particularly strong belt of this axial deformation forms the boundary, not only between the two compositionally and temporally distinct zones, but also between country rocks of different metamorphic character. I envisage this as the trace of a major, deep-reaching transcurrent fault system marking both the edge of the old continent and the controlling structural lineament for magma emplacement.

These contrasts between the eastern and western zones of the Peninsular Ranges Batholith address one of the central problems of Cordilleran petrology – that is, the compositional change that occurs whenever the locus of plutonism of any of the Cordilleran batholithic complexes moves eastwards into the ever-thickening lip of the old continent or the sedimentary aprons derived from it.

We are fortunate in having a wealth of compositional data to illustrate this. Thus, between them, Silver, Taylor, Chappell, Gromet, Gastil, Walawender and their colleagues have established a general, rather subdued, eastward increase in aluminium, sodium and potassium, contrasting with marked gradients among the trace elements, as well illustrated by the REEs. Further, oxygen becomes systematically enriched in $\delta^{18}O$ from west to east, and the strontium, neodymium and lead isotopic systems change progressively from mantle values in the west to a more evolved character in the east. Changes in the original oxygen fugacity of the magma are marked by the easterly exchange of magnetite for ilmenite, as detailed by Gastil and co-workers in 1986. In all these variations the isopleths parallel the structural grain of the batholith, clearly relating the changes to the increasing influence of the underlying old continental crust. But was this due to the composition of the latter or to its increasing thickness?

On the basis of the regional nature of these variations, largely independent of rock type, Silver and Chappell argued convincingly for a deep source, compositional control. Processes such as convection, fractional crystallization and local assimilation are only of minor importance at high levels. They envisage that the western batholith derived its magmas from the oceanic lithosphere rooting the fringing volcanic arc, and that the eastern batholith derived its magmas from a deeper, subcrustal source of basaltic composition.

Particularly pertinent to these discussions is their well-substantiated opinion that no single two- or three-component system involving model mantle, crust or ocean sediments can be invoked to explain the remarkable transverse and longitudinal compositional variations of this multiple batholith. Nevertheless, Silver and Chappell concluded that, overall, the Peninsular Ranges Batholith is a complex, but ordered, template of the character of a source region which they provisionally model in the form of a plagioclase-rich (gabbroic) residual assemblage giving way laterally and downwards to a garnet-bearing, eclogitic residual assemblage.

At this point it is of great interest to refer to the 1990 studies of Walawender and co-workers on a superunit of the eastern zone represented by the zoned plutons of La Posta-type. It is surely intriguing that within a single pluton there is a range of modal and chemical parameters matching those representing the eastern batholith as a whole, with values of Sr_i ranging from 0.7040 to 0.7070, and $\delta^{18}O$ from 9.0 to 12.0‰.

THE SIERRA NEVADA BATHOLITH

In following the great Cordilleran belt northwards (Figures 14.1 and 14.12) it is the Sierra Nevada Batholith that provides the most extended example of this asymmetric compositional zonation. As summarized by Paul Bateman in 1983 and 1992, this vast intrusive complex is really composed of two batholiths of significantly different ages spread out across a continental margin. From a study of the great roof pendants Saleeby and co-workers documented the presence of coeval volcanic rocks which were intensely deformed during the medial phases of the growth of the Cretaceous batholith. Clearly plutonism was associated spatially, although alternating in time, with a precursor volcanic arc and temporally with an episode of ductile shearing.

The Sierra Nevada Batholith provides a classic example of this continent-ward increase in compositional parameters such as aluminium and potassium, here coupled with strong gradients in Sr_i and ε_{Nd} values as reported by DePaolo (1981a). Although, for the most part, exposed at a deeper level than in Peru, the constituent granitoids have many features which mirror those of both the Coastal and the Peninsular Ranges Batholiths, especially the clear grouping of the rock units into consanguineous suites (equivalent to superunits) having elongate outcrops along the axis of the batholith, each representing the differentiation of a common parent, modelled as being produced during a single equilibrium fusion event. Furthermore, according to Bateman and his colleagues such differentiation processes continued to operate even to the final level of emplacement to produce the zoned plutons (Bateman, 1992).

More specifically, it is the rock assemblage in the western flank of the Sierra Nevada that compares best with that of the Coastal Batholith, though with the important difference that it is only the gabbros, quartz diorites and tonalites that are represented in the former locality, in contrast to the full gamut of rock types present in the latter. As in Peru, this western part of the Sierra Nevada Batholith lies on the flank of a downwarp in the Moho, and its root can be geophysically modelled as metaigneous rocks passing down into a distinctly mafic lower half of the 55 km thick crust. Remarkably, as Dodge and Bateman showed, the latter can be directly identified as composed, at least in part, of metamorphosed and deformed mafic and ultramafic rocks overlying mantle, a deduction based on the preservation of appropriate lithologies as xenoliths in volcanic pipes piercing the batholith. Whether these deep crustal rocks are of the same age as the batholith and represent, perhaps, a residuum of the source is debatable, but I am tempted to interpret them as representing new crust from which the tonalites and granodiorites originated by remelting. It is a contention fortified by the isotopic and dating studies of Leventhal and colleagues of the mafic and ultramafic xenoliths entrapped in the lavas of the Cima volcanic field east of the Sierra Nevada. The xenoliths were derived from a MORB-like source of Mesozoic age, clearly indicating invasion of the crust by asthenosphere-derived magmas at that time.

Both in California and Peru the granitic rocks as seen in outcrop would seem to occupy only the top 15–20 km of the crust, so that, in the view adopted by Hamilton and Myers in 1967, the batholiths are relatively thin. However, I regard both the multiple array of granitic plutons and their dyked understorey of altered gabbros, peridotites and pyroxenites as essentially parts of a single crustal edifice within which there are no easily definable boundaries.

AN INTERIM CONCLUSION

There are many features common to the circum-Pacific batholiths, including the lineamental control, association with a volcanic arc, the presence of precursor gabbros, the sequential nature of intrusion, and the eastward migration with compositional changes. From the studies already cited and many more, as recorded in two important *Memoirs*, Nos 159 and 174, of the Geological Society of America, it is also easy to see an overall similarity in the possible source rock composition of all these linear batholiths. Whatever complex interplay of ocean–continent subduction and transform faulting is needed to provide the energy for melting, the bulk of the Cordilleran granitic rocks were clearly part of a Mesozoic–Cenozoic crust-making process, their source materials being largely and ultimately derived from the mantle, probably the suboceanic mantle in the examples of the docked oceanic arc of western Colombia and perhaps the fringing arc of Baja California, otherwise the subcontinental mantle in the case of the marginal back-arcs such as those of the central and southern Andes. However, a crustal component became increasingly important whenever the locus of magmatism moved into the continent.

As a general model for Cordilleran-type batholiths we can envisage a two-stage process whereby a basaltic–andesitic underplate was first extracted from a compositionally layered mantle during a long period of extension and arc volcanicity at a continental lip. This was terminated by a short period of compression, immediately after which, possibly as a result of crustal thickening, there was remelting of the underplate. The new magmas were held in residence long enough for differentiation to occur, being released into the upper crust whenever corridors were created by pull-aparts between the great Andean-trending transcurrent faults – that is, the batholiths are essentially of transtensional origin. I like to believe that these fundamental crustal flaws might be sequentially stitched and sealed by the consolidation of the early magma pulses, thus forcing the locus of magmatism to move sideways and preferentially into the pristine continental lip. And overall the interplay of compression and extension represents the response to Cretaceous variations in the vectors and rates of motion between the plates.

To rationalize Daly's law, the fact that crustal mobility and magmatism are both related to an active plate boundary is hardly surprising. However, I have found many of the conclusions of modern studies rather unexpected – that is, that the plate edge batholiths were derived largely as a mantle extract, not from old crust, that they were born in an extensional tectonic environment with all the processes of differentiation taking place at depth and near to the source, and that such batholiths were not the true roots of volcanic arcs. That the siliceous magmas were so voluminous in this plate edge environment must have been due to the melting and refining processes having a sufficient residence time as a result of being blanketed beneath a thickened new crust accreting on to the old continent.

Finally, it is important to appreciate that Cordilleran magmatism represents, overall, a specific assemblage of geological features, all the components of which need to be identified before a plate edge can be identified, as many geologists have attempted to do in more ancient rock sequences.

And whether or not the subduction of oceanic crust plays an essential part is still, I believe, a matter of discussion.

SELECTED REFERENCES

Atherton, M.P. (1990) The Coastal Batholith of Peru: the product of rapid recycling of 'New' Crust formed within rifted continental margin. *Geological Journal*, **25**, 337–349.

Bateman, P.C. (1983) A summary of the critical relationships in the central part of the Sierra Nevada batholith, California, U.S.A. In: Roddick, J.A. (ed.), *Circum-Pacific Plutonic Terranes, Geological Society of America Memoir No. 159*, 241–254.

Pitcher, W.S. (1978) The anatomy of a batholith. *Presidential address. Journal of the Geological Society of London*, **135**, 157–182.

Pitcher, W.S., Atherton, M.P., Cobbing, E.J. and Beckinsale, R.D. (1985). *Magmatism at a Plate Edge: The Peruvian Andes*. Blackie-Halsted Press, Glasgow.

Silver, L.T. and Chappell, B.W. (1988) The Peninsular Ranges Batholith: an insight into the Cordilleran batholiths of southwestern North America. *Transactions of the Royal Society of Edinburgh: Earth Sciences*, **79**, 105–121.

Intraplate, rift-related magmatism: mainly the A-type, alkali feldspar granites

15

... it is difficult to escape the conclusion that the alkali rocks are more often associated with stable regions or regions of crustal foundering and are relatively inconspicuous among rocks developing in the igneous activity of orogenic belts.

C.E. Tilley *(1957)* Quarterly Journal of the Geological Society of London, *113, p.324.*

THE NATURE OF INTRACRATONIC MAGMATISM IN GENERAL

Contrary to general opinion, a distinctly bimodal magmatism involving rather ordinary biotite granites and coeval basalts has been a feature of the earth's cratons since the Early Proterozoic. Although I shall concern myself here mainly with the alkalic granites of this same environment, it is important to emphasize that they are really of small volume in relation to their calc-alkaline and essentially I-type analogues. To emphasize this point and to set in contrast the two compositional types I turn briefly to a 1988 study by Wyborn and co-workers of the 1800–1840 Ma magmatism of Australia.

These particular Proterozoic granites are associated in space and time with a polygonal network of failed rift zones that wrap around older cratonic nucleii (Figure 15.1) – zones of extension that also support coeval volcano-sedimentary basins. There is little geochemical evidence of any involvement of an older, recycled crust, so that the source of these I-type granites is modelled as that of a subcrustal, mafic underplate: a now familiar explanation. Interestingly, Nd model ages suggest that this underplated protolith was developed about 100 Ma before intrusion, confirming that the latter, and not the overlying old crust, was indeed the source involved. As there is no evidence of the Wilsonian cycle with its subduction mechanism, the magmatism is attributed to the establishment

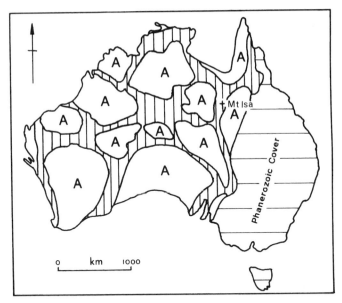

Figure 15.1 Postulated location of zones of 2300–2000 Ma underplating (vertical stipple) which are thought to represent relatively small scale mantle convection cells. (A) Archean nuclei. Adapted from Wyborn, L. (1988).

of relatively small-scale mantle convection beneath the thinning crust of the zones of extension.

Despite the predominance of calc-alkaline granites in the Australian example there are, in the region of Mount Isa, some fluorite-bearing pink to red coloured granites with rapakivi textures, so that there is, even in that particular environment, an approach to the typical alkalic, anorogenic, A-type granite that I turn to consider below.

IS THERE A UNIQUE A-TYPE?

When Loiselle and Wones introduced the term 'A-type granite' in 1979 they wanted to highlight the contrast between those potassium feldspar-rich and calcium-poor granites, often mildly peralkaline, which are common to the anorogenic cratons, and those much more abundant calc-alkaline granites lying within the orogenic belts. Although they placed an emphasis on the tectonic environment, they were well aware that there was also a unique compositional signature. Indeed, this acronym could equally well signify 'alkaline', 'anhydrous' or even 'alkali feldspar', because all of these features prove to be central attributes of the granites and quartz syenites of the plate interiors. But this is probably too broad a church, associating rocks with very different origins.

Nevertheless I propose to review the special attributes of the rift-related felsitic magmatism in both continental crust and mid-oceanic volcanic crust, and discuss the proposition that it is a manifestation of the thermal events related to crustal extension. With some reservations I include those apparently similar rocks associated with late-stage faulting of the crustal welts surviving orogenesis. If there is an affinity between granites belonging to such contrasted geological contexts, then this is a matter of great significance in petrogenesis.

Whatever their categorization there is no doubt that the potassium feldspar-rich leucogranites are often set apart by their distinctive chemistry, mineralogy, texture and mode of occurrence. The rather special rapakivi granites must be included because they not only comply with the A-type criteria summarized in the following, but their type occurrences in Fennoscandia provide a vital clue to the genetic process.

GEOCHEMICAL CHARACTERISTICS

A key factor in understanding the special features and origin of these A-type granites is the higher than average abundances of fluorine, chlorine and often boron. This has dramatic consequences because fluorine, in alliance with water, is much more effective than water alone in fluxing the magma, changing the form of the melt equilibria and lowering the temperature of the liquidi. Moreover, such polymer destabilization allows the entry of highly charged ions with the consequent formation of melt-soluble, alkali fluoride complexes in which the heavy metals can be carried forward into the ultimate melt residua. It is just such an enhanced solubility of GaF_6 over that of AlF_6 that explains why the Ga/Al ratio is so effective a discriminator of A-type granites.

From the experimental studies of Tuttle and Bowen, and also Clemens and his colleagues, it seems that A-type granitic magmas represent non-minimum, rather high temperature melts ($>830°C$), a point confirmed by the common occurrence of high temperature quartz and fayalite – though the stability of the latter is as much a function of f_{O_2} and water activity as temperature. This relatively high temperature, coupled with the fluxing effect of the halogens, not only promotes liquidity and fluidity, thereby facilitating fractionation, but also explains why A-type melts are almost entirely restite-free, generally lacking in phenocrysts, why they mix relatively easily with coexisting basaltic melts, and why they are able to rise high into the crust to fill the ring fractures and collapse cauldrons underpinning central volcanoes.

This is certainly one situation where the plutono-volcanic interface is wholly bridged by a variety of rapidly quenched, hypabyssal rocks. Preserved within the surface calderas are lavas and ignimbrites of equivalent composition, such as represented by the alkaline comendites and the fluorine-rich, topaz rhyolites. The latter are particularly good evidence of primary fluorine-bearing magmas (see the reviews by Christiansen and his colleagues in 1983, and Johannes and Holtz in 1996, p. 154), though we need to be aware that the topaz in these rocks more often occurs as cavity fills than as phenocrysts.

It is likely that the richness in fluorine has a bearing on the other compositional traits of the A-type granites. Thus in the early stages of magma evolution the presence of the halogens should promote the early crystallization of hornblende, when the precipitation and removal of the latter would explain the relatively low abundances of calcium, magnesium and aluminium, and probably also the corresponding high ratios in Fe_t/Mg. Furthermore, in the late stages, the presence of fluorine and water together greatly enhances element transport and redistribution, particularly of the alkalis, again with dramatic results. In addition, the relatively high K/Na ratio suggests that there may have been an early separation of calcium plagioclase, though, according to Johannes and Holtz in 1992, this may equally have resulted from the low water content of the magma. As for the total alkalis, these are often of sufficiently high abundance to identify these rocks as being mildly peralkaline, although it is important to point out that peralkaline, metaluminous and peraluminous suites

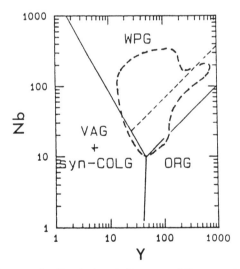

Figure 15.2 Y/Nb tectonic discriminant diagram of Pearce *et al.* (1984) with the field of A-type granites as plotted by Eby (1990). Concentrations in ppm. (ORG) ocean ridge granites; (VAG) volcanic arc granites; (syn-COLG) syn-collision granites; (WPG) within-plate granites. The broken line is the upper boundary for ORG from anomalous ridge segments. The great variety of locations is reported in Eby (1990).

often coexist in the same igneous complex. This is why alkali feldspar granite is a more appropriate general title.

The concentrations of the trace elements provide an even better discriminant. The high field strength elements such as zirconium, niobium, yttrium and the REEs (barring europium) are relatively high, as also are gallium and zinc, whereas scandium, chromium, cobalt, nickel, barium and strontium are low, having been early scavenged. That the former elements preponderate is the reason why compositional ratios based on niobium, yttrium and tantalum, delineate a unique, within-plate or, otherwise, A-type field, as proposed by Pearce and co-workers in 1984, and also by Whalen and his colleagues in 1987 (Figure 15.2).

MINERALOGICAL CHARACTERISTICS

These chemical peculiarities are everywhere reflected in a distinctive mineralogy and mode (Figure 15.3). Relatively simple two-feldspar–biotite granites closely relate with one-feldspar granites, quartz syenites and monzonites in two suites, one bearing the reaction assemblage fayalite, ferrohedenbergite–ferrohastingsite–annite, the other, more alkaline, carrying riebeckite, arfvedsonite and aegirite. These latter suites represent the typical hypersolvus granites, the epigranites, in which orthoclase cryptoperthite predominates as the pristine felsic mineral, though it is

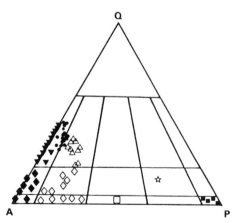

Figure 15.3 Representative modal data collated for the Niger–Nigeria anorogenic province. Note the distinct groupings of syenite modal compositions (two groups). (▼) Peralkaline granites; (◆) alkali syenites; (●) biotite granites; (△) fayalite granites and hornblende granites; (◇) syenites; (□) monzonite; (■) gabbros; (☆) hybrid compositions and igneous breccias. (Q) Quartz; (A) alkali feldspar; (P) plagioclase plotted in the Streckeisen diagram. Reproduced from Bowden and Kinnaird (1984), with the permission of the authors.

almost universally reconstructed to microcline microperthite during subsequent subsolvus recrystallization and metasomatism. Albite may or
may not accompany the potassium feldspar, and whether it is magmatic
or exsolved is rarely easy to decide. Primary quartz occurs in the
bipyramidal, high temperature form, but again this rarely survives
unreconstructed.

The incompatible elements combine with fluorine in abundant accessories such as fluorapatite, pyrochlore, cryolite, fluorite, topaz and
astrophyllite, along with zircon, allanite, cassiterite and fergusonite. So
profuse are these minerals, especially within the miarolitic cavities and
the pegmatitic fringes of contact zones, that the late phases of A-type
granites constitute a major source of the rare metals. They represent the
metallo-genetically specialized granites of Tischendorf.

The paucity of the anorthite molecule means that these granites can be
represented by the simple, haplogranite system Or–Ab–SiO$_2$–H$_2$O. In the
latter orthoclase cryptoperthite is the preferred early precipitate, with or
without albite, especially at the low pressures of the subvolcanic environment. Also, in such a magmatic system, it is likely that crystallization
differentiation will operate and be dominated by the precipitation of a
single alkali feldspar, an assumption confirmed by the geochemical
studies of Eby, who manipulated the various indicators of magmatic evolution, such as Y/Nb versus Sc/Nb (which assesses the role of
clinopyroxene and amphibole) and Y/Nb versus Ba/La (which assesses
the role of the feldspar) to show that the mafic minerals play but a minor
part, at least in the later stages of the petrogenetic history.

TEXTURE: THE RESULT OF SUBSOLIDUS REACTIONS

The subsequent evolution of the texture is dominated by the exsolution of
the alkali feldspar, when albite appears in the form of perthite lamellae,
replacement rims and new, unzoned crystals. To complicate the texture
further, the intergranular seepage of residual fluids rich in fluorine, boron
and alkalis results in a dramatic overprinting of the first phase, magmatic
fabric – a process representing the endogenetic, hydrothermal metasomatism so well documented by Bowden and Kinnaird in their studies of
the central complexes of Nigeria and Namibia. In this process extensive
second stage albitization is accompanied by the rimming of both the
fayalite and hedenbergite by amphibole and biotite, together representing
textural modifications that can advance to the extreme of fenitization
and result in the wholesale replacement of the original rock along zones
of microfracturing. It would be easy to overemphasize the effects of
a process that may represent no more than a 'stewing in its own juices',
but it seems to me unlikely that the magmatic chemistry would often
survive such a major reorganization unmodified, which surely questions

the significance of isotopic ratios based on strontium, rubidium, lead and uranium.

A RANGE OF PETROGENETIC MODELS

It may well be that there can be no unique, all-embracing model for the petrogenesis of these alkali feldspar epigranites, and certainly there is no lack of alternative proposals. The latter group into three main categories – that is, the fractionation of a mantle-derived, basaltic magma as originally proposed by Loiselle and Wones in 1979; the late stage interaction of an alkali, fluorine or chlorine-rich residual fluid with either a crystal-laden melt, as suggested by Bonin in 1982, or the subsolidus crystallate, as envisaged by Bowden and Kinnaird in 1984; and the remelting of 're-worked' crustal rocks, as proposed by Collins and others in 1982. Of course, none of these models is exclusive, but whatever their combination the major goal must be to account for the special abundances of the incompatible and high field strength elements, the relatively dry, yet often halogen-rich character of the magmas, and the association with a tensional, or at least, non-compressional, tectonic regime.

In what follows I examine examples of these A-type granite suites in their separate environmental niches, bearing in mind that Eby, in 1990, in a wide ranging survey of their chemistry using such 'elemental screens' as the Y/Nb and Yb/Ta ratios, was able to distinguish between two suites, one embracing the oceanic islands and the continental rifts and swells, and the other special to the immediately post-orogenic environments, at first sight a rather surprising subdivision.

ALKALI FELDSPAR GRANITES WITHIN THE CRATONS

The belt of mildly peralkaline alkali feldspar granites that extends from Niger to Nigeria is an excellent example of this A-type magmatism. The constituent central complexes represent the remains of a line of great volcanoes situated on the crest of continental swells and along the protorifts identified by regional magnetic surveys. Particularly interesting is the recognition of a general southward migration, though discontinuous and step-like, of these centres, stretching in time from the Silurian to the Jurassic. It is impossible simply to equate this with the movement of the continent over a mantle hot-spot because the rate is an order of magnitude less than that deduced from palaeomagnetic studies, but this does not deny a slowly fluctuating mantle plume beneath a long-standing protorift system.

But a structural control is paramount. According to a review by Bowden and his colleagues in 1987, changes in the direction of plate movement provoked reactivation of lithospheric N–S megashears. Where these intersected oblique faults the latter became tensional, opening to provide the channels for volatiles and magma generated at depth in response to pressure release: yet another example of the tectonic control of magmatism.

The roots of these eroded volcanoes of North Africa are revealed as complexes of ring dykes and cauldrons filled with a multipulse assemblage of A-type granites. Splendid accounts provided by Jacobson, Macleod and Black in 1958, by Turner in 1963, and by Bowden and Kinnaird in 1984 draw a classic picture of centred complex construction with fracture polygons, averaging 16.6 km in diameter, showing a precise lineamental control, both of their outlines and in the migration of their separate centres: I remember the Sara-Fier Complex as a particularly fine example (Figure 15.4). Space was clearly created for the multipulse intrusions by the downfaulting of dissected blocks of the old, Pan-African reworked crystalline basement, and there are examples of the infilling of the collapsed calderas by mildly peralkaline, comagmatic lavas and ignimbrites, all closely associated with vent agglomerates and, in one example, intercalated with lake sediments.

In general terms the sequence of igneous activity begins with the intrusion of a basaltic dyke swarm and continues with the extrusion of felsic volcanics from vents around a ring fracture. The latter is then filled with a great dyke of fayalite granite and this is followed by the foundering of the central block into a cauldron of biotite granite. The igneous cycle is terminated by the intrusion of a range of hypersolvus peralkaline granites in the form of bell-jar-shaped ring dykes and plutons.

What is especially important from the point of view of petrogenesis is that mafic rocks, albeit of small outcrop area, are present in a precursor role in the form of gabbroic intrusions, basaltic dykes and some basaltic lavas. Furthermore, there is much evidence of mingling and mixing as shown by net veining, the pillowed form of the basaltic enclaves, the composite nature of some dykes and the occurrence of hybrid rocks. Once again we are presented with a familiar space–time–volume relationship whereby the basaltic magmas were first on the scene but overlapped in time of intrusion and so were able to mix with the compositionally separate granitic magmas. Thus in outcrop it is difficult to entertain that there is a natural series extending from olivine gabbro to biotite granite, so that any apparent geochemical continuity is more likely to be due to the mixing: but the mafic magmas could have provided the heat required for crustal melting.

However, critical evidence is to be found in the similar ring complexes of the Aïr, Niger, where alkali granites and syenites are intimately associated with assemblages of anorthosites and leucogabbros disposed in the form of layered, funnel-shaped intrusions so typical of cumulates. The juxtaposition of these contrasted rocks within the same centred complexes

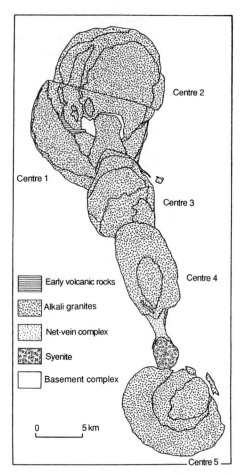

Figure 15.4 Sara-Fier Granite Complex, northern Nigeria, showing the migration of the intrusive centres and the multipulsal nature of the latter. After Turner (1963).

highlights this dilemma in explaining the link between the mafic and felsic components, especially in the face of a marked 'Daly gap' in composition. The problem was discussed in detail by Leger, in 1985, and Moreau, in 1982, the latter making the key observation that one particular monzoanorthosite member carries interstitial quartz–orthoclase intergrowths which progressively increase in amount as if separating out in response to some form of contrasted differentiation process. Nevertheless, the Daly gap is so profound in volumetric terms that it is difficult to suppose the direct derivation of haplogranite as a residuum from the cumulates. But where the isotopic data can be decoded, veiled as it often is by a late stage metasomatism, it provides a signature which is certainly not that of recycled crust.

Within the felsic rocks of this Niger–Nigerian province it is possible to identify within the felsitic rocks the two rock suites mentioned earlier, one more peralkaline than the other (Figure 15.5). The two suites may well represent differing paths of crystal differentiation, but Bowden and Kinnaird, as already noted, were so impressed with the patent evidence for extensive albitization and microclinization, especially marked in the rocks of the more peralkaline series, that they attributed many of the differences, including not only the complex textures and the special mineralogy but also the geochemical changes, even the desilication, to subsolidus reactions. No one who has studied these rocks microscopically can doubt the importance of this hydrothermal metasomatism, yet, as hinted earlier, I am convinced from my own observations that the arfvedsonite–riebeckite granites, in particular, also had a previous mag-

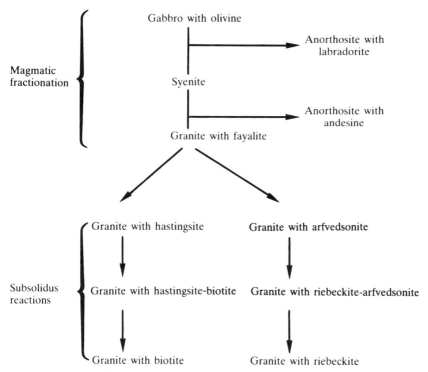

Figure 15.5 Genetic relationships of the alkalic granites according to Bowden and Kinnaird (1978). Only the primary part of this divided series is certainly generated from the melt; the later evolution may be the result of subsolidus reactions in the presence of hydrothermal fluids. After Bonin (1982).

matic history involving the precipitation of the alkaline mafic minerals, if only because the latter are so splendidly displayed in the marginal hanging-wall pegmatites fringing the ring dykes and plutons.

These rocks, with all their special features and unique geological context, are so ubiquitous that their separate description could easily become a mere catalogue. However, there is an important point to be made in following this Nigerian association across the Pan-African terrane of North Africa and into Arabia. In The Sudan and Saudi Arabia the ages of identical alkali granites disposed in identical centred complexes range down to 650 Ma, to a period of major transcurrent faulting which, in the Arabian Shield, marks the terminal phase of the Nabilah mobile belt. The latter represents but one of the criss-cross of sutures along which the miniplates slid and jostled during the period of Pan-African remobilization of the African basement. Thus, although the Saudi complexes can be regarded as immediately post-orogenic, they are only the initial phase of a 400 Ma period of specialized magmatism occurring whenever there was a resurgence of extension, rifting, deep melting and the provision of conduits for the mantle heat and mantle magmas to rise into and melt the crust.

In a wide ranging review in 1993 of such Proterozoic A-type magmatism, Windley held that the latter only appears to be anorogenetic, on the supposition that there must be a time lag after the peak of crustal thickening before the crust can be gravitationally unstable and extension can start. It may not, therefore, be possible to distinguish geologically between this alkalic magmatism in so-called intracratonic and immediately post-orogenic settings. We must look further.

ALKALI FELDSPAR GRANITES WITHIN THE RIFTS: THE OSLO GRABEN

The Permian igneous activity within the Oslo Graben has been intensively studied, beginning with the classic work of Brögger in the 1890s and later developed by Oftedahl, Barth, Holtedahl, and Ramberg and Larsen (in 1953, 1954, 1963 and 1978, respectively). From such a well-established database we can appreciate to the full the nature of the A-type granite association with both basic and silica undersaturated rocks.

The alkali granites occupy collapse cauldrons cut out from a volcanic plateau built from basaltic flows and felsitic welded tuffs: the latter are the famous rhomb-porphyries. Marvellously the amount of central subsidence can here be measured – from 500 m to more than 3000 m – and the association of vent agglomerates and ring faults draws a clear picture of the surface calderas.

In what is in effect the type area of the one-feldspar, hypersolvus, alkali feldspar granites, these are again seen to coexist with two-feldspar biotite granites. Furthermore, despite the evidence of mixing, the felsic and mafic rocks have so clearly a separate identity in outcrop that it has long been obvious to Norwegian petrologists that the two could not have been produced, the one from the other, by crystal fractionation alone. Thus an alternative thesis of separate derivations from the mantle and the lower crust has been current, the two derivative magmas diverging in evolution. Indeed, in 1954 and with great prescience, Barth estimated that only small chemical changes were necessary to produce the mildly alkaline magmas of Oslo from the granodiorites of the subcrust. And now that the presence of large masses of dense rock below the Oslo Graben has been established by reference to gravity data we can be confident that basaltic magmas could have provided the heat for the remelting of such a pre-existing granitic crust. Furthermore, Ramberg and Larsen considered that these mafic magmas were themselves remelts of the lithospheric mantle mobilized by the upwelling of asthenospheric mantle into the stretched and thinned crust of the Oslo Palaeorift, along which resurgent faulting localized intrusion and provided the conduits for magma: a very modern thesis indeed! It is, however, only proper to report that these authors believed that these mafic magmas provided, by contrasted differentiation, the syenites and alkali granites, an alternative supported by the mantle-like isotopic signature of the latter – though this could have been inherited from an igneous, crustal protolith, perhaps represented by Barth's 'granodiorites'?

In view of these different possibilities it is of the greatest interest that, in 1995, Lundqvist reported from southeastern Sweden an outcrop example of the melting of deep crustal rocks by underplated mafic magmas. The latter form a component of a very old anorogenic plutonism, now revealed by deep erosion. At intrusive contacts between gabbro and the granitic country rock the latter can be seen to have been melted, the resulting anatectic leucogranite mingling in with the mafic magma!

THE TERTIARY CENTRAL COMPLEXES OF THE BRITISH ISLES

We are led on naturally to consider here the Tertiary central complexes of western Scotland and Northern Ireland in which alkali feldspar granites are subordinate in volume to gabbros, probably greatly so because geophysical studies again reveal a substrate of dense rock. It is therefore not surprising that these particular epigranites were long considered to result from contrasted differentiation, a process invoked in 1934 by Nockolds on the assumption that the efficient precipitation of plagioclase and pyroxene from a tholeiitic magma would provide a final 10% residuum having the highly contrasted composition of a granophyre. However, despite the fact that mafic magmas were seen to be capable of generating granitic magmas in areas where continental crust was lacking, such as Iceland, doubts

concerning the volumes involved and the efficacy of the separation, coupled with certain critical field evidence, led to the view that the granitic
rocks of the Tertiary centres of Scotland were more likely to have been
generated by the fusion of crustal rocks, the newly generated felsic magmas then mixing with the mafic invader (Wager et al., 1965; Emeleus, 1991,
pp. 485–7). Certainly the mixing of magmas is everywhere in evidence in
the Tertiary Igneous Province, even to the extent of the selective extraction
of early potassium feldspar phenocrysts from a felsic magma into a contiguous mafic magma.

In the past 30 years the evidence provided by the strontium, lead and
neodymium isotope ratios has substantiated this involvement of both
mantle and crustal sources with estimates ranging from a mild degree of
contamination of a basalt melt to almost pure crustal melting, when the
mafic magma simply provides a source of heat. Diverse crustal sources
have been identified, their contributions being greatest in the earliest
magmatic pulse of any particular central complex. Current is a model
involving various degrees of assimilation of a range of available crustal
components, coupled with fractional crystallization, mixing and hybridization, which is surely flexible enough to accommodate explanations of
both the isotope abundances and the variations recorded between several
rock suites.

As an example, oxygen isotope studies of the Skye granites in 1971 by
Taylor and Forester revealed that the magmas were depleted in ^{18}O, which
they modelled by envisaging these granites as originating by the partial
fusion of hydrothermally altered, ^{18}O-depleted Torridonian arkose, blocks
of which, appropriately partially melted, occur with these epigranites.
However, such a special origin can only be one of several possibilities
because this evidence, recorded from both Skye and Mull, of the strong
interaction with meteoric water, is lacking in the similar epigranites of the
Mourne Mountains in Northern Ireland. Neither does the crust in this
latter region include Torridonian arkoses! Furthermore, according to
Meighan and his colleagues there is good geochemical evidence for a
basaltic parentage for this Mourne suite, though with a modest crustal
component identified from the isotope data. It is true that there is no
geochemical hiatus but, nevertheless, the Daly gap is quite evident in
volumetric terms.

In accepting a central role for the mafic component we have to also
accept the findings of Thompson and Morrison, in their 1988 geochemical
study of the latter, that the mafic melts themselves underwent a complex
history of derivation from the mantle before interacting with the crustal
rocks. Of course, we are now well aware that earth processes are commonly multifactorial, but, stated this generally, the model takes on a form
of 'all things to all men', and the challenge is to provide specific explanations for specific intrusive complexes, with the added expectation that the

component processes will each vary in their effects throughout the life history of such a complex. With this caution in mind I turn to the other tectonic environments featuring alkali feldspar granites, if only to find whether it is possible to constrain this rather open-ended model.

ALKALI FELDSPAR GRANITES IN AN IMMEDIATELY POST-OROGENIC SETTING: CORSICA

It might not be expected that alkali feldspar granites would feature among the granites of the collisional orogens and, indeed, it is true that when present they are of very small comparative volume. Nevertheless, in 1974 Bonin identified an important cluster of such rocks forming the latest stage of granitic magmatism associated with the Hercynian of Corsica, where alkali feldspar granites intrude a basement of 60 Ma older, calc-alkaline granites.

Once again the tectonic context and the structural control are plain to see in the location of the ring complexes at fault intersections. Once again hypersolvus granites form the roots of eroded calderas in which are preserved comagmatic ignimbrites, breccias and lahars. Here, too, is the familiar bimodal relationship between the felsic and mafic rocks, and also the evidence of both mixing and subsequent subsolvus reaction.

From the detail of this study we find a partial resolution of at least one of the problems raised in the previous discussions – that is, whether two distinct rock series existed and separately evolved by crystal differentiation. Bonin's comprehensive studies (1982, 1990), particularly the geochemical plots, as for example that comparing strontium with rubidium, show rather well not only that there is in this case a hiatus between the mafic and felsic rocks, but also that there are two distinct trends within the latter. This is best explained as due to the differing efficacy in the precipitation of an early plagioclase, by which barium and strontium are removed with the calcium, and a later precipitation of orthoclase when rubidium follows the potassium, all these extractions taking place without the necessity of involving a late stage fluid phase.

However, it is very likely that both hypersolvus fractionation and subsolvus recrystallization are involved, when the gradual build-up of fluorine in the magma and its final concentration in the late stage fluids will ensure that there will be a continuum of processes from the melt to the solid.

ALKALI FELDSPAR GRANITES IN SOUTHEAST AUSTRALIA

I turn now to the highly illuminating example of possible A-type granites in the Lachlan region, where they complete a trilogy with the I- and S-types. Although these rocks are not obviously peralkaline or hypersolvus or, for that matter, particularly abundant, they do show the main chemically diagnostic features of the type, that is, other than fluorine enrichment, and have been so exhaustively studied that the geochemical data provide important clues to the origin.

The key contextural facts are that these particular potassium feldspar-rich rocks are substantially younger than the I-type granites of the same belt, that is to say, they are of Upper Devonian age as distinct from Lower Devonian, with one suite cutting a graben-infill of Upper Devonian rhyolites. Moreover, their composition is sufficiently distinct from that of even the most evolved of the pre-existing I-types to preclude their evolution by crystal differentiation from I-type magma systems. Furthermore, comagmatic mafic rocks are but poorly represented. Search for a petrogenetic model then centres on the possibility that these A-type granites represent the hotter, vapour-absent partial melting of pre-existing, relatively anhydrous lower crustal source rocks.

The nature of the source material has again engendered much discussion. One view held by the Australian school of Chappell, White, Collins, Whalen and others (Whalen et al., 1987) is that this is represented by a mafic granulitic assemblage already depleted by the earlier extraction of water-bearing, minimum melts of I-type composition. It is a thesis that could also provide an explanation not only for the anhydrous nature of A-type melts in general but also for the abundance of fluorine, on the basis that the presence of this halogen so enhances the thermal stability of micas and amphiboles in a source rock that fluorine-bearing minerals will be preferentially retained in the residues from remelting processes. This in turn means that any further remelts at higher temperatures will necessarily be enriched in these halogens.

This may well be so, but Creaser and co-workers argued persuasively, in 1991, that a dry, mafic granulite source would be unlikely to yield melts with compositions appropriate to A-types, so they favoured, instead, a potentially more fertile source such as that provided by pre-existing tonalites and granodiorites. That this is a possible scenario is confirmed by the experiments of Skjerlie and Johnston, which showed that the high temperature, vapour-absent partial melting of a biotite–hornblende tonalitic gneiss, carrying a fluorine-rich biotite, yields an A-type felsitic melt, the fluorine being concentrated in the latter.

As I have already noted, the essence of this model was proposed by Barth for the origin of the felsitic rocks of Oslo 40 years ago, and although I am impressed with the valid point made by Whalen, Currie and Chappell that a primary diversity in the source rocks might well explain the individuality of A-type suites from different regions, I am greatly attracted to this view that granites early assembled in the crust as a direct consequence of orogenesis should provide a potent source of new magma if remelted at a new, higher temperature.

I now turn to consider how such a model would fit with current ideas on the formation of the rapakivi granites.

RAPAKIVI: A GENUINE A-TYPE

In reviewing this aspect of the A-type granite story I am fortunate to have at hand excellent reviews by Rämö and Haapala, in 1991, of the rapakivi granites of eastern Fennoscandia. These granites constitute the great batholiths of Wiborg, Aland, Laitila and Salmi, in the form of high level, subcaldera complexes, 1.65–1.54 Ga in age, intruded into the early Proterozoic Svecofennian basement of Finland.

The enigmatic texture involving the formation of the mantled alkali feldspar ovoids has excited much interest in the past, the very name arising from the fact that the ovoids weather out from their finer-grained matrix to provide a crumbly rock known as 'rapakivi' in Finnish. However, I avoid a discussion of this fascinating texture and turn instead to the general A-type character of these granites as a whole, recalling that many outcrops in the type areas are even-grained and normally porphyritic (but see p. 87).

There are the precursor swarms of basaltic dykes that we have noted elsewhere, also the early intrusions of gabbro and anorthositic gabbro which, though small in relative volume, again show the familiar contrasted association with alkali feldspar granites carrying fayalite and iron-rich biotite and hornblende. Particularly revealing is the abundance of the characteristic fluorine-bearing accessories together with niobium and tantalum-rich cassiterite and columbite in the late stage phase. Indeed, some of the most evolved rocks show many of the characteristics of tin granites.

It is almost a century since Sederholm and Hackman first established the chemical identity of the rapakivi granites of Fennoscandia by drawing attention to the heightened abundance of silicon, potassium and iron, and increased ratios of K/Na and Fe/Mg, but it is only relatively recently that the geochemical work of Sahama, also Vorma, and Rämö and Haapala have firmly established their analogy with the fluorine-bearing A-type granites the world over.

Trace element abundances coupled with the lead and samarium isotopic data confirm this identification, although Emslie considered that the rapakivi granite–anorthosite association forms a distinctive subset of this A-type category. Not surprisingly such data also place the rapakivi granites in the within-plate field, the Sm/Nd isotopic values, in particular, revealing a well-defined pattern of evolution involving the melting of crustal materials, as shown in Figure 15.6. This is wholly consistent with a model whereby the Wiborg and Laitila batholiths were generated by the partial melting of a low Sm/Nd crustal source like that represented by the

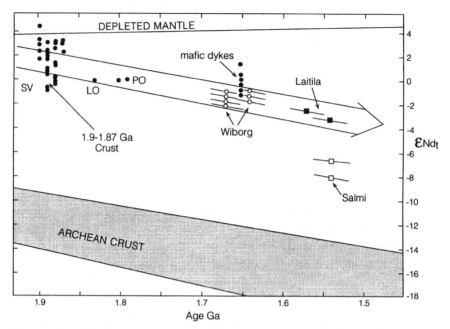

Figure 15.6 Nd isotopic data (ε_{Nd} versus age plot) of the rapakivi granites and porphyry dykes of the Wiborg (open circles), Laitila (filled squares) and Salmi (open squares) batholiths. Evolution lines for the rapakivi granite samples are marked with thin lines and a common evolution path for the Wiborg and Laitila batholiths is denoted. LO = late-orogenic granitoids; PO = post-orogenic granitoids. The Nd isotopic data for the Svecofennian and Karelian (1.9 to 1.8 Ga) igneous rocks (SV) and the evolution path of the Archean crust are from Huhma (1986) and Patchett and Kouvo (1986). Reproduced from Rämö and Haapala (1991) with the permission of the authors and the Geological Association of Canada.

Proterozoic Svecofennian crustal rocks. In contrast, the isotopic composition of the Salmi batholith is indicative of a mixed source with a major input of Archean derivation joining with Proterozoic material (Figure 15.6), a conclusion wholly consonant with the siting of the Salmi complex at the tectonic contact of the Proterozoic and Archean basements in eastern Fennoscandia. Once again the melting – possibly of a granodiorite protolith – is thought to be due to thermal input of the mantle-derived magmas represented by the isotopically primitive gabbros (Figure 15.7). Surely we can now have some geological confidence in a theoretically and experimentally controlled model for the deep-seated crustal melting by basaltic magmas, such as that proposed by Huppert and Sparks.

By switching attention from Finland to Greenland, though remaining in the same Early Proterozoic terrane, we can further appreciate the nature

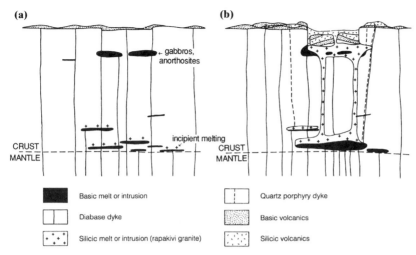

Figure 15.7 Model of the relations between diabase dykes, basic plutonic rocks, rapakivi granites and quartz porphyry dykes; the scheme is based on Huppert and Sparks (1988). Ascending mantle-derived basic magmas are in part trapped in the lower and middle crust and partly reach the surface forming fields of volcanic deposits (a). The thermal effect of the basic magmas (about 1200°C) slowly caused extensive melting of the granulitic (?) lower crust, forming silicic magmas. The latter intruded episodically into the middle/upper crust forming rapakivi granite complexes (by cauldron subsidence) and quartz porphyry dykes, or extruded to the surface forming pyroclastic deposits and rhyolite domes (b). The basic and silicic magmas locally use the same channelways and may have intruded and extruded contemporaneously, forming composite dykes and lava flows. Reproduced from Rämö and Haapala (1991) with permission.

of this generative process. There, Brown, Dempster, Harrison and Hutton reinterpreted the superbly exposed, fjord-side, three-dimensional exposures to show how the rapakivi crystal mushes rose as sheet-like bodies along ductile extensional shear zones. Moreover, these can actually be seen to have arisen out of a deep-seated, high temperature metamorphic environment of metasedimentary migmatites, with which they were in thermal equilibrium, into higher crustal levels where they produced contact aureoles.

Once again major and trace element modelling, by Harrison and others in 1985, indicated a metasedimentary source, specifically by a 35–40% batch melting of metamorphic rocks equivalent in composition to andesitic or dacitic greywacke. However, even in this favourable terrane we cannot directly observe the evidence of bulk melting, though it appears that the necessary thermal conditions were approached, the temperature estimations (850–870°C) lying close to or even within the conditions required for vapour-absent melt production involving the breakdown of biotite. Such temperatures could only have been achieved in the crust by the synorogenic emplacement of mafic magmas and, appropriately, we find that the rapakivi granites are accompanied by precursor norites.

We may speculate that these norites are representatives of a major underplating by mafic magma resulting from lithospheric extension and that this underplate provided the dominant heat source for the whole metamorphic and melt-producing event of the Ketilidian. However, it would seem unwarranted to extend this hypothesis to imply that there was a natural granite series extending from melt production associated with high grade regional metamorphism to magma intruding into the roots of post-orogenic volcanoes. Nevertheless, the model of crustal melting is real enough. But before I wax enthusiastic over this model of magma evolution at the base of a thick sialic crust involving a substantial crustal contribution, I ought first to consider the mid-plate, oceanic environment to juxtapose, so to speak, two apparently diverse hypotheses concerning the origin of A-type felsitic magmas.

ALKALI FELDSPAR SYENITES AND GRANITES IN THE OCEANIC ISLANDS

That felsic rocks with appropriate A-type characteristics occur in oceanic islands such as Reunion and Ascension raises fundamental questions not only about the origin of A-type granites in particular, but also about how granite can be generated outside the continental crust.

I choose here to examine the well-documented example of the volcanic Kerguelen Islands, which now lie within the Antarctic plate but may well have originated athwart the mid-oceanic ridge as Iceland does today. Interestingly, Giret in 1990, and also Lameyre in his 1983 review, did not favour a hot-spot thesis but compared the growth and construction of the islands to that of Iceland.

The occurrence of syenites and granites is spectacular enough to have led Watkins and co-workers, in 1974, to entertain the possibility that these islands might be situated above a mini-continent, and recent geophysical investigations by Operto and Charvis, in 1995, although confirming the relatively great thickness, 21–25 km, of the Kerguelen crust, have done nothing to dispel this view. Nevertheless, all who have worked in detail on the islands' geology and geochemistry are convinced that they represent a relatively young pile of oceanic basalts intruded by a cluster of shallow-seated, volcano-plutonic central complexes dating from the mid-Tertiary to the Recent. Here, to a degree much greater than the other environmental niches considered earlier, gabbro, peridotite and pyroxenite, the latter clearly representing cumulates, predominate among the intrusives. The subordinate felsitic rocks occur in two suites: one, oversaturated, is represented by gabbro, monzogabbro, quartz monzonite, quartz syenite and granite; the other, undersaturated, by gabbro, monzogabbro and nepheline-bearing monzonite and syenite.

Despite the contrast in geological environment and the proportions of the contrasting rock types, Lameyre and Giret liken the granite-bearing central complex of the Peninsula of Rallier du Baty, in all respects, to those of Oslo and Nigeria. It possesses a ring structure, the rocks show the hypabyssal features of quickly quenched, high temperature magmas, and their mineralogy is characterized by either perthite or orthoclase and albite, together with aegirine and riebeckite–arfvedsonite. The characteristic accessories are present, carrying an abundance of REEs and incompatible elements and, furthermore, fluorite is ubiquitous – a particularly important point.

The felsic rocks, gabbros and basalts all have a geochemical unity with no significant differences in the standard isotopic ratios, the latter clearly identifying a mantle source rich in radioactive elements for both the volcanic and plutonic rocks. These ratios are $0.70460 \leqslant {}^{87}Sr/{}^{86}Sr \leqslant 0.70588$ and $0.51233 \leqslant {}^{143}Nd/{}^{144}Nd \leqslant 0.51270$, which represent the general signature of the ocean crust.

Bearing in mind the predominance of mafic rocks with substantial cumulates coexisting with relatively small volumes of felsitic rocks, both the geological and chemical evidence seem decisive in identifying these particular A-type granites as derived from a mantle source, albeit indirectly via a basaltic underplate. Moreover, the origin of the fluorine must lie in this same primitive protolith. It is but a complication that undersaturated and saturated series here occur side by side, requiring either that there was a special (?depth) control of melting at source, or that the path of crystal differentiation during the upwelling depended on the early or late precipitation of calcium-rich amphibole, for this does not revise the main conclusion that the A-type felsic magmas can be generated from mantle-derived basaltic magmas without a contribution from the crust and that, equally, fluorine in some abundance can be released during this process, perhaps by the remelting of phlogopite.

A KIND OF CONCLUSION

From this rather extended discussion I accept that granites with certain common features identified as of A-type can appear within both the continental and oceanic interiors of plates. I also accept that such rocks are unlikely to represent highly evolved end-members of calc-alkali suites, though a contrary view has been expressed by Rogers and Greenberg.

From the tectonic point of view the key is the presence of a thick crust and the association of crustal extension and rifting with the high heat flow. The latter is attributed by many workers to the rise of a mantle plume, which would reasonably account for the high temperature of A-type magmas and their ability to remelt residues. Furthermore, it is logical to accept that remelting of mantle mineral assemblages is an anhydrous process capable of releasing fluorine and chlorine from minerals such as phlogopite. The resulting halogen-enriched basaltic melts might well underplate and carry heat up into the base of the crust, there to promote remelting of a variety of crustal materials that could as easily include the granulitic residues from earlier granite-producing processes as the first-stage granitic rocks themselves.

Though such a thesis is poorly constrained from the experimental point of view, Clemens and his colleagues, in 1986, extrapolated certain of their experimental results to support this thesis that peralkaline granites are produced by the second generation melting of already depleted igneous source rocks carrying halogen-enriched micas and amphiboles. But noting that Skjerlie and Johnston obtained A-type melts by direct melting of a tonalite bearing a fluorine-rich biotite, this does seem to offer the simplest thesis.

That magmas are generated by the extension and thinning of the continental lithosphere is supported by the modelling, in 1988, of McKenzie and Bickle, who also pointed to the likelihood of such basaltic melts being alkaline in the early stages of magma generation. I am further fascinated by the possibility that rapid extension allows immediate access of magma, whereas slow extension, or mere faulting, retains melt for a sufficient time to allow crustal melting and magma differentiation.

There are two especially disparate situations. Within the oceanic sectors of the plates, remelting associated with mantle plumes can only provide felsitic magmas by some form of extreme differentiation. This is possible to envisage in such fluidal magmas, but even so a thick carapace of basaltic lava flows would still be needed to blanket an underplated melt reservoir in which the processes of differentiation would have time to operate. That A-type felsitic rocks are produced in this oceanic environmental niche suggests that a mantle source always makes a contribution to the composition as well as providing the heat. Thus it is of particular interest to note that the elemental screens used by Eby in his review are effective even for rock suites in which there is evidence of a significant crustal contribution. This is possibly a result of the high absolute abundance of many of the incompatible elements inherited from the mantle source, because this effectively buffers the elemental ratios against changes due to crustal additions. What additions there may be are introduced, according to Bonin in 1982, by the hydrothermal systems and not through the agency of crustal anatexis.

The second extreme concerns the occurrence of alkali feldspar granites in some post-orogenic situations, especially those examples in which the direct mantle contribution is negligible, such as in southeastern Australia. It may be, as Eby thinks, that the Lachlan rocks are sufficiently unique to constitute a separate lithospheric type of the 'A' category. They are certainly not very obviously associated with rifting and maybe it would be wise to divorce them altogether. It is not surprising that granites of this particular environment show a certain uniqueness, for they are generated in a thick crustal welt where they are likely to be produced by the adiabatic melting of deep crustal rocks during the late stage uplift of dehydrated hot rocks, and when mafic magmas may or may not be involved. Whether the crustal source is to be sought in the granulitic residue from an earlier remelting episode, or in granodiorites earlier derived from this residue, or in new mantle displacing this old complex (as suggested by Bonin), is not easily resolved and must depend on particular geological contexts.

Even from the limited number of examples of intra-plate, bimodal magmatism referred to above we can see that there is a continuous spectrum of subordinate environments depending largely on the degree of extension and the thickness of the crust, so that we must accept that rigid boundaries are a philosophical luxury. Nevertheless, I believe that Loiselle and Wones were right to draw attention to this particular magma-tectonic connection, which not only directs the search for a geological answer to a geochemical problem but provides a working hypothesis stimulating comparative studies such as those presented here.

SELECTED REFERENCES

Bonin, B. (1982) *Les Granites des Complexes Annulaires. Bureau de recherches géologiques et minières, Manuels et méthodes, No. 4.*

Eby, G.N. (1990) The A-type granitoids: a review of their occurrence and chemical characteristics and speculations on their petrogenesis. *Lithos*, **26**, 115–134.

Rämö, O.T. and Haapala, I. (1991) The rapakivi granites of eastern Fennoscandia: a review with insights into their origin in the light of new Sm-Nd isotopic data. In: Gower, C.F. *et al.* (eds), *Mid-Proterozoic Laurentia-Baltica. Geological Association of Canada Special Paper No. 38*, pp. 401–415.

Whalen, J.B., Currie, K.L. and Chappell, B.W. (1987) A-type granites: geochemical characteristics, discrimination and petrogenesis. *Contributions to Mineralogy and Petrology*, **95**, 407–419.

Migmatites: are they a source of granitic plutons?

16

When we follow rocks into higher metamorphic grades, we finally end in a granitic core ... this cannot be accidental; the association of metamorphites, migmatites and granites must mean something.

Herbert Harold Read (1951) Transactions of the Geological Society of South Africa, Annex **54**, 5.

PREAMBLE

In accepting that many granites are produced by partial melting in the crust we need to search for outcrop examples of the production of melt in the volumes required to fill sizeable plutons. A likely candidate is a migmatitic gneiss. Certainly to the early workers in France and Fennoscandia the close connection between migmatization and granites seemed so obvious in outcrop that Sederholm considered migmatites to be intermediate between igneous and metamorphic rocks. He thought that the granitic *lits* of the banded gneisses originated through the agency of either melt or a tenuous fluid, the ichor, both of which derived from the nearby granites. The opposite view, promulgated by Holmquist, was that the granitic material came from the adjacent country rock, not the granites, and that it was segregated by fluid transport. Holmquist believed that such replacive migmatites were produced during metamorphism at a relatively low metamorphic grade, with partial melting only intervening at high grade.

As noted by Olsen the ensuing controversy encapsulated all the elements of a debate which has continued until today, centred on whether the process represents an open system in which magma is either injected or the rock metasomatized, or a closed system involving partial melting or metamorphic differentiation. In the event of much research neither of these alternatives proves to be exclusive and it seems that more than one can operate simultaneously. However, a problem remains about the degree of mobility of the new granitic component, and whether or not it contributes in a significant way to the large volume of granite at a high crustal level.

(a) METATEXIS AND GRANITE MAGMA

THE NATURE OF MIGMATITES

For the most part migmatites are metamorphic rocks made up of alternating bands of a granitic leucosome and a complementary mafic selvedge or melanosome. A third intervening layer of surviving country rock is known as either the mesosome or palaeosome (Figure 16.1). The ensemble is a *metatexite*.

It is usual to infer a sequence from patchy, poorly segregated, to layered, well-segregated migmatites; the stromatic gneisses. The importance of synchronous deformation in deciding the geometric form is shown by the way the leucosomic patches and layers occur along shear zones or infill the necks of boudins. What is particularly important in the context of the central theme is that such migmatites may grade into rocks in which the structural continuity is lost as the material of the leucosomes becomes increasingly predominant, even to the extent of the *anatexite* becoming indistinguishable from a restite-bearing granite (Figure 16.2).

Such an obviously progressive series has been widely interpreted as representing an increasing degree of partial melting, labelled in its advancing stages as metatexis, diatexis and complete anatexis. This progression, coupled with a final clearing or homogenization of the restitic fraction, represents the granite series of Read.

The experimental justification for the partial melting of natural rock compositions by Tuttle and Bowen and also by Winkler convinced many petrologists that the leucosomes of migmatites represented squeezed-out

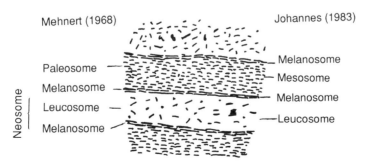

Figure 16.1 Nomenclature of a stromatic migmatite according to Mehnert (1968) and Johannes (1983).

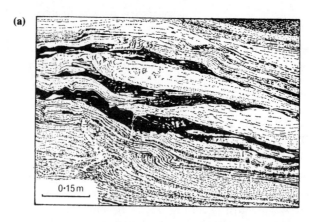

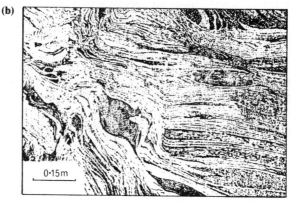

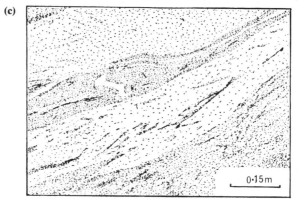

Figure 16.2 (a) Banded migmatite with biotite-rich layers (melanosomes) affected by deformation in which the biotitic layers behave incompetently. The more competent folds in the leucocratic material are often isolated in mobile and refoliated biotite schists. Gwenora Dam, west of Selukwe, Zimbabwe. (b) Banded migmatite involved in oblique shearing and synkinematic migmatization producing agmatitic structures (left). Rusfontein, Zimbabwe. (c) Mobilized diatexite; relict structures provide evidence of mode of derivation. Rusfontein, Somabula, Zimbabwe. Drawings by B.C. King.

minimum melts. As we have seen, the experiments showed that, in the presence of sufficient water, melting begins in pelitic rocks at about 650°C at 5 kbar (with $P_{total} = P_{H_2O}$). According to Mehnert, Büsch and Schneider such a melt is initially located at the three-fold contacts of the grains of quartz, potassium feldspar and plagioclase, when it is probably of very small amount, with fluid-aided recrystallization of the bulk rock the only result. Only with water steadily becoming available from dehydration reactions as the temperature increases, and its preferential partition into the melt, will the latter appear in substantial volume. However, Vielzeuf and Holloway, in experiments involving rather simple systems, showed that, even under fluid-absent conditions, large volumes of S-type granitic liquids could be produced deep in the crust at temperatures buffered at around 850°C – that is, at the biotite melting stage. In other experiments using natural rocks of more appropriate composition, Le Breton and Thompson demonstrated that metapelites undergo two stages of fluid-absent melting between 760 and 800°C at 10 kbar, as first muscovite and then biotite react. At 850°C melting is extensive.

Concerning those possible protoliths of igneous origin, we have already seen that substantial melting of amphibolite requires somewhat higher temperatures (p. 44). Even though intergrain melting begins at 650°C at 10 kbar, the amount of this low temperature melt is very small and not of the composition of real tonalites until temperatures reach or exceed 900°C (Johannes and Holtz, 1996, p. 282). Again, in the case of tonalitic lithologies, Rutter and Wyllie found that although dehydration melting begins at 825°C at 10 kbar, a melt fraction large enough to generate a mobile magma could only be achieved at temperatures greater than 950°C at 10 kbar. Thus, while we can certainly expect partial melting to be a feature of the highest grades of regional metamorphism, especially in pelitic rocks, whether or not magma is produced in bulk must often be in the balance, only to be tipped decisively in favour of magmatism by the introduction of extra heat, with mafic magmas often the carrier. But unless this contribution be substantial we must expect that only granitic melts in relatively small volume will be the normal product of regional metamorphism, certainly not granodiorites and tonalites.

Be that as it may, continuing work on real rocks revealed that leucosomes may not have minimum melt compositions. When attempts were made to remelt samples of leucosomes, Winkler and Breitbart, for example, found 'with surprise' that many did not completely liquefy under the simulated pressure, temperature and water concentrations of high grade metamorphism. Furthermore, regional migmatites more often than not show all the features of metamorphic rocks, even within the leucosomes, including blastic textures, relict mineral orientations, lack of zoning in plagioclase, and an overall mineral equilibration. Again, there is a need for caution because Vernon and Collins found otherwise, as we shall see. There are exceptions, as Sawyer told us, to the generally accepted fact that there are none of the differences that we might expect between the composition of plagioclase in the palaeosome and leucosome, or the patterns of trace elements, if melts were involved. Then, of course, the metamorphic imprint may well have been superimposed because magmatism and metamorphism are rarely separable processes.

Once again uncertainty rules. The origin of migmatites is not so easily solved as was once thought. However, we are fortunate in having some easily available reviews on which to base a revised opinion (King, 1965; McLellan, 1984; Ashworth, 1985; Olsen, 1985; Mehnert, 1987; and Brown, 1993).

OPEN OR CLOSED SYSTEMS

As a result of the difficulty in identifying with certainty the unchanged equivalent or correlative of a particular migmatite, it is not easy to determine the extent of the mass redistribution of the materials involved or, indeed, whether the system is essentially closed or open. However, we can now be reasonably assured that, on the regional scale, metamorphism is largely an isochemical process – that is, aside from dehydration. The simple fact is that although metasomatism, in the form of feldspathization, occurs locally at some granite contacts and affects some pelitic xenoliths, no one has convincingly demonstrated regional metasomatism involving the long-range migrations and the geochemical fronts on the scale early advocated by Perrin and Roubault in 1949, and by Lapadu-Harques in 1945. Although often conveniently forgotten, the thesis of metasomatic granitization is decisively refuted by the simple geological observation that existing heterogeneities such as stratification and clast form and size in metaconglomerates are preserved into the very highest grades of metamorphism. I refer the reader to Kenneth Mehnert's 1987 essay on 'The granitization problem – revisited', for a robust rejection of large-scale metasomatic granitization in general. If granites are produced from crustal rocks, then it is by the thoroughgoing partial melting of the latter.

I cannot but be amused by Eskola's approving comment in 1955 on Bowen's critical contribution to the 1948 symposium which reads, 'The Frenchmen M. Roubault and R. Perrin, the first ones to let atoms jump headlong into granite, are knocked out with an easy hand, and H. Ramberg's reasonings on diffusion are dismissed as light-minded. The soaks are confronted with some questions that they have never been able to answer.' As I was educated as a 'soak', I am acutely aware of the shortcomings of the granitization case.

Nevertheless, there is an exception to this general denial, as represented by the exchange of silica for potassium, water and oxygen in certain major shear zones, as reported by Beach and Fyfe in 1972, and this reference is important in introducing a theme on which I will have much to say later – that is, the crucial role of deformation in metamorphic and metasomatic processes.

THE ORIGIN OF THE GRANITIC *LITS*: MASS BALANCE IN MIGMATITES

The inner aureoles of some granites are characterized by vein complexes which are clearly the result of magmatic injection, but these are no real guide to the origin of those abundant regional migmatites that core the major metamorphic complexes. In the latter the field evidence is often decisive in showing that the thin, impersistent leucosomes and the accompanying anastomizing vein systems do not join back into a mother granite. Clearly the source of the granitic material must lie, for the most part, in the country rocks, as proposed originally by Holmquist.

The nature of the segregation has been particularly well established by the mass balance studies carried out by both Olsen, in 1985, and Johannes, in 1983. These show that planimetric analyses across the layering of stromatic migmatites most often yield a net composition of leucosome plus melanosome that closely approaches that of the palaeosome, indicating that little has been gained or lost overall.

From his particular studies Johannes favoured a largely isochemical layer by layer transformation of paragneiss into migmatite with increasing grade. The leucosome compositions did, in some of his examples, approach minimum melt compositions, though Johannes argued that this does not necessarily have to be the case to support a thesis of partial melting; only that the mesosome layers need to be further removed from the cotectic composition than the leucosomes.

By comparison of migmatized and unmigmatized rocks Johannes observed that the finer-grained layers are preferentially transformed into leucosomes, the melanosomic selvedges marking the former borders of the layers. Although this suggests that the migmatizing process may have been controlled simply by grain size and compositional differences, I suspect that these particular fine-grained layers actually represent ductile shear zones, and that these provided the loci for the metasomatizing fluids – a concept I will return to later.

In parallel studies Olsen identified both closed and open system migmatites, with the latter showing an excess mass in the leucosome, sometimes of a sufficient degree to require injection of new melt (Figure 16.3). Sometimes there is also evidence of minor metasomatism in the form of either an introduction of new potassium feldspar components or a K–Na exchange. Olsen's general conclusion from a wide range of examples was that the migmatization process is in reality very complex, ranging widely in mechanism and degree of departure from the isochemical norm. She further considered that the major driving force is primarily

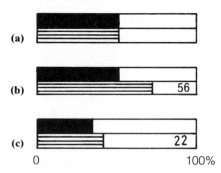

Figure 16.3 Selected examples of mass balance calculations for migmatites in the Front Range, Colorado. After Olsen (1985). Estimated (solid) and measured (lined) volume proportions of the leucosome. (a) Closed system migmatites; (b) migmatite formed by apparent injection; and (c) migmatites formed, at least in part, by metasomatism. Numbers are volume % excess mass in the leucosome. Full explanation in Olsen (1985), p.149 *et seq.*

partial melting, which produces gradients along which fluids then transport material, as in the process of metamorphic segregation.

MIGMATITES AND DEFORMATION

As a field example of the multifactorial nature of the migmatization processes I quote the conclusions of Sawyer and Barnes for the Quetico Metamorphic Belt of Canada, that certain layer-parallel leucosomes formed early on by a subsolidus process akin to pressure solution, but that these are distinct from discordant anatectic leucosomes that formed after an intervening deformation event when metamorphic temperatures reached a peak. In this example, as in many others, a near synchronous deformation was clearly a factor in the production of the migmatites.

It is a general finding that in outcrop the most obvious feature of many migmatites is their structural complexity. As a particularly well documented example I refer to the migmatites of the Moine Supergroup of northwest Scotland studied in 1985 by Barr , who was critical of the fact that much confusion had arisen in the past by seeking a single mechanism when it was obvious to him that several phases of coeval deformation and segregation were involved.

With respect to the genetic history of these particular migmatites, Barr found that the earliest stage of migmatization in the metasedimentary Moine rocks was associated with the peak of metamorphism, giving rise to subsolidus, trondhjemitic, layer-parallel leucosomes. Several phases of near synchronous deformation can be identified. Within certain structurally defined zones these early leucosomes were strongly deformed before a renewal of migmatization in which actual melting occurred within originally feldspar-bearing horizons, the resulting streaky and partially mobilized migmatites – that is to say, diatexites – being interpreted as representing crystal mushes. In outcrop there are good examples of segregation of melt into and along shear zones, along contacts with concordant amphibolites and into the necks of boudins, with agmatites appearing whenever the amphibolitic bands were disrupted (compare Figure 16.2b).

In explanation Barr proposed a model whereby partial melting is controlled by the local availability of water, which is sucked, along with any interstitial melt, into deforming shear zones by dilational pumping. The consequent weakening of the shear zones promotes the continuing deformation with the developing leucosomes responding differently from the palaeosomes. It seems that when the melt fraction increases beyond the critical melt fraction the strength of the mush falls rapidly, at which point melt is presumably expelled by filter pressing into discrete veins. To complicate the genetic history of these Moine migmatites further, Barr found evidence for even further subsolidus segregation during a final phase of ductile thrusting.

The details of all the textural changes involved make fascinating reading but sufficient has been reported here to emphasize the multifactorial nature of the migmatization processes, their different operation in different country rock lithologies and, above all, the importance of synchronous deformation in the segregation and mobilization of the leucosome. But the most important conclusion is that actual melt is often present.

DEFORMATION AND EXTRACTION

On a more general point it seems that the extraction of this melt itself requires that the rock is synchronously strained. As McKenzie and Bickle deduced from their theoretical studies, in 1988, of the fluid dynamic processes involved in extraction, it is likely that felsic melts can move with respect to their matrices only if the viscosity is lowered by a high concentration of volatiles. Otherwise the mechanical strength of the interlocking crystal framework inhibits the expulsion of the interstitial melt. Surface effects may further impede the migration of fluid, as indicated by Jurewicz and Watson's finding that felsitic partial melts show so little tendency to wet quartz grain boundaries that they remain *in situ* unless mechanically extracted. It seems certain that extraction requires some measure of stress-assisted compaction to squeeze melt into the networks of coalescing tension fractures which locally feed potential voids created in shear zones and boudin necks (Figure 16.2) – synchronous deformation features that we actually observe in most migmatite terrains. But even when melt can be seen to have coalesced in this way it only accumulates as relatively small bodies, metatects, of pegmatitic granite. In reality, as

Holmquist observed, most migmatitic venites are blind, representing exudates *in situ*: bulk extraction is not the norm. Here I introduce my central theme: that granitic magmas in bulk are born by bulk melting in the very deep crust and are but rarely extracted from layered migmatites.

My prejudice about the importance of deformation in the migmatizing process is shared by Ashworth, who, in his 1985 review, also opines that high strains are required to expel melt from granitic systems at low degrees of melting, whereas low strains will suffice at high degrees.

THE TEXTURAL CONFIRMATION OF THE PRESENCE OF MELT

Rock textures offer another approach to the solution of these problems, in particular by providing important indicators of the former presence or otherwise of a melt fraction in any specific example.

By analogy with the growth of pegmatite and aplite Yardley early suggested that coarse-grained leucosomes may have grown from aqueous fluids and fine-grained leucosomes from water-saturated melts. Conversely Flinn has provided strong textural evidence that certain granoblastic gneisses in Shetland were produced by grain-growth recrystallization presumably caused by the passage of a fluid through the rock during regional metamorphism of a grade too low for melting to have occurred. The sophisticated studies of Dougan, and also of Ashworth and McLellan, reveal something of the overall complexity of the textural evolution in such migmatitic rocks, yet there is some consensus that the common coarsening into the leucosome layers is due to growth in the presence of a dispersed melt, with the larger grains growing at the expense of the smaller ones, which is especially likely in the case of the porphyroblastic migmatites – the ophthalmites.

That the minerals of the coarse leucosome crystallized from melt is strongly suggested whenever there is idiomorphic zoning in plagioclase, but almost all other mineral growth or replacement features would seem to me to be equivocal in discriminating between crystal–melt and subsolidus reactions. More promising in this respect are the quantitative studies of microstructures – grain size, grain shape and grain interrelationship – especially when coupled with a consideration of the petrofabric (Vernon and Collins, 1988).

Carefully measured and statistically evaluated, grain size may well help to distinguish leucosome growth from either melt or fluids. In addition, the evolution in grain size appears to be sensitive to the degree of melt separation. Grain shape was utilized by McLellan in 1984 to discriminate between certain anatectic and non-anatectic leucosomes on the basis that the aspect ratio of plagioclase will be lowest in partially melted rock.

However, of all the grain-based techniques, that of recording the between-grain contact frequencies shows particular promise in detecting

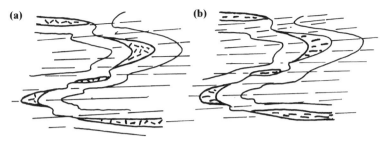

Figure 16.4 Comparison of stromatic leucosomes showing different time relationships with respect to deformation. (a) Mimetic, with melt segregation post- or syn-
deformation; (b) melt segregation pre-deformation. Any field geologist will
appreciate how schematic is this representation.

the former presence or otherwise of melt, even though there is a need to be
aware that recrystallization will often have obliterated the early formed
textures. Thus Ashworth and McLellan found that aggregate distributions, in which contacts between grains of the same species are more
common than expected of a random distribution, characterize solid-state
differentiation, whereas random distributions result from melting. In
similar studies Ashworth contrasts the highly regular spatial distribution
of the crystal grains in certain palaeosomes with the much weaker regularity in the associated leucosomes, interpreting this as indicating an
anatectic origin for the latter.

Finally, petrofabric studies are crucial in establishing the timing of
migmatization in relation to deformation, though here we must accept
that mimetic growth can easily mimic the fabrics of deformation (Figure
16.4).

Clearly methodologies exist for discovering the genetic history of particular migmatite terranes, and so far as they have been used they provide
strong evidence that there are migmatites of different origins, both
anatectic and non-anatectic, just as surmised by Olsen. So it seems proved
that melt can exist, but is unlikely to be extracted in bulk quantities. I now
turn to the situation where partial melting advances the crustal rocks to
the stage of mobilization.

(b) ANATEXIS AND GRANITE MAGMA

A GRANITE SERIES?

We have seen how mass balance studies of migmatites largely deny the
bulk drainage of melt from leucosomes. If anything, material is added not
subtracted. Furthermore, exhausted residues in bulk are but rarely *directly*

associated with migmatites, though it is possible that they may not be easily recognized as distinct from original metabasics of one type or another. Of course, there are examples of granulitic lithologies within the same orogenic zone as migmatites, and these may indeed represent residua, though whether from extraction of leucosomic melt or the differentiation of bulk partial melt is not easily decided.

Once again I turn for guidance to field examples of the relationship between migmatite and restitic granite. In a classic work based on the rocks of the Black Forest, summarized in his 1968 review, Mehnert documented a transition from metatexite to diatexite, in other words from stromatic migmatite to anatexitic quartz diorite. The core of one dome-like structure is a near homogeneous rock indistinguishable from a normal intrusive, and the dome structure itself is attributed to the bulk flow of melt moving away. According to Mehnert (1968, p. 269) this 'granite series' simply represents an increasing degree of melting and mobilization more or less *in situ*: the final diatexite did not result from the *extraction* from the leucosomes of the peripheral migmatite. This is an excellent example of the entire spectrum of the partial melting process, yet even so there is some suggestion that the observed core material had moved up from a higher energy level (Mehnert, 1968, p. 262).

In a study by D'Lemos, Brown and Strachan, in 1992, outcrops of the Cadomian Belt of northwest France were interpreted as separately representing middle and upper levels of the crust. A middle level is represented in the St Malo region by heterogeneously deformed migmatites and anatectic granites, tectonically interleaved with medium grade metasediments, whereas an upper level is represented in the Mancellian region by the same kind of metasediments but in a much lower state of regional metamorphism, and intruded by undeformed granitic plutons.

Within the former region Brown and D'Lemos described a gradual transition with increasing metamorphic grade involving a coarsening of grain size, the appearance of quartz segregations and then granitic leucosomes, leading finally to stromatic migmatites interpreted as products of low to moderate degrees of partial melting. The progression is continued with the stromatic structure becoming progressively disrupted, when schlieric and nebulitic migmatites result. Note that it was not suggested that the leucosomic material was extracted in bulk, simply that the whole rock became mobile when the degree of partial melting had reached the critical melt percentage point.

Locally, sizeable bodies of partly homogenized restitic granites are so closely associated with these schlieric migmatites that they can be confidently interpreted as anatexitic granites marking the most advanced stage of melting and partial separation of melt from restite. Interestingly, Peucat found that the homogenization of the strontium isotope systems occurs at just this step in the progression.

Throughout the zone of migmatization there is abundant evidence of a synchronous, polyphase deformation (compare p. 287). Three phases of this deformation and two of metamorphic growth are recognized, the second of the latter being associated with the peak of metamorphism and the attendant migmatization. There are also well-defined ductile shear belts and it is along these that the anatectic granites are thought to have been syntectonically emplaced.

Although there is some evidence for the local interaction of mafic magmas in the St Malo region, Brown and D'Lemos argue that the volumes were too small to provide the heat necessary for melting, the conditions for which were more likely to have been generated during transpressional overthickening and radiogenic self-heating of the metasediments.

In contrast with the situation around St Malo the granitic plutons of the Mancellian region were passively emplaced into a brittle upper crustal regime, presumably in response to faulting. They are wholly post-tectonic in relation to their greenschist or even lower grade country rocks, in which they produce static, contact aureoles. These biotite granites are peraluminous and bear minor cordierite, muscovite and small biotite schlieren which are likely to be of restitic origin. As the granites share many petrographic, geochemical and isotopic characteristics with the anatectic granites within the St Malo migmatites, D'Lemos, Brown and Strachan reasonably concluded that they are genetically related.

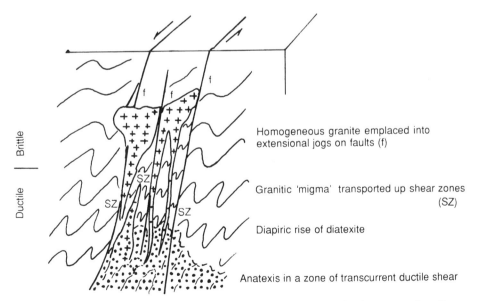

Homogeneous granite emplaced into extensional jogs on faults (f)

Granitic 'migma' transported up shear zones (SZ)

Diapiric rise of diatexite

Anatexis in a zone of transcurrent ductile shear

Figure 16.5 Model of the evolution of the granite series as in the crust of north-west France. Adapted from D'Lemos *et al.* (1992), with the permission of the authors. (After D'Lemos *et al.*, 1992.)

Assembling all the compositional, structural and age data, these workers modelled the Mancellian granites as the final arrival, in the brittle upper crust, of the anatectic granites (Figure 16.5). They suggested that the shear zones in the St Malo region represent the conduits up which granites rose from their site of generation, the driving force being the alternation of dilation and compression within the anastomizing shear zone system, with its extensional jogs and compressional shears. Of particular interest is a finding that cordierite is extensively developed adjacent to some of those shear zones in which only a limited amount of granite is present, implying, perhaps, that larger volumes of magma than are presently observed formerly passed along the shears.

This rather convincing model, with its implication that the granites moved out from a deforming understorey into a passive canopy, in the sense of both space and time, is of particular interest in that Read (1949, see 1957, p. 337) drew attention to this very example in support of his granite series long before the detailed studies reported here, commenting that the 'rise of the magma lasts longer than the tectonic events and many granites break through the structures and appear to be post-orogenic'. Read also considered that the granites were genetically unconnected with mafic magma.

A DISJOINTED GRANITE SERIES

While there are many examples of this kind of granite series, there are many more in which melting has only advanced to the stage of leucosome segregation at the end-stage of high-grade metamorphism. It would seem that only where there is exceptional heat flux does melting advance to the stage of wholesale melting sufficient to result in the disruption of original structure and the production of an anatectic granite capable of flow – a diatexite. Even so I suspect that the apparent end-stage diatexite of many migmatite terranes has arrived from some deeper, higher energy level. I will ask my reader to be patient while I develop this theme.

Referring back to the well-studied example presented by Brown and D'Lemos, I believe that the frequency of microdioritic enclaves in the high-level Mancellian granites not only suggests synchronous recharge by mafic magmas, but also generation at a higher energy level than that represented by the migmatites at St Malo, though not necessarily in a very different protolith. Indeed Power has reported quite significant geochemical differences which throw some doubt on the direct connection of the several members of the 'granite series' as proposed in the general model, so that I think it possible that the Mancellian granites were generated at an even deeper crustal level than the diatexitic granites of the present outcrop: a level where mafic magmas would be more likely to be involved.

Following this theme I am reminded of the studies of Vielzeuf and Holloway on melting relations in the pelitic system, particularly of their conclusions about an application of their data to the production of granitic melts at different levels in the Hercynian crust of the Pyrenees – that is, within the dry lower crust and within the relatively wet middle crust. The intrusion of mafic magmas into the deep crust, now observed as layered complexes, would certainly have ensured the higher temperatures required for bulk melting.

The removal of the melt would leave behind a mafic residue which Vielzeuf thought was actually represented in the Pyrenees by certain granulitic gneisses, the composition of which suggested that 40% of the granitic fluid had been extracted from an original metapelite.

Vielzeuf and Holloway considered that at the rather lower pressures and temperatures pertaining closer to the surface, the initial vapour-present melting of metapelites at the muscovite melting stage would have generated the existing thick layer of migmatites, but these would have been unable to contribute mobile granite to the system because of their near water-saturation. This scenario implies that the granites produced at depth intruded through the shallower seated migmatites, and may well have been contaminated in the process. If so, the timescale involved implies that the cycle of deep melting and final arrival of granite in the upper crust lasted 20–40 Ma. This would have provided adequate time for the clearing of restite from the granites, but I doubt that stagnant melt remained around for so long in the crust. Granite emplacement is commonly a distinctly episodic process involving repeated batch melting, so I suspect that several cycles of reactivation were involved.

The concept of bulk melting at deep levels of the crust and limited melting at shallower levels fits the pattern observed in many migmatite–granite environments. I believe it to be supported, for example, by the situation in the Coast Range Complex of British Columbia, described so vividly by Hutchison in 1982. Here, tadpole-shaped, tonalitic plutons lie within a migmatitic gneiss, the product of a major plutonic event at about 90 Ma. The cross-cutting heads of these plutons prove that they are allochthonous, but the tails are so conformable with their enveloping gneiss as to appear to root in the migmatite. From my own observations I doubt whether any parts of the tails are truly autochthonous, as the overall structure of these plutons is highly characteristic of model diapirs ballooning into a shear zone. Moreover, the plutons date at 80–70 Ma but, nevertheless, the abundance of trondhjemitic migmatites of a bulk composition not unlike that of the plutons suggests that the plutons were ultimately derived from material akin to that of their country rocks. But not, I think, at their present level. Despite the abundance of included material the plutons have obviously been homogenized to a considerable degree compared with the gneisses. Furthermore, many of the dark microdioritic enclaves are much more likely to represent the break-up of synplutonic basaltic dykes than inclusions of amphibolite derived from the envelope.

These are all matters of interpretation, but I believe that these tadpole plutons of northern British Columbia originated deep in the crust, perhaps even at the crust–mantle interface, by wholesale bulk melting brought about by heat introduced by the injection of hot basaltic magmas. I envisage the batches of granitic magma rising up shear zones into the still cooling and ductile migmatites during one of the pulses of deformation.

I am attracted to this extended model because the compositions of diapiric plutons intersecting migmatite terrains rarely correspond exactly with that of the anatexitic parts of the host migmatites, and are even more rarely in physical continuity with them. This is even true when the geochemistry indicates a common source. A simple explanation is that mobility is bound to create some kind of hiatus, when the newly created magma moves away under the influence of buoyancy and deformation. The degree of unsticking is likely to extend from a mere surging and pulsing, analogous to the mobilization of the still fluidal centres of plutons, to an upwelling of the new melt well away from the generative source, when a hiatus appears as a strong compositional contrast across intrusive contacts – even though these are often disguised by a new conformity of diapir and envelope imposed by forceful emplacement!

ANATEXIS: THE REMOVAL OF RESTITE AND ITS FATE

There is a general concensus, based on the results of experimental and analytical geochemistry, coupled with the outcrop evidence of the relative chronology of the granite suites, that granitic magmas can often result from the deep-seated, partial melting of a mafic underplate or old crust – itself predominantly of amphibolitic character. That the partial melting of such source rock should yield tonalitic magmas is wholly in accord with experimental work (Johannes and Holtz, 1996, p. 286 *et seq.*), but we also have the sure proof of the outcrop, as so well demonstrated by the Black Forest studies (Büsch, 1966; Mehnert, 1968). In the latter locality, as already noted, the most advanced term in the melting progression is a dioritic mobilizate – a diatexite – which overlaps in composition the parent amphibolitic gneisses, and which only begins to diverge with advancing mobilization, when the heterogeneous diatexite transforms to a more igneous-looking quartz diorite (Figure 16.6). It is most significant that the dense concentrations of biotite in the original melanosome layers of the gneiss end up as the dark mafic schlieren of the quartz diorite, showing that what is to become restite begins to be segregated from the felsic melt during the early stages of melting, a process of separation continuing even more effectively during progressive mobilization.

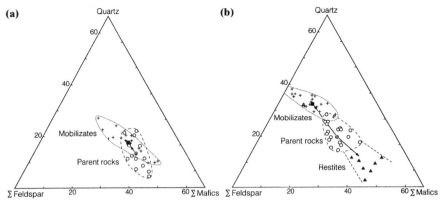

Figure 16.6 (a) Average mineral composition of biotite–hornblende–gneisses (parent rocks, O) and associated dioritic mobilizates (+). Initial stage of development as shown in a roadcut near Todtnauberg, southern Black Forest. (b) More advanced stage of mobilization. The parent rocks (O) separate into dioritic mobilizates (+), associated with the corresponding restites (▲). From outcrops at Hirschbachtal, near Schapbach, central Black Forest. After Büsch (1966) and Mehnert (1968). Note that ⊙ denotes average of all parent rocks, ■ all mobilizates.

It is important to realize that this transformation involves continuous reaction and recrystallization, producing major changes in the mineralogy, and even a degree of local metasomatism as frequently revealed by K-feldspar blastesis. That this is so is entirely in accord with the theoretical studies of Huppert and Sparks reported above (p. 128), which indicated that a significant fraction of the refractory components would be resorbed (or recrystallized), though reprecipitation might rapidly ensue as the new magma moved away. As a result many crystals will be of genuine igneous origin, though nucleated on to unresorbed residue. If this is, indeed, the expected scenario, then it is unlikely that much of the original restite mineralogy will survive.

Another significant observation is that such anatexitic granites commonly show strong mineral fabrics, which is not surprising considering the obvious syntectonic and kinematic nature of the whole process. I believe this may hold a vital clue as to both the mechanism of the continuing disengagement of melt from restite and the homogenization of the former, because it is in this environment of deep-seated melting and synchronous deformation that mechanical sorting is likely to be important. According to Kumar the action of shear on a viscous suspension of crystals produces dispersive pressures that lift and deflect crystals into zones of less intense shear, and I surmise that this is a central process in the clearing of the granitic mush, and is especially effective if the mush is pumped away through the multiple conduits provided by active shear zones.

But what of the collected residue, how can we recognize this material resulting from 'degranitization' in the outcrop?

MIGMATITES AND GRANULITES

It has long been recognized that magmatic and high-grade metamorphic conditions overlap in the granulite facies. However, I am not qualified to comment usefully on the part mafic granulites may play as the exhausted residues of granite extraction, or to enter the debate on whether the associated charnockites – the orthopyroxene-bearing granites – represent metamorphosed igneous rocks or high temperature melts derived from the wholesale fusion of even deeper-seated sources.

That the nature of the residue might be recognized in hornblende granulite facies rocks has often been discussed in terms of 'degranitization', with a search being made for rocks of appropriate composition – that is, rich in mafic minerals or their reaction products, such as biotite, cordierite, garnet, sillimanite, and even corundum. These are the 'kinzigites' (significantly enough defined from the Black Forest locality of Kleine Kinzig!), rocks which Mehnert tells us (1968, p.337) are closely associated with granitic mobilizites (but see Büsch et al., 1980). Such a relationship is well shown in the ancient basement of Quebec and Labrador, where Percival studied the interface between granulite facies metamorphism and crustal magmatism. There, a high degree of fusion of a source akin to a granulite facies paragneiss is thought to have given rise to intrusions of a garnet- and orthopyroxene-bearing peraluminous 'diatexite', better labelled, I suppose, as a restite-rich granodiorite representing a primitive stage of granite production in the lowermost crust.

Some further insight as to this kind of relationship is provided by a study of a mafic migmatite terrane in East Antarctica reported by Tait and Harley in 1988. As appropriate to the high metamorphic grade, the palaeosomes of these granulitic migmatites consist largely of garnet, orthopyroxene and a strongly zoned plagioclase, with lesser biotite and apatite. The garnet and orthopyroxene are enriched in selvedges against leucosomes consisting of weakly zoned plagioclase, lesser orthopyroxene, some biotite, rare quartz and accessory ilmenite and apatite. These leucosomes, which make up about 15–30% of the rock, are either layer-parallel or vein-like, when they link, intersect and feed into dioritic pools replete with schlieric restite, in the manner of melt. Despite this appearance of intrusion, geochemical mass balance calculations strongly suggest the local extraction of melt, though certain inconsistencies about potassium, rubidium and barium indicate an accompanying weak metasomatism, probably associated with the growth of the biotite, and yet another minor inconsistency concerning yttrium and the REEs seems to be related to a partial dissolution of the garnet.

These workers do not discuss the source of the high temperatures that would be required and neither do they identify exact experimental analogues on which to model the melting process, but I join them in their contention that this is an example of the early stages of the partial melting, under high pressure, of a deep-seated tholeiitic protolith, yielding a metaluminous I-type quartz diorite, the compositional equivalent of a basaltic andesite and a likely candidate for a primary member of a granitic suite of rocks.

ANOTHER INTERIM GEOLOGICAL CONCLUSION

In the development of ideas on the origin of granites the popularity of the thesis of metasomatic granitization waned under the combined assault of quantitative geochemical and structural studies. Nevertheless, one of the central tenets of the thesis, that granites were of crustal derivation, has been strengthened by the finding that many granites image their source rocks. The essential difference in modelling is that it is now believed that the crustal source needs to be partially melted to provide what is essentially a magma. New problems then arise as to what represents the residue at depth, and if this is entrained in part as restite, how the mush is clarified?

We have seen that partial melting is not only possible but is almost demanded by a variety of evidence, experimental and field based. Nevertheless, I express yet another prejudice in doubting that the partial melting, as represented by the migmatization associated with metamorphism, will *often* yield granite magma in the bulk proportions indicated by the relatively rapid arrival of voluminous batches of magma in the upper crust. As we have also seen, there are good enough examples of the complete anatectic melting process to show that extraction does not provide bulk melt, and that only wholesale melting to a degree sufficient for mobilization will suffice. For partial melting to occur on so grand a scale requires that extra heat be available, probably at a very deep level in the crust, and as a result of mantle diapirism and the abyssal injection of mafic magmas. Furthermore, according to the calculations of Bergantz in 1989, whereas the deep-seated injection of basaltic magma proves to be an efficient transporter of heat, it requires repeated intrusion, coupled with turbulent convection at the melt face, to produce large volumes of mobile new magma.

It may be the final paradox of the granite problem that although the migmatite terranes we observe represent an essential step in the granite-producing process, they, in their turn, need to be remelted in bulk to supply plutons! It is an evolutionary process well exemplified in Southwick's interpretation of the late stage generation of granite plutons by the bulk remelting of earlier formed migmatites in the Archaean of the Superior Province of Canada.

The intervention of extra heat may occur at all stages of the plutonic process, so that the various melts, either mantle-derived, lower crustal-derived, or even directly derived from the intracrustal metamorphic migmatites, are all possible contributors to the plutons exposed at the present surface. It is when these separate contributions mix and mingle that the real complexities of natural processes become apparent!

SELECTED REFERENCES

Ashworth, J.R. (1985) Introduction. In: Ashworth, J.R. (ed.), *Migmatites*, Blackie, Glasgow.

D'Lemos, R.S., Brown, M. and Strachan, R.A. (1992) Granite magma generation, ascent and emplacement within a transpressional orogen. *Journal of the Geological Society of London*, **149**, 487–490.

Mehnert, K.R. (1987) The granitization problem – revisited. *Fortschritte der Mineralogie*, **65**, 285–306.

The waning stages: the role of volatiles in the genesis of pegmatites and metal ores

17

To many petrologists a volatile component is exactly like a Maxwell demon; it does just what one may wish it to do.

Norman L. Bowen (1928)
The Evolution of the Igneous Rocks, *p. 282.*

INTRODUCTION

A great deal has been written on the nature and influence of the volatile components of magmas and I only touch on the subject in so much as it throws light on the role of granitic magmas as host of these volatiles, on how the latter affect the course of crystallization during the waning stages of granitic magmatism, and also on their part in the production of granitic pegmatites and aplites and in facilitating granite-associated mineralization.

THE ORIGIN OF THE FUGITIVE CONSTITUENTS

As Fyfe has always emphasized there is no shortage of combined water in crustal rocks. It is released in such vast amounts as a result of the dehydration reactions involved in the melting and contact metamorphism of pelites and altered volcanic rocks that we can be sure that the intrusion of granitic magma often takes place within a great plume of vapour. The effect on the more distant country rocks is rarely commented on, but I know that, at least in Donegal, assemblages of plutons lie within retrograde aureoles of almost regional dimensions.

Many studies show how contrasting fluid compositions evolve from different protoliths during this prograde metamorphism. Thus Wilkinson, in extraction studies involving the pelitic envelope of the Cornish granites, identified water-dominated fluids containing up to 25 mol.% CO_2, 1.7 mol.% N_2 and 0.1 mol.% CH_4. Semiquantitative mass balance calculations support the intuitive thesis that most of these materials were derived from the metamorphic devolatilization of pelites containing small amounts of organic or elemental carbon and nitrogen. Wilkinson refers to similar studies, and estimates that the total fluid yield may often be of the order of 1.1 mole of fluid per kilogram of pelite. He concluded that there is therefore no need to appeal to an alternative source for the CO_2 and N_2 components of the fluid inclusions trapped in syn-intrusion quartz veins, and further suggested that such contact metamorphic vapours played an important part in the mobilization and redistribution of ore-forming components during emplacement of the Cornish granites.

I believe that this is likely to be generally true for granites for which, as in the Cornish example, there is much evidence that these components were derived from a sedimentary protolith. Indeed, there is a popular and widely held view that the polymetallic mineralization in the Cornish region was the product of just such a large-scale convective meteoric–hydrothermal fluid circulating around the margins of a cooling batholith, a view sustained by the early studies of the $^2H(D)$ and ^{18}O isotopes by Sheppard. However, in more recent studies, Alderton and Harmon showed that the system is more complex, with fluids of likely magmatic derivation being diluted with meteoric waters and basinal brines. It seems that early tin- and tungsten-bearing magmatic fluids were replaced by leaches of copper, lead and zinc from the metamorphic aureole, and then diluted by an influx of calcium-rich basinal brines associated with a lead–zinc–fluorine mineralization.

Such an interplay of differently sourced fluids was envisaged by Hanson in his theoretical treatment of the migrations set up during contact metamorphism. The initial expulsion of fluids from the magma is modelled as being followed and gradually superseded by those derived from the country rock. Together these circulate at lithostatic pressures, permeating along grain boundaries and through the microfractures produced by fluid expansion. As the system cools and the pressure drops to hydrostatic pressures, surface-derived fluids infiltrate the fracture system and interface with the derived fluids along a diffuse boundary which migrates inwards with time. Clearly a mixed source is the rule.

Furthermore it seems that the consequential mineralization is often intimately related to the erosion process. Thus Samms and Thomas-Betts envisaged the history of mineralization as representing a cooling sequence connected with high level emplacement and the progressive erosion of the cover rocks above a bell-jar-shaped pluton. As erosion bites deeper, the pattern of flow of the upwelling fluid migrates away from the peak of the pluton, so providing a possible explanation for the overall zonation of the mineral deposits.

I am ill-equipped to tread the minefield of fluid inclusion studies and mineralization, but I can appreciate how difficult it is becoming to label fluids as juvenile, magmatic or meteoric. Furthermore, we must expect that because there are different protoliths and different country rock acceptors there will be multiple origins for the volatile constituents. As an illustration I quote a study by Laouar and co-workers on the sulphur isotopes within the sulphide phases of British and Irish granites. On the basis of $\delta^{34}S$ content these granites divide into two groups, one with $\delta^{34}S$ ranging from $-4.5\%o$ to $+4.4\%o$ and another ranging from $+6.2\%o$ to $+16\%o$. As sulphides in the relevant country rocks are characterized by unusually high $\delta^{34}S$ values, the Laouar team suggested that, on the basis of simple comparison, this latter group of granites contains a significant proportion of the sulphur derived from such sedimentary rocks. This is consonant with the $\delta^{18}O$ and Sr_i evidence that there was significant incorporation into these granites of envelope material which had previously undergone low temperature fluid–rock interaction. This is an apparently simple illustration, perhaps not wholly unambiguous if we are to believe Coleman, but it points the way.

THE ISOTOPES OF HYDROGEN AND OXYGEN

From the work of Hugh Taylor, summarized in 1988, we have learnt that the δD values of the minerals of granites are relatively constant and identical to those in most other crustal rocks, both continental and oceanic. Therefore even the primary magmatic water in these granites is unlikely to be juvenile, but is derived at source by dehydration or partial melting of crustal rocks or subducted oceanic crust. Furthermore, the fact that the high $\delta^{18}O$ values of many granites contrast so sharply with the low values of basalts requires that the parent material of the former had originally been weathered or hydrothermally altered. This is yet another indication of the originally meteoric derivation of much of the water. Taylor and his colleagues went on to show that the stable isotope signatures provide an excellent way to characterize hydrothermal fluids of different origins. Apparently fluid–rock interaction effects are most clear cut when low $\delta^{18}O$, low δD meteoric waters are involved, but the effects of sea water, sedimentary formation water, metamorphic dehydration water and truly magmatic water can also be separately distinguished.

Particularly illustrative of the principles involved are the well authenticated examples of abrupt changes in the isotopic values across batholithic assemblages of plutons straddling a hidden junction between crustal blocks of contrasted composition. From such situations, as represented in the Peninsular Ranges Batholith and the plutons of north-central Idaho, we can easily appreciate the separate contributions of mantle and crust to the water component of the magmas (Figure 17.1) and, to an extent, discuss the specific nature of the source rock.

However, a considerable complication is introduced, especially rel-

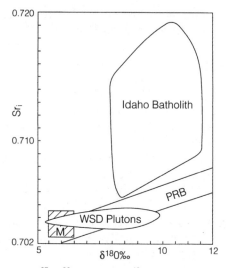

Figure 17.1 Plot of initial $^{87}Sr/^{86}Sr$ versus $\delta^{18}O$ adapted from Taylor (1988). The field of the Idaho Batholith is contrasted with those of the plutons of the Wallowa–Seven Devils terrane (WSD), the Peninsula Ranges Batholith (PRB) and mantle (M). The Idaho Batholith (Bitter-root Lobe) is emplaced within the ancient continental crust, the WSD plutons within an accreted island arc complex and the PRB within a marginal continental arc.

evant to the granite problem, by the continuity of aqueous fluid interactions, which means that the rocks that ultimately undergo partial melting may well have had a prior history of fluid exchange and thus carry, on melting, isotopic signatures different from those with which they started. The effect of the almost ubiquitous burial metamorphism is very much a case in point.

According to Taylor the impact of these interactions is a complex function of the water to rock ratio, temperature and the duration of the interaction. Viewed in the simplistic terms of increasing crustal depth, involving a change from hydrostatic to lithostatic pressures, an increase in temperature and a longer potential life of interactive systems, whole rock $\delta^{18}O$ values change from being widely variable to relatively uniform, whereas $^{18}O/^{16}O$ ratios evolve from showing extreme disequilibria between coexisting minerals to the reverse; all this with but slight change in the whole rock chemical composition or other isotopic parameters. The inherent complexity of rock systems is manifest when we appreciate that such fluid interactions may occur before, during or after melting!

Clearly, such complex interactions have the capacity to modify the original $\delta^{18}O$ signature, especially of those granites derived from materials of mantle origin which show low $\delta^{18}O$ values. Perhaps, as Taylor noted, this is the reason for the apparent rarity of low $\delta^{18}O$ granites in the geological record, when other isotopic data suggest that there is indeed a fair proportion of 'primitive' granites. Such a disguise resulting from hydrothermal melting processes may be common but, despite the problem this introduces, it most certainly reinforces the point that granite magmas are but a part of gigantic meteoric–hydrothermal convective circulation systems deriving their volatile constituents from both source rock and envelope.

But if the volatiles in large part emanate from crustal rocks and are scavenged by the granites, how are they finally concentrated and from whence came the ore-forming metals that they carry?

THE CONCENTRATION PROCESS

As we have already seen, both the extent and type of the differentiation process are largely controlled by the relative abundances of the volatile constituents, namely water, hydrogen sulphide, hydrogen fluoride and hydrogen chloride, as they are progressively concentrated by the crystallization of the feldspars and quartz and insufficiently taken up into the hydrous micas and amphiboles. Such constituents depolymerize the melt with dramatic results, the most influential of which are the enhancement of the fractionation process by a combination of lowering the melt viscosity, extending the range of crystallization and increasing the diffusivities of the melt components.

As crystallization continues, the vapour pressure needed to keep water in solution in the melt rises until it exceeds the hydrostatic head, when water vapour exsolves as an immiscible phase at the so-called second boiling point of the magma. This delightfully simple explanation of the origin of the late stage fluids, originally proposed by Bowen, was later amplified into an all-embracing model by Jahns and Burnham, by which it is not only possible to explain the concentration of the metals but also the formation of pegmatites and aplites and even account for explosive volcanism. Whether a second boiling occurs must depend on the initial water content of the magma and the degree to which water is consumed in the formation of hornblende and biotite, and these factors therefore determine whether or not a granite evolves to the pegmatitic and hydrothermal stages.

ON VOLATILE CONCENTRATION AND THE ORIGIN OF PEGMATITES AND APLITES

Pegmatitic assemblages of feldspar, quartz and a mica are a common feature of granitic rocks, taking the form of irregular patches within the margin of a pluton, of fringes along contacts and within discrete sag-sheets, and forming veins and dykes concentrated in roof zones. Com-

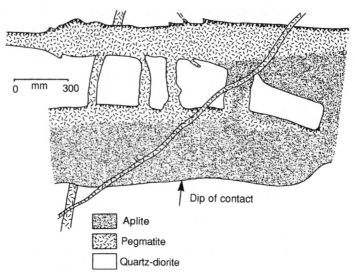

0   mm   300

Dip of contact

Aplite

Pegmatite

Quartz-diorite

Figure 17.2 Pegmatite–aplite sheet of the type associated with the Rosses Complex. The concentration of pegmatite on the hanging wall is thought to be due to the collection and trapping of the volatile constituents in this position. Crovehy, Donegal. After Pitcher (1953).

monly, giant feldspars, interposed with plumose mica and quartz lobes, grow on a contact surface, fanning out to end abruptly against an inner zone of aplite or aplogranite (Figure 17.2). Such fringes are best displayed on hanging walls where they underlie projections and fill re-entrants in a way suggestive of the rise and collection of volatiles with the consequent impoverishment of the rest magma. Equally it is easy to find an explanation of the resulting textural contrast between pegmatite and adjacent aplite in differences in nucleation density and growth rate.

There are many variations on this theme. Sometimes the order pegmatite–aplite is reversed, especially in vertical veins, pegmatite lying central to aplitic rims, or there is an interbanding of pegmatite or pegmatite and aplite, the latter representing cyclic precipitation as a vein is opened by the pulsatory pumping in of vapour-rich fluids. More complex situations arise where a measure of mineral replacement is evident.

In the laboratory the growth of such textural variants has been closely simulated in experiments involving slightly water-undersaturated melts. For example, Jahns and Burnham were able to replicate the various relationships simply by changing the water concentration and pressure, and in one experiment simulate the example of the aplitic dyke with a pegmatitic centre. Thus the underlying mechanism is broadly understood and the theoretical basis of it has been brilliantly explored by Burnham (1979 and 1982). As an illustration of the application of this central hypothesis London ascribed most of the features of pegmatites to the effects of an aqueous vapour phase coexisting with silicate melt plus crystals. In a long series of experiments he was able to determine the exact phase relationships within, for example, lithium aluminosilicate systems and he used the fluid inclusions in their natural analogues to evaluate the pressure–temperature conditions and the changing fluid composition during the transition from the magmatic through the pegmatitic to the hydrothermal state.

According to London the crystallization path involves the disequilibrium growth of feldspars and quartz through the supersaturation caused by increasing concentrates of water and other fluxing components in the melt. This is not likely to be due to undercooling, but is the consequence of the disruption of the alkali-aluminosilicate framework of the melt owing to the incompatibility of water, boron, fluorine and phosphorus in feldspar. The probability of forming crystal nuclei is consequently decreased and hence the crystallization of feldspar delayed until a significant degree of supersaturation is reached with respect to feldspar. Perhaps surprisingly, London and his colleagues found that vapour saturation may not always be attained and it is thus significant that graphic quartz-feldspar intergrowths are only produced in vapour undersaturated conditions. This is an important finding because they went on to show that melt fractionation under such vapour undersaturated conditions leads to both silica depletion and increasing alkalinity, factors which promote greater melt solubilities of the incompatible, high-charge elements.

In his review of the genesis of granite pegmatites London also touched on the problem of the regional chemical zonation in granite pegmatite fields. Along with Cerny he suggested that such zoning is inherited from an original vertical zoning in the cupola of the parental pluton, the highly fluxed materials of which were then injected as a single pulse into the capping country rocks. In reality, of course, there are many possible scenarios. Thus Möller contended that a combination of processes is often involved whereby the residual melt, resulting from fractional crystallization deep within a pluton, mixes with intergranular fluid within the largely crystallized periphery (Figure 17.3).

At this point I would remind the reader that the detailed study of the crystal-rich fluids so often retained in inclusions has been the special province of our Russian colleagues. They have long debated the exact nature of these silicate-rich hydrous melts whose compositions lie between typical silicate magmas and aqueous fluids (reviews by Sobolev, 1991a, b, 1992). Indeed, quite recently Marakushev and Shapovalov questioned the traditional view that ore deposits accompanying granites are related to aqueous solutions released by residual melts. They argued otherwise in that the experimental data demonstrate that ore metals are preferentially concentrated in the melt! Effective extraction of the ore

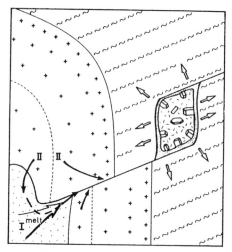

Figure 17.3 Schematic two-component model of granitic pegmatite formation after Möller (1989), reproduced in Lehmann (1990). Fluid I is a late residual hydrous granitic melt; fluid II represents the intergranular liquid from crystallized granite which mixes continuously with fluid I. Reproduced with permission of Springer-Verlag and the authors.

metals can only result by the immiscible separation of dense salt phases, which act as extractors. Of these the fluorides were found to be particularly effective in extracting tungsten and the REEs, yet relatively ineffective in the case of tin, niobium and tantalum. We have much to learn.

THE GEOCHEMICAL CONSEQUENCES OF THE ADDITION OF A VAPOUR PHASE

Once a fluid exsolves from the melt each volatile phase is variously partitioned between crystal and fluid; for example, chlorine behaves very differently from fluorine. This affects the phase equilibria and the partitioning of the elements between vapour and melt as well as between crystal and melt, so that the overall process of concentration and deposition becomes extremely complex.

In his attempt to place constraints on the chemistry of this system Candela, in 1992, although admitting that our understanding is in its infancy, nevertheless presents a valuable quantitative model for vapour evolution in granite systems and a discussion of the partitioning between vapour and melt. He explained, for example, how extreme loss of europium to the vapour phase leads to pronounced europium anomalies in the resultant rock, and how direct vapour phase crystallization in fluorine-rich magmas produces extreme enrichment of the incompatible elements and thus an overall flattening in REE patterns. Further, the preferential solubility of the heavy REEs over the light REEs explains the tendency for these patterns to develop a U-shape on the appropriate plots. All this reinforces the point that the presence of a vapour phase introduces many complexities which I do not think have yet been taken into account in many geochemical and radiometric models.

The presence of high volatile concentrations is amply confirmed by the capture of the fugitive and otherwise incompatible lithophile elements in borosilicates, phosphates, fluorides, sulphides and arsenides, and in pristine form within fluid inclusions. Of all the representative minerals tourmaline is the most common, its content of iron indicating the likelihood of metasomatic interchanges between fluid and country rock – a view confirmed by the experimental finding that iron is leached into boron-rich and iron-poor pegmatitic melts with the consequent precipitation of tourmaline. Continuing permeation by vapours and fluids, along with decreasing temperature, leads to widespread alteration of the primary precipitates and to complex mineral parageneses involving a myriad of secondary phases. So much so that there is often a debate on whether a particular mineral phase is of primary or secondary origin, especially in the complex pegmatites of rare element ore bodies where the crystallization of cleavelandite is a case in point.

LATE STAGE METASOMATISM

Often such permeations produce even more drastic effects on the main body of a granitic intrusion. Greisens are a familiar example where the alkalis are leached out, but there are also cases of fairly profound desilication leading to syenitic rocks of metasomatic origin, processes described in detail by Cathelineau and by Fonteilles and Pascal. A particularly fine example is that represented by the pipe-like bodies of alkalifeldspar intersecting the Barnesmore pluton of Donegal and described by Dempsey, Meighan and Fallick. Desilication and microclinization are very real deuteric processes and it is with some reluctance I avoid following up this fascinating trail and turn to the perhaps more significant matter of mineralization.

ON THE TYPICALITY OF METALLIZATION

The most dramatic end result of efficient fractionation in the presence of volatiles is the concentration of metals to the extent of becoming ore bodies. It is a very wide field of study and I only intend to touch on just one aspect, the typicality of metallization.

The association between metal concentration and silicic magmatism has been the special concern of our Russian colleagues, in particular this matter of typicality. In joining in this discussion I find it useful, as Plant and co-workers recommended, to distinguish between metalliferous and mineralized granites because even though a magma may acquire ore-forming elements from its source, these are not necessarily concentrated without the intervention of the volatiles and brines that derive as end-products of particularly efficient and specialized differentiation processes. Furthermore, the processes of scavenging and hosting, and the final release of these volatile carriers can be very complex, as is well documented by Eugster and Wilson.

So complicated, in fact, that Hannah and Stein were right to question the applicability of relating the type of mineralization with the 'alphabet soup' of granitic types, as they rather disparagingly label the crude attempts of myself and others to establish a genetic classification. It is indeed likely that the type and degree of metal enrichment are determined as much or more by the nature of the fractionation and fluid-release than by the overall chemistry of the bulk magmas and, certainly, Hannah and Stein provided convincing examples showing how the multifactorial nature of these concentration processes confound simple schemes based on magmatectonic environment and original source input. Furthermore, I remember Sobolev telling me that tungsten is associated with M- and I-type granites in the form of scheelite-bearing metamorphic skarns, whereas the metal occurs with S-type granites in the hydrothermal context of pegmatites and greisens.

However, this general reservation does not deny a role for the source rocks, and Hannah and Stein themselves conceded that certain tectonic settings do correlate, albeit loosely, with both granite composition and type of metallization. This is because the global tectonic setting is a vital factor in determining the source rock composition, plutonic history and the availability of volatiles at the site of magmatism.

There are some well-established examples of this correlation. Thus the copper porphyry deposits are closely related to arc magmatism with its I- and M-type tonalites and granodiorites, with gold a common associate, often joined by molybdenum in the marginal continental arcs. However, in the specific case of the Mesozoic Andean we shall see how complex was the process of recycling and concentration of copper and other metals which were derived first from exhalative deposits within the volcaniclastic host rocks of batholiths, and then concentrated in the copper porphyry systems that form the latest stage of a longstanding magmatic event.

The relationship between granitic magmatism and tin and tungsten deposits is very different, as has been especially well demonstrated within the 'tin girdle' of southeast Asia. The current model for their origin involves the metals being originally garnered from clastic sediments, this time during a cycle of partial melting and the subsequent volatile-promoted fractional crystallization that led to the intrusion of ilmenite-bearing, S-type granites during a continental collision-motivated orogeny. However, this is not the only environment of tin deposits, which also occur in association with the alkali granites of the anorogenic ring complexes, with their highly evolved magmas concentrating niobium, tantalum and cassiterite. These are particularly well exemplified in the Nigerian complexes with their tin–zinc–niobium mineralization and alkali–fluorine metasomatism.

In contrast with these rocks, those I-type granodiorites and granites that characterize the post-tectonic uplift and fault regimes are much less highly mineralized. According to Plant and her colleagues this is not because these granites are non-metalliferous, but because they fail to maintain a sufficiently high heat flow to sustain late stage fluid circulation. This may be because of a paucity of radiogenic minerals in such primarily mantle-derived magmas, coupled with the fact that they are divorced from the high heat flux associated with subduction processes. Certainly it is within those granites with a high heat production due to higher than average contents of uranium and thorium that the concentration processes are the most efficient and long lasting. The Cornubian Batholith is a case in point, hosting a major polymetallic ore field in which activity continued, albeit abated, for 100 Ma.

UPLIFT AND EROSION: A COMPLICATION OR A CAUSE?

Any attempt to survey the association of ore systems with granite type through time is complicated by the fact that the roofs of the associated granites very often lie close to an original land surface (possibly due to their ascent being terminated at the base of the meteoric water-table). Furthermore, granitic plutons, with their cupolas of satellitic tin veins, copper porphyries, and epithermal systems in general, tend to be associated with near contemporaneous uplift and are therefore sensitive to erosion, a point early emphasized by Meyer. This is probably why the ages of such deposits are strongly biased towards the Phanerozoic, and why they only survive at early times within intra- and pericratonic depressions.

In fact the correlation between granite emplacement, faulting and uplift has often been commented on, more especially by Leake, who ascribed post-consolidation uplift to the buoyancy effect of the concentration of relatively low density granite in the upper crust. The effects are often dramatic, involving the stripping off of the sedimentary and volcanic cover and the formation of a red bed molasse, often with the progressive exposure of the apical cap of plutons at that critical phase of the cooling history when ore concentration processes are just beginning to wane. This is the explanation of the often noted connection between granite mineralization, its zonation, and peneplanation; they are but parts of this integrated process of intrusion and uplift.

Paradoxically the granite-filled crust then becomes isostatically stabilized, with the result that the tops of many granitic plutons continue to lie near the general erosion level for hundreds of millions of years, as in the example of the 400–470 Ma Caledonian plutons of northern Britain and Ireland (Leake and Cobbing, 1993). We may ask what processes of delamination or extension in the lowermost crust are responsible for this remarkable situation? I wish I knew.

THE COPPER PORPHYRIES: GARNERING ORE-FORMING METALS FROM THE MANTLE

The world-wide association of copper and molybdenum porphyries with calc-alkaline magmatism led Sillitoe (1973; 1981) and many other authors to couple this type of mineralization with Cordilleran-type batholiths and to postulate that the metal-bearing fluids and magmas were a product of the partial melting of metalliferous oceanic sediments and basalts subducted beneath the arc.

Within the crust the porphyries are seen as filling volcanic conduits that channelled the magmas and associated volatiles, locating major hydrothermal systems that carried in the copper and molybdenum. The nature of the hydrothermal processes involved has been deduced from modern analogues in regions of active volcanism, from field studies in areas of rapid erosion such as the Andes, from restorations of complexly faulted deposits, and from deep drill-hole information, all matters reviewed in depth by Norton. I content myself with comments on the copper porphyry associated with the Coastal Batholith of Peru, the magmas of which were generated, as we have seen, during a longstanding cyclic process involving materials ultimately derived from the mantle. Its complex history is closely paralleled by that of the mineralization, which, as C.E. Vidal showed, began with the formation of exhalative stratiform deposits of barite and sphalerite in response to the submarine volcanicity within a 'back-arc' basin (Figure 17.4).

The low temperature static metamorphism that accompanied the burial of the volcanic fill of this basin motivated a second metasomatic phase of mineralization involving the local replacement of the early gabbros and diorites to form mantos of amphibole–magnetite–chalcopyrite–apatite. Then, after the inversion of the basin and during a long history of plutonic intrusion, further mineralizing phases took four different forms: (1) quartz–specularite–chalcopyrite–tourmaline–(potassium feldspar) veins adjacent to the early potassium-rich diorites and monzonites, in the genesis of which high salinity 'second boiling' fluids were involved; (2) chalcopyrite–molybdenite–scheelite skarn deposits, which were clearly of

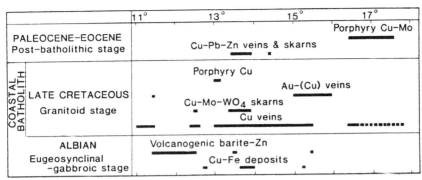

		11°	13°	15°	17°
	PALEOCENE–EOCENE Post-batholithic stage		Cu-Pb-Zn veins & skarns		Porphyry Cu-Mo ━━
COASTAL BATHOLITH	LATE CRETACEOUS Granitoid stage	•	Porphyry Cu ━ Cu-Mo-WO₄ skarns • Cu veins ━	Au-(Cu) veins ━━━	━━•••••••••
	ALBIAN Eugeosynclinal –gabbroic stage		Volcanogenic barite-Zn ━━ Cu-Fe deposits •	•	•

Figure 17.4 Metallogenic stages between 11° and 18°S in the Western Cordillera of Peru, showing the approximate distribution of known mineral deposit types. After Vidal (1985).

contact metasomatic origin; (3) quartz–calcite–ankerite–auriferous-pyrite-bearing veins associated with various calc-alkaline plutonic suites; and finally, late in the history of the batholith, (4) the important porphyry coppers in which chalcopyrite–molybdenite–pyrite disseminations and stock-work fillings formed in and around small intrusions and breccia bodies enclosed within a halo of strong potassic, phyllic and propylitic alterations (Figure 17.5).

We owe to Le Bel a detailed study of the examples of the Cerro Verde and Santa Rosa porphyry copper deposits in southern Peru. From an examination embracing the mineralogy of the micas, the stable isotopes calculated as δD, $\delta^{18}O$ and $\delta^{34}S$, coupled with extensive fluid inclusion studies, Le Bel was able spatially, temporally and thus genetically to relate the copper mineralization to stock-like bodies of porphyritic monzonite and breccia that evidently filled volcanic conduits; the magmas involved represent a very late stage in the evolution of a long-lasting plutonic system. Within these magmas Le Bel envisaged fractional crystallization, involving plagioclase and amphibole, controlling the primary evolution of the partial melts derived from a subcrustal source. However, as we have already seen in the discussion of the within-continent-margin arc of the Mesozoic Andes, there is no need to call upon subducted oceanic crust; the new Andean crust is itself a complex assemblage of submarine volcanic rocks, basaltic dykes, gabbros and early formed quartz diorites. Within this, copper and other metals had already been concentrated be-fore the sequence of remelting events that supplied the bulk of the tonalites and granodiorites and so were readily available to enter into the final act of concentration.

According to Le Bel the particular magma representing the porphyries evolved under moderately high f_{H_2O} and, migrating upwards along ten-sional faults, produced a separate aqueous volatile phase as it became

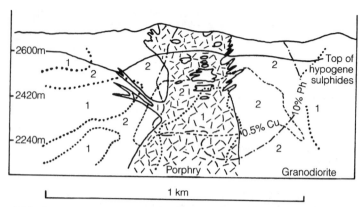

Figure 17.5 Cross-section of the Santa Rosa porphyry copper deposit near Arequipa, Peru, according to Le Bel (1985). Shows hypogene alteration zones: (1) propylitic; (2) phyllitic. Also 0.5% copper isograd and 10% isopleth of the phengite substitution ($Si^{IV} \leftrightarrow Al^{IV}$) in muscovite.

oversaturated. The alkalis and metals such as copper, iron and manganese partitioned into this aqueous phase as chloride complexes, after which a dramatic decrease in pressure from 1500 to 300–500 bar induced boiling at 400°C, which led to ore deposition by an increase in f_{H_2S}. All these events and the ambient pressure, temperature and concentrations are recorded in the fluid inclusions, as are those later events that produced the phyllic alteration zone.

But how do we explain how such extensive aureoles of mineralization come to be associated with such small stock-like bodies of porphyritic granite lying high in the crust, when metal concentration and transport must require the through flow of liquid or gas on an immense scale? This has been frequently discussed, most recently by Shinohara and his colleagues in a review of Climax-type Mo porphyry deposits. Of the two major hypotheses, volatile transfer and transport in a continuously convecting magma column, these authors inclined towards the latter, envisaging a volatile-rich magma ascending, degassing with consequent loss of buoyancy, and thus descending – a very likely process in such a volatile-charged system with conduits probably extending down to a voluminous feeder pluton.

There are many other excellent examples of the complex recycling, refining and concentration of chalcophile elements to the point where they become ore-forming. The problem as I see it is not so much in accounting for the metallization that exists, but in understanding why little or no mineralization occurred in essentially analogous, favourable magma-structural sites, such as, for example, in northern Peru, still along the axis of the Coastal Batholith. It is a problem that has occupied other workers who, like Baldwin and Pearce, have sought with limited success to discriminate between productive and non-productive porphyritic intrusions in the Chilean Andes.

One obvious possibility in Peru is the nature of the country rocks. Although everywhere volcaniclastic lithologies are predominant, interbeds of saline deposits are far more abundant in the vicinity of the porphyry coppers of southern Peru, suggesting that the greater availability of derived brines was the critical factor. That this is the likely explanation is confirmed by the higher saline concentrations in the fluid inclusions in associated plutonic rocks compared with those in other segments of the batholith.

Whatever the complexities of the concentration process the eventual source of the copper and molybdenum must lie in the mantle, from which the metals were directly extracted during arc-related magmatism. We can be sure of this because in the continental margin arc of the Mesozoic Andes the contribution from the old crust was minimal, and in the younger oceanic island arcs it was clearly non-existent.

If we turn now to the natural history of tin we find it to be in complete contrast, involving not only a crustal source but one established early in earth history.

GARNERING ORE-FORMING METALS FROM THE CRUST: A GEOCHEMICAL HERITAGE

Tin granites are geochemically specialized. The systematic geochemical campaigns in Russia provided Barsukov and his colleagues with a sufficient database to show conclusively that granites associated with tin deposits had a considerably higher intrinsic tin content than those that were not, even in those facies not altered by post-magmatic processes. Furthermore, tin granites and tin deposits have a distinctly provincial character, with tin enrichment and ore formation often temporally repetitive, as expressed by Routhier in his 'étagement temporel', which provided the basis of his concept of the geochemical heritage of tin. For example, in the Bolivian tin province a Precambrian tin mineralization was apparently regenerated by both Permo-Triassic and Tertiary magmatism. An even clearer example is provided by the South China tin province, which has been discussed by Liu Yingjun and co-workers at Nanjing, in terms of a heritage which envisages synsedimentary tin and tungsten being enriched by magmatic processes at various times from the Precambrian up to the Devonian.

Such specialization and heritage strongly suggests that tin (and to a lesser extent tungsten) was very early concentrated in the crust and has been continually recycled during the production of crustally derived granites. However, the conventional form of proof of crustal source, involving specific signatures of the strontium and neodymium isotope ratios, is complicated by the extreme fractionation of the magmas involved in the concentration of tin.

Tin-associated granites are characterized by other specific factors, including anomalously high boron – as in southwest England – or fluorine – as in southeast Asia – and also by the exceptional enrichment in the lithophile elements such as rubidium, caesium, lithium, thorium, uranium, niobium, tantalum and tungsten. I have already commented on the depolymerizing role of water, fluorine and boron, which promote a more efficient fractional crystallization, but it is surely significant that Al^{3+} plays a similar part in peraluminous melts. The complexing of tin in the presence of hydroxyl, fluoride, chloride and boride ensures that the natural sequestration of the metal into the mafic minerals is progressively overridden as differentiation proceeds. Not surprisingly the concentration process is complex in detail, but it appears from much work that the tin-bearing capacity of fluids is largely dependent not only on the temperature, but also on the chloride concentration, pH and oxygen fugacity.

In outcrops of the tin granites the evidence of a former fugitive component is ubiquitous in the form of highly evolved granites with heterogeneous textures indicative of fluidization processes, the frequency of pegmatitic phases and well-authenticated examples of extreme fractionation. Thus the Tregonning cupola in Cornwall shows an upward transition from biotite granite through lithium mica granite to a lithium mica leucogranite and aplite – the country rock cap is intruded with lithium mica pegmatites – a progression marked by an increasing tin concentration. However, as Stone showed, a decided complication is introduced into this neat story by the finding that the mineral lodes were formed about 25 Ma later. Stone considered that this can only be explained by an enduring meteoric convection system, genetically related and energized by continuing heat flow from the granite, which redistributed tin previously concentrated in both the apical part of the pluton and the roof rocks into the lode fractures. Such recycling is probably common, especially in high heat production granites, as suggested by Fehn. Indeed, in Cornwall, the long-lasting nature of such hydrothermal processes provides the only logical explanation for the formation of hydrothermal deposits of kaolin, which are most distinctly related to the erosional peneplains resulting from the uplift and unroofing of the plutons.

Leaving aside such complications, most of these matters are dealt with in detail in a comprehensive review by Lehmann, who concluded that in tin granites world-wide crustal material forms the dominant component. What is particularly significant is the continental crustal location of many tin granite provinces. Thus the three dominant tectonic environments of such granites are: post-orogenic magmatism in continent collision belts, as in the case of the Hercynian tin province of Europe; intracontinental rift zones, as in the example of the Cretaceous tin granites of Nigeria; and the active fault zones immediately inside the continent margin arcs, as in southeast China. I fully agree with Lehmann in his suggestion that it is the thick crust in all these environments that allows sufficient process space and arrival time for a high degree of melting and extended fractionation, and that magma intrusion and ponding at upper crustal levels is favoured by brittle fracture in extensional zones.

The origin of tin is surely a good example of the interrelationship of source, tectonic history, magma type and magmatic processes, none of which operates to the exclusion of the others. However, as Lehmann wrote, 'The geotectonic stability of host environments seems to have generally a first-order control on the time distribution of ore deposits' (1990, p.12). This is surely a statement applicable to magmatism in general.

SELECTED REFERENCES

Jahns, R.H. and Burnham, C.W. (1969) Experimental studies of pegmatite genesis: 1. A model for the derivation and crystallization of granitic pegmatites. *Economic Geology*, **64**, 843–864.

Lehmann, B. (1990) *Metallogeny of Tin. Lecture Notes in Earth Sciences No. 32.* Springer-Verlag, Berlin.

Stein, H.J. and Hannah, J.L. (1990) (eds) *Ore-bearing Granite Systems; Petrogenesis and Mineralizing Processes. Geological Society of America Special Paper No. 246.*

The sources of granitic magmas in their various global tectonic niches

18

It is of course difficult and in many cases it will always be impossible to tell exactly to what extent truly juvenile ichors, derived by squeezing out or by some other way of differentiation from the sima, have been added to the palingenetic magma in any particular case. An igneous complex including all members of the differentiation series, down to peridotites, should in all probability contain much juvenile material, while granites connected with only small amounts of more basic differentiates could be presumed to be largely palingenetic.

Pentti Eskola (1932) Zeitschrift für Kristallographie,
Abt. B, *42*, 475.

INTRODUCTION

One important theme of the previous chapters has been that granitic rocks image their sources, the nature of which differs in response to the global tectonic environment. This theme is now expanded using chemical characteristics, set in the frame of geological environment, to explore the range of sources. In what is now almost a conventional statement these environments are: mid-ocean ridge, oceanic island arc, continental margin arc, arc-interior thrust or megashear belt, intercontinental collision zone and continental rift zone – an order involving an ever-increasing complexity of source rock availability.

A COMMENT ON THE USE OF RADIOGENIC ISOTOPES

Together with the major and trace elements the radiogenic isotopes form an essential tool in the identification of the source rocks of granites. The use of the isotopes addresses in particular the nature and composition of the source, the extent of mantle involvement, the age of the continental crust involved, and discriminates between geotectonic settings. A prerequisite of such studies is that the rock samples should be of the same age, magmatically consanguineous and have behaved as a closed system since formation. One may wonder how many studies have fulfilled these rigorous conditions!

As is well known it is the presumed lack of isotopic fractionation during magmatic processes that offers a window through to the original composition, and it is the comparative plot of $^{143}Nd/^{144}Nd$ versus $^{87}Sr/^{86}Sr$ that has proved most significant in source identification (Figure 18.1), with the added potential of providing the Sr/Nd ratio of a crustal component. There are, as might be expected, certain pitfalls inherent in these techniques, and Vidal provided, in 1987, a useful review of their use and misuse, inserting a note of caution into the interpretation of such data,

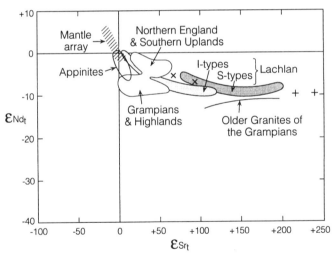

Figure 18.1 ε_{Nd}–ε_{Sr} diagram as devised by DePaolo and Wasserburg (1979), with $^{143}Nd/^{144}Nd$ and $^{87}Sr/^{86}Sr$ normalized in relation to chondritic initial ratios at the present. The so-called 'bulk earth' reference point is where $\varepsilon_{Nd} = 0$, $\varepsilon_{Sr} = 0$. Granitic rocks produced by simple mixing of mantle-derived and crustal materials will be defined by a downward curved array on this diagram because the Sm/Nd ratio in sediments is less than that in most mantle-derived rocks. Some compositional fields of granitic rocks described in the text are depicted. (×) Sediments of the Southern Uplands of Scotland; (+) sediments of the Moinian of the northern Highlands of Scotland. Data from Harmon *et al.* (1984).

even that generated by the now popular use of Sm/Nd systematics. We must always bear in mind that isotopic signatures may be third- or fourth-hand in origin. Nevertheless, the simple geologist can only marvel at the manipulation of the Sm/Nd isotope ratios, originally by DePaolo and Wasserburg, to provide model ages which are estimates of the original mantle extraction, and at the finding that the shale Sm/Nd ratio constrains the granitic sample evolution line because shales exhibit remarkably constant Sm/Nd ratios through space and time. In quoting this kind of data in much of what follows I am in the hands of the experts!

THE PLAGIOGRANITES OF THE MID-OCEAN RIDGES

There is no other viable possibility than that the plagiogranites are the differentiates of mid-oceanic ridge basalts, albeit altered by extensive leaching. We scarcely need the support of the primitive geochemical signature to prove that the source of plagiogranites lies ultimately in the mantle, though their evolution involves first an extraction of basaltic material and then a subsequent fractionation – a two-stage process that forms the basis of most current models of the genesis of granitic rocks.

THE M-TYPE QUARTZ DIORITES OF THE OCEANIC ISLAND ARCS

Quartz diorites and tonalites occur in oceanic island-arc magmatic sequences, though in small volume relative to their mafic volcanic and gabbroic associates. As an example, in the Aleutian arc, where gabbros form the peripheries of zoned, quartz diorite plutons, Perfit and co-workers (1979, 1980) reported that the geochemistry of these augite-bearing plagioclase-rich quartz diorites, with their low Sr_i (0.70299–0.70377) and low $\delta^{18}O$ values (≤ 6.0), was entirely comparable with those of the associated volcanic rocks. It is thus reasonable to conclude that these rocks were derived by fractional crystallization of the same high aluminium basaltic parental material which provided the comagmatic lavas. For such a geochemically primitive parent a mantle source is most likely, even though subducted oceanic crust may have formed its intermediary residence. Of great interest is the finding that even in this situation the initial ratios vary within a single zoned pluton.

Similar associations of primitive quartz diorites and tonalites are found wherever there is sufficiently deep erosion in ocean arcs, as in the Kuril Islands, the Caribbean, the Philippines and the Solomon Islands, in many cases associated with gold-bearing porphyry copper deposits. As suggested in an introductory chapter, such granitoids of oceanic arcs might usefully be distinguished as an M(mantle)-type from the I-type of continental margins, even though the modal and geochemical parameters overlap; the M-types are, however, uniformly low in silica and comparatively impoverished in the large ion lithophile elements. This apparent transition is well demonstrated by Mason and McDonald's comparison of the clearly identifiable I-type plutonic rocks of the New Guinea mobile belt, with the geochemically more primitive assemblage of the adjacent New Guinea–New Britain arc.

Genetically these quartz diorites and tonalite rocks of the oceanic arcs are important in showing that intermediate granitic rocks can be generated without the intervention of continental crust. Furthermore, there is no need to stress the importance of subduction in this and other arc-type environments. As Fyfe in 1988 asserted, most workers agree that it is the injection of volatiles derived from the dehydrating slab into the presumed mantle wedge that initiates the magmatic cycle. Such volatile material, perhaps tenuous magma, enriches the wedge in silica and some of the incompatible elements, including potassium, rubidium, strontium, thorium and uranium. This is the stuff of textbooks, but my personal difficulty is in recognizing compositional characteristics that unequivocally identify contributions from the ocean crust, enriched mantle and mantle extracts in general.

TONALITES AND GRANODIORITES OF THE CONTINENT MARGIN ARCS

As we have seen, the special relationship between the type of granite and its tectonic environment is well illustrated in the geological setting of the Mesozoic batholiths of the circum-Pacific regions. This is because they provide a young, and therefore near actualistic basis for the discussion of their characteristics, and also because they have an especially distinctive compositional and source signature. Furthermore, the temporal sequence, namely, crustal extension, basin subsidence, volcanicity and basin fill, burial metamorphism, intrusion of gabbros and basalt dykes, basin inversion and finally batholith intrusion, clearly reflects a major cycle of crustal growth. These are reasons enough for returning to discuss further Cordilleran batholiths. Maybe they hold the clue to crustal growth in general.

THE ANDES

Without doubt the great linear batholiths of the Western Cordillera of the Central Andes represent a stage in the growth of the South American continent. Although the long-standing subduction of the oceanic crust of the Pacific Nazca plate is thought to have provided the energetics of the episodic production of the vast volumes of magma, the essential tectonic control was an extension of the continental lip, which led to the thinning of the continental crust above an up-arch of the mantle. I have supposed above that it was the adiabatic partial melting of the mantle arch that produced the basaltic underplate that, in its turn, formed the source of the diorite–granite assemblage.

Specifically, the most important genetic features of the Peruvian example, represented by the Coastal Batholith, are first the preponderance of tonalite and granodiorite, with granite in the strict sense occupying less than 10% of the outcrop, and second the lack of continuity of composition between gabbro and tonalite. Direct fractional crystallization from basalt magma is ruled out by the temporal and compositional hiatus coupled with a lack of any outcrop evidence of cumulate rocks. Of course, it is possible to argue to the contrary, but it seems that, once again, the partial melting of basaltic rocks seems to provide the best-fit model, the residue being left in the upper mantle. Long residence in the thick, new crust ensured sufficient time for the fractionation and evolution of these remelts to form compositionally expanded granite suites. The conclusion of many years of work on the Coastal Batholith of Peru is that it does not represent the root of the precursor arc, but is a consequent arc, and I firmly believe that this is true of Cordilleran-type batholiths in general.

The great length of the Coastal Batholith allows isotopic examination in axial segments either with or lacking an ancient crustal substrate. In the latter case the rocks have characteristic signatures suggestive of derivation from an isotopically homogeneous reservoir that had never been in contact with the ancient continental crust, and with a composition akin to enriched subcontinental mantle. Only where the old crust is in obvious contact with the batholithic intrusions does this isotopic evidence indicate the crustal contamination of the originally mantle-derived magmas, as is very well demonstrated by the findings of Mukasa and Tilton (Figure 18.2).

This and many other studies of Cordilleran batholiths conclusively prove, yet again, that an expanded spectrum of granitic rocks can derive from modified mantle extracts without any contribution from older recycled continental crust. Nevertheless, crustal contamination is always possible and often important. Always it seems that such magmas arise by a two-stage process of extraction from the mantle.

BAJA CALIFORNIA

We have already seen that the Peninsular Ranges Batholith of California and Baja California bears many similarities to the Coastal Batholith of Peru, especially in its relatively simple history of emplacement, at least on

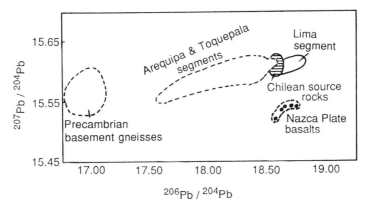

Figure 18.2 Lead isotope composition fields of the Arequipa and Lima Segments of the Coastal Batholith of Peru compared with those of the gneisses of the envelope of the Arequipa intrusives. Also shown are values from Nazca plate basalts and the isotopically homogeneous source of Chilean volcanics as determined by Barreiro and Stern (1982). After Mukasa and Tilton (1985).

its western flank, as a single, continuous intrusive cycle without the overprint on to older plutonic arcs such as occurs in the Sierra Nevada. Gromet and Silver identified some remarkable regularities of longitudinal compositional trends, but they also reported persistent regional asymmetries in geochronological, petrological, geochemical and isotopic characteristics that are only a subdued feature of the Peruvian Batholith.

These asymmetries have transverse compositional gradients independent of the individual intrusions, so that they must be imaging the underlying source rocks. The sense of the consistent REE trends cannot be explained by any combination of crystal fractionation, assimilation or mixing processes, but only by assuming that a difference already existed within the mafic source rocks – for example, in the abundances of the light REEs and the relationship between Sr_i and $\delta^{18}O$. This original heterogeneity can be best explained on the basis of a compositionally layered mantle in which plagioclase-bearing mineral assemblages abruptly give way, both laterally and downwards, to garnet-bearing assemblages, presumably in response to increasing pressure, a model wholly consistent with the experimental results of Wiley and his associates reported earlier (p. 38).

Thus we begin to see the outlines of an overall model in which the felsic character of the crust is continuously reinforced and preserved by extracts from the mantle, with the mafic residues sinking back into the latter. It is a model that directly addresses the central problem of the creation of continental crust.

SIERRA NEVADA

Isotopic studies of the Sierra Nevada Batholith carry the evolutionary story a stage further. In the western USA it has long been known that there was an easterly change in the bulk composition of the granitic rocks, and in 1959 Moore drew a quartz diorite boundary line crudely dividing a westerly belt of gabbros, diorites and tonalites from an easterly zone of granodiorites and granites, a change marked by the increasing potassium content of the intrusive rocks (Figure 18.3). Then, as early as 1967, Doe and his associates began to use lead isotopes to explore the root causes of this progression. Later, in 1973, Kistler and Peterman not only confirmed this division on the basis that the Sr_i values increased from 0.704 in the west to 0.708 in the east, largely independently of the relative age of the constituent plutons, but they also interpreted these values as a reflection of the composition of the source rocks. From this they hypothesized that, at depth, an oceanic lower crust gave way eastwards to one formed from the Precambrian basement.

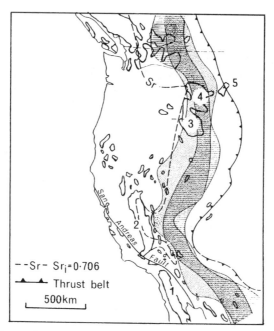

Figure 18.3 Mesozoic granitic intrusions of western North America (open outcrops) showing the relation of the belt of muscovite-bearing, peraluminous plutons (stipple) to the Phanerozoic metamorphic belt (lined), to the Cordilleran thrust belt and to the initial $^{87}Sr/^{86}Sr$ 0.706 isopleth (Sr). Batholiths: (1) Peninsular Ranges; (2) Sierra Nevada; (3) Atlanta lobe of the Idaho Batholith; (4) Bitterroot lobe of the Idaho Batholith; (5) Boulder. Adapted from Miller and Bradfish (1980).

In the early 1980s DePaolo reinforced the distinction, finding that the ε_{Nd} values showed a negative correlation with Sr_i. This he explained on the basis that the ambient crust represented various mixtures of juvenile material added during the Mesozoic – we have already seen an example of this process in action – and other material derived from the ancient craton, probably in the form of sediments, the proportion of which increased to the east. More recently Chen and Tilton have refined these source rock identifications by a sophisticated use of the lead isotopes. They found that along an easterly traverse the $^{206}Pb/^{204}Pb$ ratio first increased and then decreased, in contrast with the $^{208}Pb/^{204}Pb$ ratio, which also increased but then remained relatively constant. On the basis of the known isotopic compositions of likely crustal rocks these results are again interpreted as pointing to a derivation from a mantle-related source in the west, with the progressive eastward involvement of sediments derived from an ancient basement. The refinement comes in the interpretation of the latter as constructed of granulites and amphibolites, evidence for which exists in outcrop.

The distinction of these two different lithospheric types in the region of the Sierra Nevada Batholith can be sufficiently precise to locate a masked boundary line. Thus in 1990 Kistler used oxygen and strontium isotope data to identify the latter as representing a shear zone which is actually exposed, albeit very discontinuously. He interpreted high $\delta^{18}O$ values as indicating a significant sedimentary component in the sources of the magmas to the west of the junction, and lower values a predominantly igneous source in those lying to the east. Significantly, this boundary lies close to both the early established Sr_i 0.706 boundary and the zone of highest $^{206}Pb/^{204}Pb$ ratios. Clearly an effective technique is being refined for determining not only the composition of the sources of granitic magmas, but also their location and, with limitations, an estimation of their separate contributions to any particular magma.

Particularly interesting is the further possibility of determining the age of these source rocks; for example, Chen and Tilton used lead isotope plots to calculate an age of 1.8 Ga for the continental basement of California, an age which not only lies within the range determined from actual outcrops, but also corresponds to a regionally dominant age of 1.78–1.80 Ga determined by Bennett and DePaolo.

With all this information it becomes possible to model this particular continental edge in terms of a fore-arc, greywacke-flysch melange, with contemporary ultrabasic and basic intrusions, lying oceanwards of a continental marginal arc with its pile of volcaniclastic rocks and lava flows, itself flanking a shelf-type sequence of clastics forming an apron to the edge of the craton. Derived M- to I-type tonalite, granodiorite and granite plutons image the composition of the appropriate crustal zone, and some straddle the boundaries between them. Of course, the reality is that the arc may as well be internal or external to a continental margin, with all the separate crustal elements representing foreign terranes displaced on margin-parallel mega-shears – the very mobility that, in association with subduction, motivates magmatism!

THE GRANODIORITES AND GRANITES OF THE ACTIVE HINTERLANDS OF MARGINAL ARCS

Inside the craton edge with its volcano-plutonic arcs there is a wide zone of crustal deformation involving faulting, thrusting and differential uplift. This is associated with an intraplate magmatism which I do not think is directly related to a subduction plane, but is simply the result of compression causing block uplift and depression, and this so disturbs the thermal equilibrium that melting ensues. I examine this type of granitic magmatism by special reference to the case histories of the Idaho Batholith, the granites of southeastern Australia and the Newer Granites of the Scottish Caledonides.

IDAHO

Eastward into the Cordilleran interior and into the presumed ancient continental edge of North America, and largely east of or straddling the 0.706 Sr_i isopleth (Figure 18.3), there is a well-defined belt of two-mica granites and granodiorites extending from Sonora to southeast British Columbia. Although this plutonism was broadly synchronous with that of the Cordilleran Batholiths, it is distinctive in not being obviously motivated by continental collision or extensional tectonism, but by strike-slip fault movements, the magmas filling pull-aparts in shear zones. Furthermore, this plutonism extends over a belt hundreds of kilometres wide within strongly thrusted late Palaeozoic metamorphic rocks. Mafic intrusions are much less evident than in the Cordilleras, and both the dominant intermediate metaluminous granites and the less abundant felsic peraluminous granites show clear isotopic evidence of a major crustal input.

Miller and Barton reviewed the evidence for both magma generation and its motivation, but before commenting on these general matters I turn to Fleck's geochemical studies of part of the Idaho Batholith as an in-depth example of the type of analysis now available using a combination of neodymium and strontium isotope and trace element data. It is of the greatest interest that Fleck's traverse crosses what is now recognized as a terrane boundary separating the late Palaeozoic–Mesozoic accreted terranes from the Precambrian 'sialic' crust, which is also the location of the 'magic' 0.706 Sr_i isopleth. Fleck recorded a dramatic increase in Sr_i from 0.704 to 0.712 in less than 10 km, and this is paralleled by a leap in $\delta^{18}O$ from between 6 and 8‰ to between 9 and 12‰. There is also a clear inverse correlation between ε_{Nd} and Sr_i, the former falling from +5 to −16.

The most convincing explanation is again that this represents a mixing of magmas derived from contrasted sources. Indeed, these very constants can be recognized as typical of each of the two country rock packages of western Idaho, namely, the Proterozoic orthogneisses, pelitic schists and migmatites of an eastern block, and the late Palaeozoic–Mesozoic sediments and volcanics of a western block. But the explanation is not quite so simple, as is indicated by the synplutonic intervention of basaltic magma which can be seen in outcrop to have mingled and mixed with its felsic host, as recorded by Foster and Hyndman. Furthermore, there is little evidence of the assimilation of country rock at the presently exposed high level in the crust.

By comparing all the relevant ratios within the various plutons with those of possible country rock candidates, a simple contamination–bulk assimilation model is precluded. Neither is a simple two-member mixing adequate to explain the observed variations. Fleck concluded that the mixing involved melts derived from a number of crustal sources, but with a contribution from subcontinental lithosphere, which is surely represented by the mixing and mingling of mafic magma as recorded in the outcrop. Even more intriguingly it seems that not only are several sources required, but these need to have been partially melted and therefore already differentiated from their particular source before intermixing. These are entertaining complications indeed, but I do not find them surprising considering the heterogeneity of any observable old crustal material.

Such research joins with the mounting evidence of a multiple source, open system type of magmatic process as powerfully advocated by Arculus and Powell, and Hildreth and Moorbath under the acronym MASH, with the meaning of a combination of melting, assimilation, storage and homogenization. So it appears, after all, that granite-producing processes can be 'all things to all men'!

Returning to the regional problems of the western USA, Miller and Barton are most reluctant to dispense with the model of subduction and underplating, even when the distances involved are so great that the subduction zone would have the shallowness of a thin-skinned thrust! Also there is the fact that a proportion of the granitic rocks of the Cordilleran interior are strongly peraluminous and show clear crustal signatures, so that they can reasonably be labelled as 'S'-types. More significantly, they are most clearly associated with a continental compressional tectonism – that is, a continental 'subduction' initiated in the Late Palaeozoic and resurgent in the mid-Mesozoic. I believe that Miller and Barton have too easily rejected the alternative thesis of thermal blanketing, and have too readily relegated it to a mere facilitating mechanism. But they have a point in questioning any simple statement of the magma-tectonic concept.

THE LACHLAN EXAMPLE OF THE WAY IN WHICH GRANITES IMAGE THEIR SOURCE

In the Tasman Mobile Belt of eastern Australia a broad belt of largely Palaeozoic sedimentary rocks fringes the ancient Australian plate. Granites are especially abundant and in the Lachlan region multiple batholiths of Silurian and Early Devonian age intrude a Lower Palaeozoic slate belt. The tectonic regime is a deceptively simple one of broad open folding with discrete belts of axially cleaved, tight isoclines, and, perhaps surprisingly, there is no consensual model for the origin of the Lachlan rocks, nor satisfactory explanations for either the exceptional width of this belt (800 km), or the cause of the major episode of crustal melting represented by the granites. Nothing seems to fit the conventional picture of a destructive plate margin, certainly not the composition of the granites; neither is there any simple age polarity in the intrusives or the fold zones. Nevertheless, Collins and Vernon, in 1992, provided a model involving the closing of an early formed back-arc basin by a later, west-dipping, doubled system of subduction ('delamination' according to Collins in 1994), the granites being but part of an on-going, mid-crustal, thermal perturbation associated with the latter. On the other hand Chappell, in 1994, favoured a simpler thesis involving crustal extension leading to crustal thinning, basin formation and up-arching of the mantle, with a consequently enhanced heat flow. The evidence of a static, and largely imposed, contact metamorphism, seems to me to favour the emplacement of the majority of the shallow-seated Lachlan granites quite late in the structural history. I certainly find it difficult to connect the Lachlan granite-producing event with a subduction regime, and I would suggest instead that many of its features are reminiscent of the late, uplift-related, Caledonian granite event of the British Isles.

We have already seen how Chappell, White and their co-workers recognized distinct S- and I-type granites in Lachlan, not only on the basis of modal criteria but also on a rich assemblage of geochemical data. They established that there is a remarkably precise regional distribution on either side of a north–south line, into a western province with mainly, but not exclusively, peraluminous S-type, and an eastern province with wholly metaluminous I-type granites. The strong modal and chemical differences between these types are reflected by their respective isotopic compositions: thus the I-type and S-type granites have oxygen isotope values less and greater than 10‰, respectively, and although the former range in Sr_i from 0.704 to 0.712, and in ε_{Nd} from +3.5 to −8.9, the latter show corresponding values of 0.708 to 0.720 and −5.8 to −9.2.

Irrespective of whether the I-type granites are produced more directly by fractional crystallization of mantle-derived liquids or less directly by partial melting of mantle-derived solids, their source materials must have been broadly basaltic to andesitic in composition. In contrast, the composition of the S-type granites demands a substantial metasedimentary component, though this might have arisen by either the mixing of mantle-derived and sedimentary partial melts, as argued by Gray, or the partial melting of metasediments alone, as proposed by Chappell and White. We recall that the latter workers attribute the variations as being due to the intensity of the weathering of the source materials as recorded by the efficacy of the removal of calcium and sodium, basing their case on the strongly peraluminous nature of these Lachlan S-type granites, the lack in them of metaluminous enclaves, and the fact that the variations can be simply modelled without involving a mantle component.

Whatever the complexities and actualities of the generative processes, the Lachlan granites clearly do image their source rocks. But as the latter are not represented in time or composition within the present outcrops, a more ancient basement has to be entertained. To the west of the I–S line the deep source is envisaged as forming the margin of the continental crust, whereas to the east the source is likely to have been a mafic underplate, the line itself representing the edge of the ancient plate – possibly a terrane boundary. Following up this concept in a little more detail, we note that McCulloch and Chappell concluded that one particular I-type suite was produced from material of predominantly mantle derivation that had aged – by about 700 Ma according to Compston and Chappell – before the generation of the granites at 410 Ma. The studies of Williams and his colleagues on the inherited zircons revealed that the mafic protolith may have been assembled in two periods, one 1075–800 Ma and the other 650–450 Ma ago. Furthermore, the volcanism related to these episodes may well have yielded detritus to the early Palaeozoic flysch that then became the source of the S-type granites. What a marvellous complexity is revealed by the true study of natural things!

THE CALEDONIAN EXAMPLE

The compositional and isotopic systematics of the Caledonian intrusions of Scotland and northern England have been studied in detail by Harmon, Halliday, Clayburn and Stephens, and form the subject of a comprehensive review by Peter Brown (1991). The data indicate that an early group of granites, emplaced during the later phases of the Caledonian orogeny proper (about 470 Ma), were the products of various anatectic meltings of Late Proterozoic metasediments within the upper crust, an interpretation wholly in accord with a geological history not dissimilar to that of an interior Cordilleran setting, though the complementary arc was long ago displaced.

A later group of granites (440–390 Ma), which are post-plate closure, post-peneplanation and clearly fault-uplift related, has a more complex, multiple source, variously involving three different kinds of crust, together with a substantial contribution from the upper mantle, though whether at second- or third-hand is often difficult to decide. If by way of contemporaneous mafic magmas then these could have supplied some of the heat required for local crustal melting and assimilation.

The individual plutons are variously multipulse, with their several magma infills drawn from deeper-seated magma reservoirs. Fractional crystallization was a central process in the evolution of the magmas, but it is often modelled as following the incomplete mixing of partial melts derived from isotopically distinct sources. Furthermore, Nd–Sr studies of the microdiorite enclaves in just two plutons suggest a further diversification by basaltic recharges of the reservoirs.

To discuss even briefly the multiple-source hypothesis of this later group we need to understand something of the complexity of the crust of the northern British Isles, the different parts of which arguably represent at least five displaced terranes (Figure 18.4). In general terms the crust of the most northerly, the Northern Highlands Terrane, is constructed of an Archaean understorey of mafic to intermediate rocks of granulite and amphibolite facies, with a cover of psammitic metasediments of Proterozoic age. In the Grampian Terrane to the south of a Mid-Grampian Line, the lower crust seems to become more silicic and younger southeastwards, and in analogy to exposures in southwest Scotland this marks a change in the composition of the basement to a complex of alkalic igneous rocks of Lower Proterozoic age. The Late Proterozoic metasedimentary cover also becomes more lithologically varied. South of a terrane boundary now marked by the Highland Border Fault lies a belt of weakly deformed sediments of Lower Palaeozoic age, representing the paratectonic Caledonides, and farther south again, south of yet another terrane boundary – the Solway Line: the presumed Iapetus Suture – this lacks altogether the very ancient basement, which is replaced by the debris of a Late Proterozoic arc.

The hundred or so plutons widely scattered over this region of varying crustal composition may well be expected to image their appropriate source rocks, but as far as the major elements are concerned such a variation is not immediately apparent. There are some regionally localized suites but, even so, there is a certain geochemical homogeneity in that the granodiorites and granites are mostly relatively high potassium, calc-alkali, I-types marked by a relative enrichment in strontium and barium. There is no marked transverse compositional change in the major elements and the weak variation in some minor constituents would seem to me to correspond to a tripartite division into regional suites. There is certainly no compelling evidence for a contemporary subduction zone or

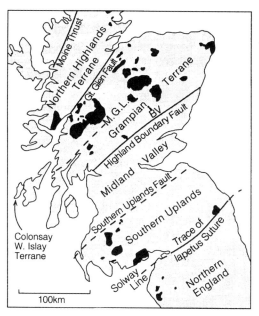

Figure 18.4 Distribution of the Late Caledonian granitic plutons showing the relation to possible terrane boundaries. The Mid-Grampian Line (MGL) marks a change in the character of the source within the Grampian Terrane. (BV) Ben Vuirich Granite.

zones. In the main these younger granites do not represent the formation of new crust, but merely the recycling of existing material.

Since 1978, when Pidgeon and Aftalion first reported the discovery of inherited zircons in those granites lying north of the Highland Boundary Fault, there has been an intensive probing of the isotopic signatures of the granites and these certainly image their appropriate source crust, but not in a simple way. Halliday, Stephens and Harmon found, for example, that in individual cases the isotopic composition suggests involvement of at least four different source regions. Always there is a contribution assigned to extracts from the upper mantle. To the north of the Highland Boundary there is good evidence of the involvement of upper crustal materials combined with contributions from a lower crust of increasing age northwestwards across the Grampian and Northern Highlands Terrane – a crust of two main types, seemingly separated along a Mid-Grampian Line (Figure 18.4) across which there is a marked northward drop in εNd values. At depth the terrane to the north almost certainly consists of ancient mafic to intermediate granulites, but that to the south possibly involves younger and more silicic rocks: findings wholly in accord with the geological expectation that there are two fundamentally different

'Lewisian' basements. Indeed, the application by Pidgeon and Compston of the advanced SHRIMP ion microprobe – which enables ages to be assigned to specific zones in inherited zircons – has confirmed the presence of Lower Proterozoic material in the Ben Vuirich granite, which outcrops in the southern Grampians – an application of wonderful potential, as we have already seen.

To the south of the Highland Boundary, in the Midland Valley and Southern Uplands, where the zircons within the granites do not carry any memory of an ancient basement, an upper mantle-derived source becomes more evident in the isotopic composition of these granites, but here mixed with sedimentary derivatives from the cover rocks. Still further south beyond the Solway Line the sedimentary input – presumably derived from the Upper Proterozoic island arc – becomes predominant.

We should not expect source identifications to be more precise than this because the deep crust is almost certainly a structural mélange. Thus, for example as depicted in Figure 18.5, seismic images of the continental lithosphere of Scotland show it to consist of large-scale, anastomosing, ductile shear zones, and we easily envisage that the components of the deep crust were structurally intermixed before any melting occurred.

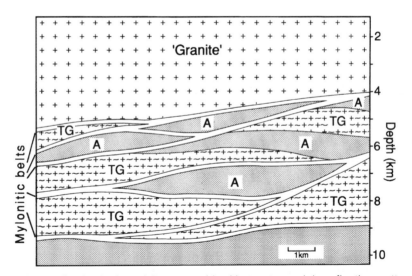

Figure 18.5 Geological model prepared by Norton to explain reflection patterns on a seismic section across the Caithness Basin, northeast Scotland. Reproduced in a modified form from Blundell (1990) with permission. Whatever the reality this model section illustrates well the tectonic complexity of the middle crust of northern Scotland. (TG) Tonalitic gneiss; (A) amphibolite.

Indeed, I think it possible that this tectonic mixing was extended to include the upper mantle. Perhaps the mixing of source material is as much a matter of tectonic displacement as is the mixing of the partial melts themselves!

Another important lesson to be learnt from this example is that the granites can take on a regional identity, despite a complexity of source, suggesting that it is the overall lithospheric signature that is indelible. Furthermore, despite the sedimentary source contributions, the younger granites, with the possible exception of the central pulse of the Loch Doon pluton and the Cairnsmore of Cairsphairn pluton, are not obviously peraluminous or immediately recognized as S-types. This is presumably duc to the sources themselves being so often of igneous derivation. Either they did not pass through a sedimentary cycle, or if they did, such sources represent the immature debris of an island arc. However, that direct mantle extracts were also involved in this particular example of granite genesis is demonstrated by the mafic enclaves mingled with some of the magmas, by the synchronous intrusion of mafic lamprophyres and appinites, and by the basaltic character of the associated late stage volcanicity.

A MULTIPLICITY OF SETTINGS AND SOURCES IN A RESURGENT OROGEN

The Caledonian example, coupled with the generalities of the MASH hypothesis, has surely prepared the reader for the real complications involved in natural processes, but I now emphasize this fact of complexity by reference to the Northern Appalachians. Here the Acadian Orogenic Belt, as presently displayed, is constructed of a number of displaced terranes that bring together disparate tectono-magmatic assemblages. Thus Sandra Barr's holistic study of this region identifies an Aspy Terrane with a protracted Silurian and Devonian plutonic history, during which a range of granites was generated sequentially in volcanic arc, post-arc, collisional and extensional tectonic settings. In contrast, the flanking Bras d'Or and Mira terranes are characterized by two Late Precambrian, expanded calc-alkaline I-type granitic suites developed in separate volcanic arc settings.

Thus, in this part of the Acadian Orogenic Belt the plutons were certainly not the result of a single tectonic event or even share a common source. They and their host terranes were amalgamated in the Late Devonian by major strike-slip displacements, and only the most detailed field based structural analysis coupled with petrological studies, and a precise geochronology, can prepare such a region for geochemical investigations; surely a cautionary tale of a 'hundred plutons'.

EXPANDED PLUTONIC SEQUENCES IN COLLISION ZONES

We may expect an even greater variety of crustal sources where continental collision brings varied cratons into juxtaposition. The Hercynian of Europe is a case in point, the granites showing a considerable range of compositional types and source signatures. In particular, the peraluminous, two-mica granites form a central facies and we can often identify them as forming the finale of the granite series of Read. However, in this penultimate chapter I cannot resist commenting on a rather different example, that of the Neogene of the High Himalayas, which offers a most remarkable insight into collision tectonics and its associated magmatism.

THE LEUCOGRANITES OF THE HIGH HIMALAYAS

The great belt of leucogranites that threads through the High Himalayas, parallel to the Indus Suture, is particularly well exposed in three dimensions. According to France-Lanord and Le Fort the crust of the Eurasian plate (the Tibetan slab) was thrust over the crust of the Indian plate, thus doubling the thickness of the crust (Figure 18.6). It seems that the underlying slab was heated by the hot base of the Eurasian plate and the resulting dehydration released fluids into the overlying crystalline nappe, thereby inducing partial melting. The new magmas were assembled as great lenticular sills of muscovite, biotite and tourmaline-bearing leucogranites, in which local modal variations attest to multipulse intrusion. Despite this physical inhomogeneity there is a considerable overall compositional homogeneity compatible with the new magmas having a

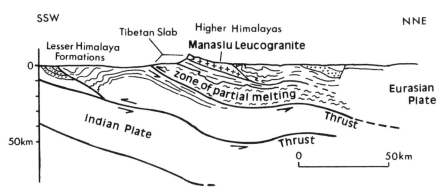

Figure 18.6 Schematic cross-section giving the present structure of the Himalayas in Central Nepal according to France-Lanord and Le Fort (1988).

minimum melt composition. The emplacement of the leucogranites was evidently controlled, in both time and place, by the regional thrusting and, appropriately, the intrusives exhibit a distinct mineral layering and a superimposed foliation.

There is thus good spatial and temporal evidence that the paragneisses below the sheets and lenses were the probable source of these leucocratic magmas, which is confirmed by this specific source being identified by the combined radiogenic and stable isotope studies reported by France-Lanord and Le Fort. For several plutons the principal isotopic characteristics are broadly similar, though marked with local heterogeneities with ranges of the order: $\delta^{18}O$ between 9 and 14‰; Sr_i between 0.730 and 0.825; ε_{Nd} between -11.5 and -17‰; $^{206}Pb/^{204}Pb$ between 18.70 and 19.91; and $^{207}Pb/^{204}Pb$ between 15.78 and 15.90; values typical of magmas derived from old crustal material involving a high proportion of recycled sediments. In the Manaslu area the gneisses of the Tibetan slab are the only formation isotopically compatible with being the source of the Manaslu leucogranite. Deniel and co-workers, and also France-Lanord and Le Fort made detailed comparisons which point to the important conclusion that the preservation of the $\delta^{18}O$ isotope variations from the protolith implies that a convecting aqueous fluid was not involved in the melting process, but nevertheless it seems to me that other volatiles were present in profusion, as shown by the abundance of tourmaline. These workers envisaged that the pluton resulted from the coalescence of numerous aliquots of magma without appreciable intermixing – a revolutionary model indeed!

This is a very remarkable example and its acceptance confounds my reluctance to countenance a sufficient extraction of melt from migmatites to provide plutonic volumes of magma. I cannot help supposing that the process of assembly is somehow more complicated than extraction from the *lits* of metamorphic migmatites and perhaps involved the partial melting of already segregated material, but, nevertheless, here is surely a realistic example of the production of crustal granite in huge volume. Furthermore, it would seem to provide a model particularly relevant to the origin of sheeted, multipulse plutons created in thrust zones, where the repeated squeezing out of incremental aliquots provides an attractive mechanism for the extraction process.

INTRA-PLATE GRANITES OF THE CONTINENT AND OCEANS

I have already commented in some detail on the possible origins of the A-type alkalic granites in their several tectonic environments, concluding that they result from a pressure release caused by extension and fracturing of thickened crust. This triggers remelting, in deep zones, of either dehydrated and depleted lower crust or the upper mantle.

Bonin (1990) believed that the change in composition characteristic of alkalic magmas appearing soon or long after orogenesis is due to the source changing in character during the first 100 Ma after an orogenic event, that is, from exhausted mixed crust and mantle material representing the disintegrating residuum of the dying orogeny, to the new mantle-derived underplate that replaced it. According to Windley this switchover implies that there is no fundamental break between orogenesis and anorogenesis or their respective magmatisms. This view is supported by the relationship of intracratonic felsitic magmatism with extensional tectonics.

Referring again to the example of the intraplate magmatism of the Australian Proterozoic described by Wyborn, this has essentially the same pattern as the ensialic extension causing Andean-type plate edge magmatism. In both instances the felsic magmas are modelled as deriving from an earlier formed mafic underplate melted out of up-arched mantle, and the reason that the intraplate magmatism is so much more obviously bimodal, with granites predominating, is possibly because of the limited degree of partial melting in the case where up-arching is subdued and the crust only fractionally thinned. In such intraplate magmatism, what was partially melted is not easily discerned when both the lower crust and the mafic underplate may represent simple extracts from the mantle, though here the determination of the neodymium model age, as reported in the Australian example, may provide a sufficient discriminant.

I do not really understand why, in these anorogenic environments, some magmas are enriched in the halogens and others not, though an explanation is attempted below. When they are, the exact identification of the source is bedevilled by the associated effects of potent volatile-rich conductive systems which undoubtedly modify the original isotopic signatures of the source. What is more, as Bonin suggested, any crustal contributions may well have been introduced by volatile interaction and need not therefore be the result of the anatexis of crustal rocks, a complication indeed!

Be that as it may, we have seen how all the relevant isotopic studies require that there should be a mantle input, albeit via an intermediary basaltic extract. That this can provide the bulk source in the case of some anorogenic granites is demonstrated by the generation of alkalic silicic magmas in the oceanic islands. Otherwise a wide variety of diverse crustal material has been identified as an additional source, the only proviso being that, in the case of the alkalic magmas, the sources had already suffered a loss of water and other components during an earlier remelting at a lower temperature. Of course, the bulk of these deep sources would have inherited a mantle signature by reason of being originally arc volcanics and intrusives.

THE ORIGIN OF THE VERY ANCIENT TONALITES

In much of this discussion on the origin of the various granite types we have been largely concerned with examples drawn from the past 1 Ga of earth time, when the tectonics of plates controlled magmatism. Such studies have provided presently acceptable models for granite generation, so we might well ask if they also apply to those granitic rocks of earlier eras, especially those born in the Archaean crust. We should also enquire whether these very ancient rocks are in any way different from their younger analogues and whether there is any fundamental change in the character of granitic magmatism back into the Archaean.

Comparisons have certainly been made between young and old orogenic belts and their magmatism, particularly in attempts to identify Cordilleran, plate edge-type phenomena in ancient terranes. As examples, McGregor and other workers identified deformed tonalites of Cordilleran type within the Late Archaean of Greenland, whereas Bridgwater, Esher and Watterson compared the Proterozoic Ketilidian mobile belt with that of the Andean with its tonalite–granodiorite batholiths intruded into a volcanic crust. Hietanen recognized a similar Cordilleran-type marginal arc situation in the mid-Proterozoic of Finland, and M.R. Wilson reported an essentially similar situation in parts of the Svecofennian of Sweden, where it is complete with copper porphyries. It seems that most workers take the conceptual step of identifying plate junctions on the basis of the type of magmatism. In so doing they join many others in the growing acceptance of models for crustal growth based on new magmatic additions similar to those envisaged for the mid-Phanerozoic evolution of the Andes.

In general terms Proterozic magmatism and its products are of much the same character as in the younger era, with the exception that there appears to have been a much greater volume of intraplate, anorogenic felsitic magmatism. There are many examples of alkalic granites of this age and I have already commented in some detail on the specificity of the rapakivi facies. We have also seen from the example of the Australian Proterozoic how voluminous was a bimodal magmatism involving a homogeneous biotite granite with I-type affinities. Overall, anorogenic magmatism was associated with rifts and failed rifts resulting from the extensional thinning of the existing thick continental crust, just as in the later era.

Within the orogenic belts of these ancient times both I- and S-type granitic rocks have been identified, as in the younger Archaean of Brazil (by Hofmann) and the younger Proterozoic of Sweden (by M.R. Wilson). Clearly the kinds of global tectonic processes and the magmas that they spawned were operating and generating magmas in a similar way far back in earth history, though not necessarily at the same rate or magnitude throughout this time.

That the origin of granitic rocks as old as 2550 Ma involved recycled crustal material is evident from the trace element and isotope studies of the 2550 Ma Q̂orqut Granite of southern west Greenland by Moorbath and co-workers in 1981. However, more generally, these ancient rocks are geochemically primitive tonalites, the volume of which increases with age, possibly for the simple reason that they evolved when a thin crust was being constructed by the growth of contiguous greenstone belts intruded by tonalite and leucotonalite in a way not dissimilar to the Mesozoic Andean. Nevertheless, the ancient tonalite–trondhjemite association is compositionally different from the tonalite–granodiorite association of the Phanerozoic. According to Tarney and Saunders in 1979, also Martin in 1987, these ancient granitic rocks differ from those of Cordilleran batholiths by having higher Rb/Sr and Na/K ratios, low abundances of the large ion lithophile elements, moderate to extreme heavy REE depletion and a tendency to develop positive europium anomalies in the residual fluids. Figure 18.7 illustrates one expression of this distinction, reflecting, according to Martin, a change of source with time.

Presumably the high degree of HREE fractionation signifies that garnet was involved in the petrogenesis and the composition of the magmas can be modelled by the partial melting of garnet-bearing rocks (Figure 18.8), but at pressures equivalent to depths greater than 25 km.

All this suggests that the conditions or mechanisms of magma generation were different in the Archaean – the convergent plate boundaries of

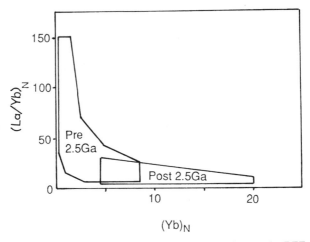

Figure 18.7 (La/Yb)$_N$ plot (normalized) showing the change in REE contents of granitic rocks with time, and claimed by Martin to reflect a change of source mineralogy. Adapted from Martin (1987).

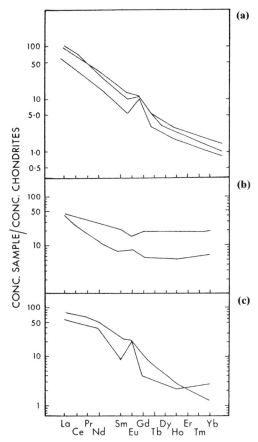

Figure 18.8 (a) REE pattern of typical Archaean tonalites with moderate to very slight positive Eu anomalies, LREE enrichment and downward sloping HREE patterns. (b, c) Calculated REE patterns of melts derived from 20% equilibrium partial melting of garnet-free amphibole (b) and garnetiferous granulite and eclogite (c). Only the garnetiferous source rocks give melts which can be matched to the Archaean tonalite patterns. After Leake (1990). Reproduced with the permission of the author and the Geological Society of London.

a different form, the generation of tonalite especially voluminous. There is no shortage of opinion on the reasons. Allègre and Ben Othman contended that the ancient tonalites represent only juvenile mantle extracts. However, Taylor, Jones and Moorbath rejected this latter explanation on the grounds that granites of any age may include a proportion of crustal components and equally tonalites of any age may be predominantly derived from mantle sources. They also contended that the primitive character indicated by initial strontium, neodymium and lead isotopic data is not related to the age of the tonalites but to their petrogenesis, which is itself closely dependent on the presence or absence of the available continental crust. Thus it is of particular interest that O'Nions and Pankhurst pointed out that the geochemistry of these ancient tonalites is more consistent with derivation from a more evolved source than the early mantle – that is, the very early greenstone volcanic belts, which bear the same relationship to their coeval granitic rocks as the arc volcanics and tonalites in the much younger Andean terrane. However, if as suggested earlier the petrogenesis involved garnet, this would seem to require a thicker crust at this early period than seems likely. Perhaps the higher heat flow was in some way responsible for a different pattern of melting at this early stage. Indeed, Martin argued that it was the oceanic volcanic crust that melted – rather than the mantle wedge – as it was subducted deep into the mantle. With garnet and hornblende as the residual phases the partial melt released from the subducted basaltic material would be impoverished in ytterbium, and the corresponding HREE patterns strongly fractionated.

I am not able to judge which is the most appropriate model, but this contrast between young and very old tonalites must be in some way connected with the global tectonic regime of the time. Intuitively I feel it likely that volcanic and volcaniclastic rocks of primitive character would have been the primary sources in the early crust, and with the melting process more vigorous I would expect M-I-types of tonalites to have been predominant, with the proportion of the more contrasted types increasing in tandem with the development of a thick and progressively more sialic crust.

Many important matters arise in the context of Archaean intrusives and their metamorphic equivalents, but I end this particular discussion with a rather *ex cathedra* statement of belief. I find myself wholly in accord with Tarney, Rox and Bartholomew, who held that although the high grade granulite terranes so common in the Precambrian were once thought to be the lower crustal refractory residue left after the extraction of the upper crustal granites, all modern geochemical studies have failed to support this refractory residua model. If this is true we have to assume that any such residue sank back into the mantle, itself a not entirely credible alternative except that it reinforces the oft-repeated thesis of these essays that most granitic magmas are born, and were always born, near or at the crust–mantle interface.

SELECTED REFERENCES

Bennett, V.C. and DePaolo, D.J. (1988) Proterozoic crustal history of the Western United States as determined by neodymium mapping. *Bulletin of the Geological Society of America*, **99**, 674–685.

Chappell, B.W. (1984) Source rocks of I- and S-type granites in the Lachlan Fold Belt, southeastern Australia. *Philosophical Transactions of the Royal Society of London*, **A310**, 693–707.

Harmon, R.S., Halliday, A.N., Clayburn, J.A.P. and Stephens, W.E. (1984) Chemical and isotopic systematics of the Caledonian intrusions of Scotland and northern England: a guide to magma source region and magma–crust interaction. *Philosophical Transactions of the Royal Society of London*, **A310**, 709–742.

A kind of conclusion: a search for order among multifactorial processes and multifarious interactions

19

Granites and their associated extrusive rocks are formed in large volumes whenever the continental crust is heated by rising hot mantle, or thickened by collision processes. The complexity of rocks of the granitic family is related to the complexity of the processes which lead to thermal perturbations.

William S. Fyfe (1988) Transactions of the Royal Society of Edinburgh, *79, 339.*

In a sense the Masters of Washington and Heidelberg were right: granite is born of basalt, but only by a complex series of partial meltings, first of mantle material to provide a basaltic underplate, then of the latter to supply the primitive granitic melts that are subsequently differentiated by the diverse processes we have studied. And concerning the mantle source there is a veritable deluge of modern literature discussing the possible processes by which basaltic melts are generated and extracted within the various global tectonic niches, either during subduction at plate margins or whenever the crust is in extension above the hot spots or ridges in the mantle. Without embarking on a thesis well outside my experience I accept that basalt is so produced and does underplate the continental crust, though often transformed into its metamorphic form of amphibolite (e.g. Bergantz, 1989). Certainly the experimental evidence for an origin by partial melting is convincing, though the fluid mechanics involved in the extraction is still a matter of research and discussion (e.g. Spiegelman, 1993).

Focusing on the origin of granites, the fact that basalt is so often an immediate precursor in the intrusion sequence, especially in the marginal arcs, strongly suggests that it is in the partial melting of such basaltic extracts, often represented by amphibolite, that we must seek the principal source of the voluminous tonalites of the Cordilleras (e.g. Rapp *et al.*, 1991). But again, although the experimental work has shown that amphibolite is a fertile source rock for tonalitic liquids, the physical processes of extraction remain poorly understood (Petford, 1995).

But just as often, and much more indirectly, granite is born of materials differentiated or otherwise derived during previous geological cycles, in which case the processes involved are endlessly repetitive, with the developing crust, either juvenile or matured in the erosion cycle, variously contributing material.

At active ocean–continent plate margins the formation of granite is an essential part of the crust-making process, but within the continental plates it involves a large measure of recycling and redistribution of the existing crust, especially when the latter is sufficiently thrust thickened to melt without any accession of heat from the mantle. All this is part of a global refining process of which granite is the magmatic end product.

While we probably know by now most of the processes that are involved, we lack certain vital information by which the existing models can be tested and refined; for example, an experimentally verified understanding of the physical nature of granitic magma, the separation process and the complete crystallization history. As well as providing such essential constraints, one of the most important remaining tasks is to assign proportionality – that is, to describe the ever-changing roles of the various processes throughout the evolution of specific granitic suites, and here the challenge is often to be directed to the geochemist. The central difficulty is that although the generative processes are concentrated in the high energy regime at the base of the crust, we are most often presented with a dead pluton after its arrival in the upper crust. Thus we see only residual traces of the principal generative processes.

Progress is more certain in determining the nature of the source rocks and the origin of the volatile elements involved in remelting processes. The key lies in further developing isotope systematics, especially in realizing the potential of the rhenium–osmium isotopic system. Already studies of B and ^{10}Be have enabled Morris and co-workers to quantify sediment recycling at convergent plate margins, while δD values have been employed by Giggenbach to trace the involvement of sea water in arc magmatism. Even the familiar ^{40}K–^{40}Ca system is yielding new information, allowing Marshall and DePaolo to propose that the potassium of many of the granites of the continental interiors is of deep crustal origin.

In other respects the challenge is geophysical on three levels: the physical state of granitic magma; the topological description of the roots of plutons and the ambient lower crust; and the nature of the heat flux. Unfortunately the root may not survive *in situ* owing to subsequent decouplings within the crust and upper mantle – yet such a delamination may answer the enigma of the elusive cumulate that geochemistry tells us ought to bottom batholiths, yet geophysics so often fails to locate. But more than anything else it is a geological challenge, requiring an understanding of the whole evolutionary history of the ambient crust. Only then can a holistic model be formulated. Granites cannot be studied in the isolation of either the laboratory or the field; they are certainly not chemical or physical numbers but essential parts of geological histories. I found it relevant, in my own studies, particularly in the Andes, to know equally the ages of the country rock ammonites, the overall gravity anomalies and the values of the initial strontium ratio for the tonalites!

I make no attempt to rehearse all the evidence and prejudices reported here but only to draw together several of the main themes that thread these essays. I believe that granitic magmas only occur in volume where a thick crust, old or new, can provide a fertile source, a thermal blanket and sufficient space for differentiation processes to operate. Furthermore, concerning the thermal state of the source terranes I am intrigued by the results of E-an Zen's modelling, which suggests that it is mantle upwelling and not subduction, or even thrust loading, that is the likely cause of anatexis. Are we not moving away from conventional views of subduction and towards models involving the uprise of mantle during crustal extension?

I have emphasized the overall tectonic control of granitic processes, in particular the location of partial melting and intrusion, and illustrated the interdependence of tectonic movements, plutonism, volcanism and sedimentation. On every scale, folds, fractures and their intersections can be seen to have focused intrusion and located contacts, whereas the shapes of the granite intrusions are largely controlled by the response of their envelopes. Furthermore, the generation and segregation of the magmas is triggered, their transport motivated and channelled, and accommodation provided by movements on deep-reaching thrusts and faults. Except perhaps in the deepest crust, granitic magmas probably migrate through dykes and sheets rather than rise in diapirs, the dykes feeding into ballooning plutons in the upper crust. True, salt dome-like piercement diapirs are probably much rarer than presently conceived, certainly in the middle and upper crust, and I suspect that many of the so-called diapirs of the world's basements represent no more than deformed plutons, sometimes even complex fold cores! Whatever the truth, the arrival form is hardly a reliable guide to the nature of the magma transport and supply.

Another major theme is that granitic magmas image their deep source, and that the various contributions can be recognized. This is despite the granite-making process showing a remarkable capacity for the homogenization of partial melts derived from a veritable structural and compositional mélange at depth. What I find particularly intriguing in view of the great range of possible sources and processes is that granite

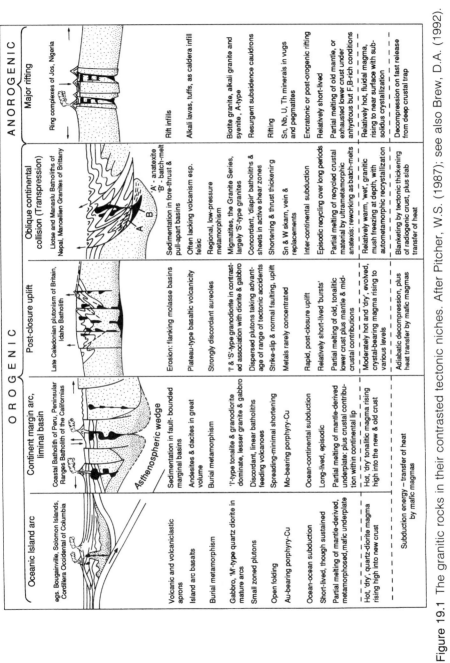

Figure 19.1 The granitic rocks in their contrasted tectonic niches. After Pitcher, W.S. (1987); see also Brew, D.A. (1992).

the world over is so often just the plain granite that one might expect of a residua system; that is, of course, until we ask geochemical questions!

Both the source rocks and the processes that melt out and segregate the granitic magmas are manifestly reflected in differences in the geological environment, so that it has been possible to discuss the origin of granites in the terms of a global tectonic context, though we must always be aware of the complexities introduced by the reworking of older material. In summarizing this geodynamic aspect I begin with the island arcs of convergent plate margins (Figure 19.1). In this situation the abundance of metamorphosed mafic igneous rocks as enclaves in the quartz diorites confirms the chemical evidence that the partial melting of a deep-seated underplate of deformed igneous rocks is a likely process for the origin of these M-type granitic rocks – a melting presumably triggered by the volatiles released from the subducted oceanic slab.

In the extensional regime of the marginal continental arcs, such as those of the southeastern Pacific margin, the batholiths also bear evidence of a basic progenitor in the form of precursor gabbro intrusions, basaltic enclaves and synplutonic dykes, and the contents of intersecting volcanic pipes. As we have seen, this mafic source is again best modelled as deriving from a new basaltic underplate that was partially melted either from a mantle wedge lying above a subduction zone or from the up-arched mantle below the pulled-apart crust of the continental margin. It was this mafic underplate that was then partially melted in its turn to provide the calcium-rich, but relatively potassium-poor, I-type tonalites and granodiorites. Compositional variations in such batholithic rocks can be explained on the basis that partial melting may have occurred at different levels of the lower crust or upper mantle, when garnet may or may not have been involved in accord with the ambient pressure. Perhaps the occasional high-potassium suites also owe their special origin to the depth factor, representing extra deep drainage of the mantle. Then, concerning this all-important matter of assigning proportionality, these arc environments are likely to represent the simplest case of a single source located in the mantle, only to be complicated where the melts came into contact with old crust during the migration of the arc, and when the combined MASH process of melting, assimilation, storage and homogenization comes into play.

Inside a marginal arc there is often an active tectonic zone in which melting is initiated in response to thrust thickening and the changing of crust–mantle relationships along both low angle thrusts and steep faults. Heat from the juxtaposed mantle and invasion by mantle-derived magmas melt the lower crust so that crust–mantle interaction is inevitable. However, the volumes of magma produced are not of the scale of the long-lived volcano-plutonic arcs, the volatiles and heat produced being far less than when generated during subduction of an oceanic slab. So much less in fact that lower temperature partial melts predominate in the form of simple granite, and also the magmatism is distinctly bimodal with granitic plutonism coeval with basaltic volcanicity.

As a result of the tectonic mayhem at the base of the continental crust the sources are varied, with mature and immature metasediments, metavolcanics and old granites partially melting to produce per-aluminous and metaluminous extracts that variously mix with the geochemically primitive magma. Either the peraluminous S-type or the metaluminous I-type granites may predominate, and they can even be juxtaposed, together with gabbro, in a single pluton. The varied granitic assemblages of the Cordilleran interior of North America and the Caledo-nian granites of the British Isles are examples, and there the assignment of proportionality of process is a challenge indeed.

The responses of a stabilized craton to global tectonics is to tend to fragment, with extension leading to crustal thinning and rifting. Again, the effect is to cause adiabatic melting of the up-arching mantle and the production of a basaltic underplate, apparently independently of any subduction mechanism. Again, simple granites predominate in a dis-tinctly bimodal magmatism. And again it seems that the contribution of mantle and crustal derivatives can be various.

A special case is the association of alkalic granites with such rift zones. Their magmas may represent not only a low degree of partial melting but melting under anhydrous conditions deep under a cover of thick crust. The causes of these special conditions are hardly understood, especially why the magmas have high concentrations of the halogens and boron, but it has been suggested that long storage of the primary partial melts, trapped under a thick lithospheric blanket, results in the preferential melting out of the volatiles from mantle minerals, the enriched magmas only being released by the rupture of rifting. As for identifying the source rocks, this is now bedevilled by the superimposed changes wrought by the volatiles. At this point I think it is necessary to emphasize that here, too, granite magmatism is unconnected with subduction mechanisms; also that a good part of the heat required for crustal melting is carried in by the basaltic magmas so evident in the bimodal magmatism of intracratonic terranes.

Turning to those granitic rocks born in the plate collisional orogens, partial melting resulting from the heat generated by progressive thrust-thickening of radiogenic crust may well be sufficient to provide all the components of a granite series, perhaps without the direct involvement of a mantle contribution, though I refer back to E-an Zen's calculations that seem to call for such an additional requirement. Nappe formation may bring very hot lower crustal slabs into higher zones of the crust and so promote such partial melting, as indeed would shear-heating and increas-ing the circulation of volatiles. In such a purely crustal environment we can expect that granites will predominate, also that these will be largely peraluminous S-types because of the abundance of recycled material in the crust. However, where immature volcaniclastic greywackes predominate we are just as likely to find I-type granitic rocks in this geological niche, and such a mixture of types is particularly evident in the Upper Palaeozoic Hercynian terrains of France and Iberia. I have already discussed the problem of the extraction mechanism of these crustal magmas and I repeat a prejudice that they are not often born of the migmatites that we now see exposed, but by the bulk partial melting of similar, very deep-level migmatites – a contradiction that may not appeal to the reader.

With multifactorial generative processes involving different source rocks, and operating in different global tectonic settings, and where there are multifarious interconnections, these can rarely be precisely defined, so that any genetic categorization must be regarded as a philosophical abstraction; some would say impeding more than advancing understanding. Certainly the specific designations, 'S' and 'I', ought to be used in their original connotation of source derivation alone, and 'M' may join them on the same basis. However, the use of 'A'-type is of a different genre. In my final summation, with its attempt to erect a genetic classification of granites, I ask the reader to accept the use of such labels as a mere device, a kind of aide mémoire. As for the categories themselves, it is probably best to avoid the alphabetical 'soup' and designate them in now conventional global tectonic terms. I rest my case.

SELECTED REFERENCES

Brown, M., Candela, P.A., Peck, D.L., Stephens, W.E., Walker, R.J. and E-an Zen (eds) (1996) The Third Hutton Symposium on the Origin of Granites and Related Rocks. *Geological Society of America Special Paper* 315, 359 pp.

Pitcher, W.S. (1982) Granite type and tectonic environment. In: Hsü, K. (ed.), *Mountain Building Processes*. Academic Press, pp. 19–40.

Bibliography

Agar, R.A. (1986) *Journal of African Earth Sciences*, **4**, pp.105–121.

Agar, R.A. and Le Bel, L. (1985) In: Pitcher, W.S., Atherton, M.P., Cobbing, E.J. and Beckinsale, R.D. (eds), *Magmatism at a Plate Edge: The Peruvian Andes*. Blackie, Glasgow, pp.119–126.

Aguirre, L. (1983) In: Roddick, J.A. (ed.), *Circum-Pacific Plutonic Terranes. Geological Society of America Memoir No. 159*, pp.293–316.

Aguirre, L. and Offler, R. (1985) In: Pitcher, W.S., Atherton, M.P., Cobbing, E.J. and Beckinsale, R.D. (eds), *Magmatism at a Plate Edge: The Peruvian Andes*. Blackie, Glasgow, pp.59–71.

Aguirre, L., Charrier, R., Davidson, J., Mpodozis, A., Rivano, S., Thiele, R., Tidy, E., Vergara, M. and Vicente, J.-C. (1974) *Pacific Geology*, **8**, pp.1–38.

Ahmed-Said, Y. and Leake, B.E. (1990) *Mineralogical Magazine*, **54**, pp.1–22.

Akaad, M.K. (1956) *Quarterly Journal of the Geological Society of London*, **112**, pp.263–288.

Alderton, D.H.M. and Harmon, R.S. (1991) *Mineralogical Magazine*, **55**, pp.605–612.

Allègre, C.J. and Ben Othman, D. (1980) *Nature*, London, **286**, pp.335–342.

Allègre, C.J. and Minster, J.F. (1978) *Earth and Planetary Science Letters*, **38**, pp.1–25.

Allen, C.M. (1992) *Transactions of the Royal Society of Edinburgh, Earth Sciences*, **83**, pp.179–190.

Anderson, E.M. (1951) *Dynamics of Faulting and Dyke Formation with Application to Britain* (2nd edn). Oliver and Boyd, Edinburgh, 206 pp.

Arculus, R.J. (1987) *Journal of Volcanology and Geothermal Research*, **32**, pp.1–12.

Arculus, R.J. and Powell, R. (1986) *Journal of Geophysical Research*, **91**, pp.5913–5926.

Arzi, A.A. (1978) *Tectonophysics*, **44**, pp.173–184.

Ashworth, J.R. (ed.) (1985) *Migmatites*. Blackie, Glasgow, 301 pp.

Ashworth, J.R. and Brown, M. (1990) *High-Temperature Metamorphism and Crustal Anatexis*. Unwin Hyman, London.

Ashworth, J.R. and McLellan, E.L. (1985) In: Ashworth, J.R. (ed.), *Migmatites*. Blackie, Glasgow, pp.180–203.

Atherton, M.P. (1981) *Journal of the Geological Society of London*, **138**, pp.343–349.

Atherton, M.P. (1990) *Geological Journal*, **25**, pp.337–349.

Atherton, M.P. (1995) Granite magmatism. In: Le Bas, M.J. (ed.), *Milestones in Geology. Geological Society of London, Memoir No. 16*, pp.221–235.

Atherton, M.P. and Plant, J.A. (1985) In: *High Heat Production (HHP) Granites, Hydrothermal Circulation and Ore Genesis*. Institution of Mining and Metallurgy, London, pp. 459–478.

Atherton, M.P. and Sanderson, L.M. (1985) In: Pitcher, W.S., Atherton, M.P., Cobbing, E.J. and Beckinsale, R.D. (eds), *Magmatism at a Plate Edge: The Peruvian Andes*. Blackie, Glasgow, pp. 208–227.

Atherton, M.P. and Sanderson, L.M. (1987) *Geologische Rundschau*, **76**, pp. 213–232.

Atherton, M.P. and Webb, S. (1989) *Journal of South American Earth Sciences*, **2**, pp. 241–261.

Augustithis, S.S. (1973) *Atlas of Textural Patterns of Granites, Gneisses and Associated Rocks*. Elsevier, Amsterdam, 378 pp.

Ayrton, S. (1988) *Schweizerische Mineralogische und Petrographische Mitteilungen*, **69**, pp. 1–19.

Ayrton, S.N. (1991) In: Didier, J. and Barbarin, B. (eds), *Enclaves and Granite Petrology*. Elsevier, Amsterdam, pp. 465–476.

Bacon, C.R. (1989) *Geochimica et Cosmochimica Acta*, **53**, pp. 1055–1066.

Bacon, C.R. and Druitt, T.H. (1988) *Contributions to Mineralogy and Petrology*, **98**, pp. 224–256.

Bailey, E.B. (1958) *Transactions of the Geological Society of Glasgow*, **23**, pp. 29–52.

Bailey, E.B. and Maufe, H.B. (1960) *The geology of Ben Nevis and Glen Coe and the surrounding country. Memoirs of the Geological Survey of Scotland, Expl. Sheet 53*, 2nd edn.

Bailey, E.B. and McCallien, W.J. (1956) *Liverpool and Manchester Geological Journal*, **1**, pp. 466–501.

Baker, D.R. and Vaillancourt, J. (1994) *Mineralogical Magazine*, **58A**, pp. 40–41.

Baldwin J.A. and Pearce J.A. (1982) *Economic Geology*, **77**, pp. 664–674.

Balk, R. (1937) *Structural Behaviour of Igneous Rocks. Geological Society of America Memoir No. 5*, 177 pp.

Barbarin, B. (1989) *Schweizerische Mineralogische und Petrographische Mitteilungen*, **69**, pp. 303–315.

Barbarin, B. (1990) *Geological Journal*, **25**, pp. 227–238.

Barbarin, B. (1996) *Geology*, **24**, pp. 295–298.

Barbieri, M., Caggianelli, A., Di Florio, M.R. and Lorenzoni, S. (1994) *Mineralogical Magazine*, **58**, pp. 553–566.

Barnes, C.G., Allen, C.M., Hoover, J.D. and Brigham, R.H. (1990) In: Anderson, J.L. (ed.), *The Nature and Origin of Cordilleran Magmatism. Geological Society of America Memoir No. 174*, pp. 331–346.

Barr, D. (1985) In: Ashworth, J.R. (ed.), *Migmatites*. Blackie, Glasgow, pp. 225–264.

Barr, S.M. (1990) *Geological Journal*, **25**, pp. 295–304.

Barreiro, B.A. and Stern, C.R. (1982) *Eos*, **63**, p. 1148.

Barrière, M. (1977) *Journal of the Geological Society of London*, **134**, pp. 311–324.

Barsukov, V.L. (1957) *Geochemistry*, **1**, pp. 41–52. (Translated from *Geokhimiyia* (1957) pp. 36–45.)

Barth, T.F.W. (1954) *Studies on the Igneous Rock Complexes of the Oslo Region XIV. Norske Videnskaps-Akademi Oslo, Skrifter Mat.-Nat. Kl*.

Barth, T.F.W. (1962) *Theoretical Petrology*. John Wiley, New York.

Bateman, P.C. (1981) In: Ernst, W.G. (ed.), *The Geotectonic Development of California*. Prentice-Hall, Englewood Cliffs, New Jersey, pp. 71–86.

Bateman, P.C. (1983) In: Roddick, J.A. (ed.), *Circum-Pacific Plutonic Terranes. Geological Society of America Memoir No. 159*, pp.241–254.

Bateman, P.C. (1989) In: Le Maitre, R.W. *et al.* (1989) *A Classification of the Igneous Rocks and Glossary of Terms.* Blackwell Scientific Publications, Oxford, 194 pp.

Bateman, P.C. (1992) *Plutonism in the Central Part of the Sierra Nevada Batholith, California. US Geological Survey Professional Paper No. 1483*, 186 pp.

Bateman, P.C. and Chappell, B.W. (1979) *Bulletin of the Geological Society of America*, **90**, pp.465–482.

Bateman, P.C. and Dodge, F.C.W. (1970) *Bulletin of the Geological Society of America*, **81**, pp.409–420.

Bateman, P.C., Clarke, L.D., Huber, N.K., Moore, J.G. and Rinehart, C.D. (1963) *A Summary of the Critical Relations in the Central Part of the Sierra Nevada Batholith, California, USA. US Geological Survey Professional Paper No. 414D*, 46 pp.

Bateman, P.C., Dodge, F.C.W. and Kistler, R.W. (1991) *Journal of Geophysical Research*, **96**, pp.19555–19568.

Bateman, R.B. (1984) *Tectonophysics*, **110**, pp.210–231.

Bateman, R.B. (1985) *Journal of Geology*, **93**, pp.293–310.

Bateman, R.B. (1989) *Journal of Geology*, **97**, pp.766–768.

Beach, A. (1976) *Philosophical Transactions of the Royal Society of London*, **A280**, pp.569–604.

Beach, A. and Fyfe, W.S. (1972) *Contributions to Mineralogy and Petrology*, **36**, pp.175–180.

Beard, J.S. and Lofgren, G.E. (1991) *Journal of Petrology*, **32**, pp.365–401.

Becker, G.F. (1897) *American Journal of Science*, **4**, p.22.

Becker, F. (1901) referenced in Harker (1909).

Beckinsale, R.D., Sanchez-Fernandez, A.W., Brook, M., Cobbing, E.J., Taylor, W.P. and Moore, N.D. (1985) In: Pitcher, W.S., Atherton, M.P., Cobbing, E.J. and Beckinsale, R.D. (eds), *Magmatism at a Plate Edge: The Peruvian Andes.* Blackie, Glasgow, pp.177–202.

Ben Othman, D., Fourcade, S. and Allègre, C.J. (1984) *Earth and Planetary Science Letters*, **69**, pp.290–300.

Bender, J.F., Hanson, G.N. and Bence, A.E. (1982) *Earth and Planetary Science Letters*, **58**, pp.330–344.

Bennett, V.C. and DePaolo, D.J. (1987) *Bulletin of the Geological Society of America*, **99**, pp.674–685.

Bergantz, G.W. (1989) *Science*, **245**, pp.1093–1095.

Bergantz, G.W. (1991) In: Kerrick, D.M. (ed.) *Contact Metamorphism. Mineralogical Society of America Reviews in Mineralogy*, **26**, pp.13–42.

Berger, A.R. (1971) *Geological Journal*, **7**, pp.437–458.

Berger, A.R. and Pitcher, W.S. (1970) *Proceedings of the Geologists' Association, London*, **81**, pp.410–462.

Bergman, L. (1986) *Acta Academiae Aboensis* (Finland), *Ser. B*, **46**(5), pp.5–74.

Billings, M.P. (1942) *Structural Geology.* Prentice-Hall, New York, 473 pp.

Bishop, A.C. (1955) *Bulletin de la Société Jersiaise*, **16**, pp.309–314.

Bishop, A.C. (1963) *Proceedings of the Geologists' Association, London*, **74**, pp.289–300.

Blake, D.H., Elwell, R.W.D., Gibson, I.L., Skelhorn, R.R. and Walker, G.P.L. (1965) *Quarterly Journal of the Geological Society of London*, **121**, pp.31–50.

Blake, S. and Campbell, I.H. (1986) *Contributions to Mineralogy and Petrology*, **94**, pp.72–86.

Blundell, D.J. (1990) *Journal of the Geological Society of London*, **147**, pp.895–913.

Bonin, B. (1974) *Bulletin de la Société géologique de France*, **19**, pp.865–871.

Bonin, B. (1982) *Les Granites des Complexes Annulaires. Bureau de recherches géologiques et minières, Manuels et méthodes, No. 4*, pp.1–183.

Bonin, B. (1990) *Geological Journal*, **25**, pp.227–238.

Boriani, A., Colomba, A., Giobbi, E.O. and Pagliani, G.P. (1974) *Rendiconti della Società Italiana di Mineralogia e Petrologia*, **30**, pp.893–917.

Boriani, A., Burlini, L., Caironi, V., Origoni, E.G., Sassi, A. and Sesana, E. (1988) *Rendiconti della Società Italiana di Mineralogia e Petrologia*, **43**(2), pp.367–384.

Bott, M.H.P. and Smithson, S.B. (1967) *Bulletin of the Geological Society of America*, **78**, pp.859–878.

Bouchez, J.L. (1995) *Terra Nova*, **7**, *Abstract Supplement No. 1*, XI-2(8).

Bouchez, J-L. and Gleizes, G. (1995) *Journal of the Geological Society of London*, **152**, pp.669–679.

Bouchez, J-L., Gleizes, G. *et al.* (1990) *Tectonophysics*, **184**, pp.157–171.

Bowden, P. and Kinnaird, J.A. (1978) *Transactions of the Institution of Mining and Metallurgy*, London, **88**, B66 and B70.

Bowden, P. and Kinnaird, J.A. (1984) *Geologisches Jahrbuch*, **B56**, pp.1–68.

Bowden, P., Batchelor, R.A., Chappell, B.W., Didier, J. and Lameyre, S. (1984) *Physics of the Earth and Planetary Interiors*, **35**, pp.1–11.

Bowden, P., Black, R., Martin, E.C., Ike, E.C., Kinnaird, J.A. and Batchelor, R.A. (1987) In: Fitton, J.C. and Upton, B.G.J. (eds) *Alkaline Rocks. Geological Society of London Special Publication No. 30*, pp.357–379.

Bowen, N.L. (1915) *Journal of Geology*, **23**, pp.1–91.

Bowen, N.L. (1928) *The Evolution of the Igneous Rocks*, Princeton University Press, Princeton, NJ, 334 pp.

Bowen, N.L. (1948) *Geological Society of America Memoir No. 28*, pp.79–90.

Bowes, D.R. and McArthur, A.C. (1976) *Krystalinikum*, **12**, pp.31–46.

Bowes, D.R. and Wright, A.E. (1967) *Transactions of the Royal Society of Edinburgh*, **67**, pp.109–143.

Brew, D.A. (1992) In: Bartholomew, M.J. *et al.* (eds), *Basement Tectonics 8*. Kluwer Academic, Dordrecht, pp.169–177.

Bridgwater, D., Esher, A. and Watterson, J. (1973) *Philosophical Transactions of the Royal Society of London*, **A273**, pp.513–533.

Brindley, J.C., Gupta, L.N. and Kennan, P.S. (1976) *Proceedings of the Royal Irish Academy*, **76B**, pp.337–348.

Brögger, W.C. (1894) *Die eruptive Gesteine des Kristiana-gebietes I. Norske Videnskaps-Akademi Oslo, Skriften Mat.-Nat. Kl.4.*

Brown, G.C. (1982) In: Thorpe, R.S. (ed.), Andesites: *Orogenic Andesites and Related Rocks*. John Wiley, Chichester, pp.437–464.

Brown, G.C. and Fyfe, W.S. (1970) *Contributions to Mineralogy and Petrology*, **28**, pp.310–318.

Brown, G.C., Thorpe, R.S. and Webb, P.C. (1984) *Journal of the Geological Society of London*, **141**, pp.413–426.

Brown, M. (1993) *Earth Science Reviews*, **36**, pp.83–130.

Brown, M. and D'Lemos, R.S. (1991) In: Haapala, I. and Condie, K.C. (eds), *Precambrian Granitoids: Petrogenesis, Geochemistry and Metallogeny. Precambrian Research*, **51**, pp.393–427.

Brown, P.E. (1991) *Caledonian and Earlier Magmatism*. In: Graig, G.Y. (ed.), *Geology of Scotland*, 3rd edn. Geological Society of London, pp. 229–295.

Brown, P.E. (1994) Book reviews, *Journal of Petrology*, **35**, pp. 1175–1177.

Brown, P.E., Dempster, T.J., Harrison, T.N. and Hutton, D.H.W. (1992) *Transactions of the Royal Society of Edinburgh: Earth Sciences*, **83**, pp. 173–178.

Brun, J-P. (1980) *Earth and Planetary Science Letters*, **47**, pp. 441–449.

Brun, J.P. and Pons, J. (1981) *Journal of Structural Geology*, **3**, pp. 219–229.

Brun, J.P., Gapais, D., Cogne, J.P., Ledru, P. and Vigneresse, J.L. (1990) *Geological Journal*, **25**, pp. 271–286.

Bryon, D. (1992) *Transactions of the Royal Society of Edinburgh: Earth Sciences*, **83**, Abstracts, p. 487.

Bryon, D., Atherton, M.P. and Hunter, R.H. (1994) *Mineralogical Magazine*, **59**, pp. 203–211.

Bryon, D., Atherton, M.P. and Hunter, R.H. (1995) *Contributions to Mineralogy and Petrology*, **117**, pp. 66–75.

Bryon, D.N., Atherton, M.P., Cheadle, M.J. and Hunter, R.H. (1996) *Mineralogical Magazine*, **60**, pp. 163–171.

Buddington, A.F. (1959) *Bulletin of the Geological Society of America*, **70**, pp. 671–747.

Burnham, C.W. (1979) In: Yoder, H.S. (ed.), *The Evolution of the Igneous Rocks*. Princeton University Press, Princeton, NJ, pp. 439–482.

Burnham, C.W. (1982) In: Rickard, D. and Wickman, F.E. (eds), *Chemistry and Geochemistry of Solutions at High Temperatures and Pressures*. Pergamon Press, Oxford, pp. 197–229.

Burnham, C.W. (1992) *Transactions of the Royal Society of Edinburgh*, **83**, pp. 387–398.

Burnham, C.W. and Nekvasil, H. (1986) *American Mineralogist*, **71**, pp. 239–263.

Büsch, W. (1966) *Neues Jarbuch für Mineralogie. Abh.*, **104**(2) pp. 190–227.

Büsch, W., Schneider, G. and Mehnert, K.R. (1974) *Neues Jahrbuch für Mineralogie*, **8**, pp. 345–370.

Büsch, W., Matthes, S., Mehnert, K.R. and Schubert, W. (1980) *Neues Jahrbuch für Mineralogie und Petrographie, Abh*, **137**, 223–256.

Bussell, M.A. (1975) *Proceedings of the Geologists' Association*, London, **87**, pp. 237–246.

Bussell, M.A. (1983) *Lithos*, **16**, pp. 169–184.

Bussell, M.A. (1985) In: Pitcher, W.S., Atherton, M.P., Cobbing, E.J. and Beckinsale R.D. (eds) *Magmatism at a Plate Edge: The Peruvian Andes*. Blackie, Glasgow, pp. 128–155.

Bussell, M.A., Pitcher, W.S. and Wilson, P.A. (1976) *Canadian Journal of Earth Science*, **13**, pp. 1020–1030.

Bussy, F. (1990) *Geological Journal*, **25**, pp. 319–324.

Caironi, V. (1985) *Rendiconti della Società Italiana di Mineralogia e Petrologia*, **40**, pp. 341–352.

Candela, P.A. (1986) *Economic Geology*, **81**, pp. 1–19.

Candela, P.A. (1990) In: Stein, H.J. and Hannah, J.L. (eds), *Ore-bearing Granite Systems. Geological Society of America Special Paper No. 246*, pp. 11–20. See also p. 7.

Candela, P.A. (1992) *Transactions of the Royal Society of Edinburgh: Earth Sciences*, **83**, pp. 317–326.

Candela, P.A. and Bouton, S.L. (1990) *Economic Geologist*, **85**, pp. 633–640.

Capais, D. and Barbarin, B. (1986) *Tectonophysics*, **125**, pp. 357–370.

Capdevila, R. and Floor, P. (1970) *Boletín de Geología y Mineralogía*, España, **81**, pp.215–225.

Capdevila, R., Corretgé, G. and Floor, P. (1973) *Bulletin de la Société géologique de France*, **15**, pp.209–228.

Carmichael, I.S.E. (1963) *Quarterly Journal of the Geological Society of London*, **119**, pp.95–131.

Carrigan, C.R. and Eichelberger, J.C. (1990) *Nature* **343**, pp.248–251.

Carroll, M.R. and Wyllie, P.J. (1990) *American Mineralogist*, **75**, pp.345–357.

Carron, J.P. (1969) *Bulletin de la Société française de minéralogie et de cristallographie*, **92**, pp.435–446.

Cashman, K.V. (1988) *Bulletin of Volcanology*, **50**, pp.194–209.

Cashman, K.V. (1990) *Mineralogical Society of America Reviews in Mineralogy*, **24**, pp.259–314.

Castro, A. (1986) *Journal of Structural Geology*, **8**, pp.633–645.

Castro, A. (1987) *Geologische Rundschau*, **76**, pp.101–124.

Cathelineau, M. (1986) *Journal of Petrology*, **27**, pp.945–965.

Cawthorn, R.C. and O'Hara, M.J. (1976) *American Journal of Science*, **276**, pp.309–329.

Cerny, P. (1982) In: Cerny, P. (ed.), *Granitic Pegmatite in Science and Industry. Mineralogical Association of Canada, Short Course Handbook 8*, pp.405–461.

Chapman, C.A. (1962) *Journal of Geology*, **70**, pp.534–564.

Chappell, B.W. (1984) *Philosophical Transactions of the Royal Society of London*, **A310**, pp.693–707.

Chappell, B.W. (1994) *Journal and Proceedings, Royal Society of New South Wales*, **127**, pp.47–59.

Chappell, B.W. (1997) *Transactions of the Royal Society of Edinburgh: Earth Sciences*, **88**, pp.159–170.

Chappell, B.W. and White, A.J.R. (1974) *Pacific Geology*, **8**, pp.173–174.

Chappell, B.W. and White, A.J.R. (1991) In: Didier, J. and Barbarin, B. (eds), *Enclaves and Granite Petrology*, Elsevier, Amsterdam, pp.375–381.

Chappell, B.W. and White, A.J.R. (1992) *Transactions of the Royal Society of Edinburgh: Earth Sciences*, **83**, pp.1–26.

Chappell, B.W., White, A.J.R. and Wyborn, D. (1987) *Journal of Petrology*, **28**, pp.1111–1138.

Chappell, B.W., White, A.J.R. and Hine, R. (1988) *Australian Journal of Earth Sciences*, **35**, pp.505–521.

Chappell, B.W., Wyborn, D., Bryant, C.J. and White, A.J.R. (1997) (*in press*).

Charrier, R. (1973) *Earth and Planetary Science Letters*, **20**, pp.242–249.

Chen, J.H. and Tilton, G.R. (1991) *Bulletin of the Geological Society of America*, **103**, pp.439–447.

Christiansen, E.H., Burt, D.M., Sheridan, M.F. and Wilson, R.J. (1983) *Contributions to Mineralogy and Petrology*, **83**, pp.16–30.

Christensen, J.N. and DePaolo, J. (1993) *Contributions to Mineralogy and Petrology*, **113**, pp.100–114.

Clarke, D.B. (1981) *Canadian Mineralogist*, **19**, pp.3–17.

Clarke, D.B. (1992) *Granitoid Rocks*. Chapman and Hall, London, 283 pp.

Clarke, D.B. (1995) *Mineralogical Magazine*, **59**, pp.311–325.

Clarke, D.B., McKenzie, C.B., Muecke, G.K. and Richardson, S.W. (1976) *Contributions to Mineralogy and Petrology*, **56**, pp.279–287.

Clemens, J.D. (1989) *Journal of Petrology*, **30**, pp.1313–1316.

Clemens, J.D. and Mawer, C.K. (1992) *Tectonophysics*, **204**, pp. 339–360.

Clemens, J.D. and Vielzeuf, D. (1987) *Earth and Planetary Science Letters*, **86**, pp. 287–306.

Clemens, J.D. and Wall, V.J. (1981) *Canadian Mineralogist*, **19**, pp. 111–132.

Clemens, J.D., Holloway, J.R. and White, A.J.R. (1986) *American Mineralogist*, **71**, pp. 317–324.

Cliff, R.A. (1985) *Journal of the Geological Society of London*, **142**, pp. 97–110.

Cliff, R.A. and Cohen, A. (1980) *Earth and Planetary Science Letters*, **50**, pp. 211–218.

Cliff, R.A., Yardley, B.W.D. and Bussy, F.R. (1996) *Journal of the Geological Society of London*, **153**, pp. 109–120.

Cloos, E. (1936) *Neues Jahrbuch für Mineralogie*, **73B**, pp. 355–450.

Cloos, H. (1925) *Einführung in die tektonische Behandlung magmatischer Erscheinungen, pt 1. Das Riesengebirge in Schlesien. Gebr.* Borntraeger, Berlin, 194 pp. (Quoted in Balk, R. (1937) *Structural Behaviour of Igneous Rocks. Geological Society of America Memoir No. 5*, p. 59.)

Cloos, H. (1936) *Einführung in die Geologie.* Borntraeger, Berlin, 503 pp.

Cloos, H. (1941) *Geologische Rundschau*, **32**, pp. 709–800.

Cobbing, E.J. and Pitcher, W.S. (1972) *Journal of the Geological Society of London*, **128**, pp. 421–460.

Cobbing, E.J., Pitcher, W.S. and Taylor, W.P. (1977) *Journal of Geology*, **85**, pp. 625–631.

Cobbing, E.J. *et al.* (1981) *The Geology of the Western Cordillera of Northern Peru, Institute of Geological Sciences, UK, Overseas Memoir*, **5**, 143 pp.

Cole, G.A.J. (1894) *Scientific Transactions of the Royal Dublin Society*, **5**, pp. 239–248.

Cole, G.A.J. (1902) *Proceedings of the Royal Irish Academy*, **24B**, pp. 203–230.

Cole, G.A.J. (1916) *Scientific Proceedings of the Royal Dublin Society*, **15**, pp. 141–158.

Coleman, M.L. (1977) *Journal of the Geological Society of London*, **133**, pp. 593–608.

Coleman, R.G. and Donato, M.M. (1979) In: Barker, F. (ed.), *Trondhjemites, Dacites, and Related Rocks.* Elsevier, Amsterdam, pp. 149–168.

Collins, L.G. (1988) *Hydrothermal Differentiation.* Theophrastus Publications, Athens, 382 pp.

Collins, W.J. (1994) *Geology*, **22**, pp. 143–146.

Collins, W.J. and Vernon, R.H. (1992) *Tectonophysics*, **214**, pp. 381–400.

Collins, W.S., Beam, S.D., White, A.J.R. and Chappell, B.W. (1982) *Contributions to Mineralogy and Petrology*, **80**, pp. 189–200.

Compston, W. and Chappell, B.W. (1979) In: McElhinny, M.W. (ed.), *The Earth: its Origin, Structure and Evolution.* Academic Press, London, pp. 377–426.

Compton, R.R. (1955) *Bulletin of the Geological Society of America*, **66**, pp. 9–44.

Corretgé, L.G., Gallastequi, G. and Cuesta, A. (1984) *Trabajos de Geología*, **14**, pp. 17–26.

Corry, C.E. (1988) *Geological Society of America Special Paper No. 220.*

Cox, R.A., Dempster, T.J., Bell, B.R. and Rogers, G. (1996) *Journal of the Geological Society, London*, **153**, pp. 625–635.

Cox, R.A. *et al.* Reported in Fowler, M.B. and Leat, P.T. (1996) *Geoscientist*, **6**, pp. 4–29.

Creaser, R.A., Price, R.C. and Wormald, R.J. (1991) *Geology*, **19**, pp. 163–166.

Cross, W., Iddings, J.P., Pirsson, L.V. and Washington, H.S. (1902) *Journal of Geology*, **10**, pp. 550–690.

Cruden, A.R. (1988) *Tectonics*, **7**, pp. 1091–1101.
Cruden, A.R. (1990) *Journal of Geology*, **98**, pp. 681–698.
Czamanske, G.K., Ishihara, S. and Atkin, S.A. (1981) *Journal of Geophysical Research*, **86**, pp. 10431–10469.

D'Amico, C., Rottura, A., Maccarrone, E. and Puglisi, G. (1981) *Rendiconti Società Italiana di Mineralogia e Petrologia*, **38**, pp. 35–52.
Daly, R.A. (1912) *Canadian Department of Mines Memoir No. 38*, Part II, pp. 547–857.
Daly, R.A. (1925) *Proceedings of the American Academy of Arts and Science*, **60**, pp. 1–80.
Dalziel, I.W.D. (1981) *Philosophical Transactions of the Royal Society of London*, **A300**, pp. 319–335.
Davies, F.B. (1982) *Journal of Geology*, **90**, pp. 467–484.
Davies, G.F. (1991) *Earth and Planetary Science Letters*, **98**, pp. 405–407.
Debon, F. and Le Fort, P. (1988) *Bulletin de Mineralogie*, **111**, pp. 493–510.
Dempsey, C.S., Meighan, I.G. and Fallick A.E. (1990) *Geological Journal*, **25**, pp. 371–380.
Dempster, T.J., Hutton, D.H.W., Harrison, T.N., Brown, P.E. and Jenkin, G.R.J. (1991) *Contributions to Mineralogy and Petrology*, **107**, pp. 459–471.
Den Tex, E. (1990) *Geological Journal*, **25**, pp. 215–219.
Deniel, C., Vidal, Ph., Fernandez, A., Le Fort, P. and Peucal, J.-J. (1987) *Contributions to Mineralogy and Petrology*, **96**, pp. 78–92.
DePaolo, D.J. (1981a) *Journal of Geophysical Research*, **86**, pp. 10470–10488.
DePaolo, D.J. (1981b) *Earth and Planetary Science Letters*, **53**, pp. 189–202.
DePaolo, D.J. and Wasserburg, G.J. (1979) *Geochimica et Cosmochimica Acta*, **43**, pp. 615–627.
Didier, J. (1964) *Bulletin du Bureau de recherches géologiques et minières*, **3**, pp. 32–48.
Didier, J. and Barbarin, B. (eds) (1991) *Enclaves and Granite Petrology*. Elsevier, Amsterdam, 625 pp.
Didier, J. and Lameyre, J. (1969) *Contributions to Mineralogy and Petrology*, **24**, pp. 219–238.
Didier, J., Duthou, J.L. and Lameyre, J. (1982) *Journal of Volcanology and Geothermal Research*, **14**, pp. 125–132.
Dixon, J.M. (1975) *Tectonophysics*, **28**, pp. 89–124.
Dixon, S. and Rutherford, M.J. (1979) *Earth and Planetary Science Letters*, **45**, pp. 45–60.
D'Lemos, R.S., Brown, M. and Strachan, R.A. (1992) *Journal of the Geological Society of London*, **149**, pp. 487–490.
Dobrethou, G.L. and Lescov, S.A. (1987) *Soviet Geology*, **10**, pp. 9–17.
Dodge, F.C.K. and Bateman, P.C. (1988) *American Journal of Science*, **288A**, pp. 341–357.
Doe, B.R. (1967) *Journal of Petrology*, **8**, pp. 51–83.
Dougan, T.W. (1979) *Contributions to Mineralogy and Petrology*, **78**, pp. 337–344.
Dowty, E. (1980) In: Hargreaves, R. (ed.) *Physics of Magmatic Processes*. Princeton University Press, Princeton, NJ, pp. 419–485.
Drescher-Kaden, F.K. (1969) *Granitprobleme*. Akademie-Verlag, Berlin, 586 pp.
Drescher-Kaden, F.K. (1982) In: Drescher-Kaden, F.K. and Augustithis, S.S. (eds), *Transformists Petrology*, Theophrastus Publications, Athens, pp. 25–36.
Druitt, T.H. and Bacon, C.R. (1989) *Contributions to Mineralogy and Petrology*, **101**, pp. 245–259.

Durrance, E.M. and Bristow, C.M. (1986) *Proceedings of the Ussher Society*, **6**, pp.318–322.

Eberz, G.W., Nicholls, I.A. Maas, R., McCulloch, M.T. and Whitford, D.J. (1990) *Chemical Geology*, **85**, pp.119–134.

Eby, G.N. (1990) *Lithos*, **26**, pp.115–134.

Egeler, C.G. and DeBooy, T. (1956) *Verhandelingen Koninklijk Nederlandsch geologisch Sur.*, **17**, pp.1–86.

Egeler, D.H. and Burnham, C.W. (1973) *Geological Society of America Bulletin*, **84**, pp.2517–2532.

Elwell, R.W.D. (1958) *Journal of Geology*, **66**, pp.57–71.

Emeleus, C.H. (1991) *Tertiary Igneous Activity.* In: Craig, G.Y. (ed.), *Geology of Scotland*, 3rd edn. Geological Society of London, pp.455–502.

Emslie, R.F. (1978) *Precambrian Research*, **7**, pp.61–98.

England, P.C. and Thompson, A. (1986) In: Coward, M.P. and Ries, A.C. (eds), *Collision Tectonics. Geological Society of London Special Publication No. 19*, pp.83–94.

England, R.W. (1990) *Journal of the Geological Society of London*, **147**, pp.931–933.

Enrique, P. (1983) *Revista D'Investigacions Geologiques, Universitat de Barcelona*, **36**, pp.25–38.

Erdmannsdörffer, O.H. (1942) *Heidelberger Academie der Wissenschaften, Math. Nat. Kl. 2.*

Eskola, P. (1932) *Mineralogie und Petrologie Mitteilungen*, **42**, pp.455–481.

Eskola, P. (1955) *Bulletin Société géologique de Finlande*, **28**, pp.117–130.

Eugster, H.P. and Wilson, G.A. (1985) In: *High Heat Production (HHP) Granites, Hydrothermal Circulation and Ore Genesis.* Institution of Mining and Metallurgy, London, pp.87–98.

Evernden, J.F. and Kistler, R.W. (1970) *US Geological Survey Professional Paper No. 623*, 42 pp.

Fehn, U. (1985) In: *High Heat Production (HHP) Granites, Hydrothermal Circulation and Ore Genesis.* Institution of Mining and Metallurgy, London, pp.99–112.

Fenn, P.M. (1977) *Canadian Mineralogist*, **15**, pp.135–161.

Fernandez, A.N. and Barbarin, B. (1991) In: Didier, J. and Barbarin, B. (eds), *Enclaves and Granite Petrology.* Elsevier, Amsterdam, pp. 263–273.

Fernandez, A. and Tempier, P. (1971) *Bulletin du Bureau de recherches géologiques et minières*, France, **2**(IV) pp.357–366.

Fernandez, A.N. and Gasquet, D.R. (1994) *Contributions to Mineralogy and Petrology.* **116**, pp.316–326.

Fernandez, A.N., Feybesse, J. and Mezure, J. (1983) *Bulletin de la Société géologique de France*, **3**, pp.319–326.

Flagler, P.A. and Spray, J.G. (1991) *Geology*, **19**, pp.70–73.

Fleck, R.J. (1990) In: Anderson, J.L. (ed.), *The Nature and Origin of Cordilleran Magmatism. Geological Society of America Memoir No. 174*, pp.359–374.

Flinn, D. (1995) *Geological Journal*, **30**, pp.415–422.

Flood, R.H. and Shaw, S.E. (1975) *Contributions to Mineralogy and Petrology*, **52**, pp.157–164.

Flood, R.H. and Shaw, S.E. (1979) *Journal of Geology*, **87**, pp.417–425.

Flood, R.H. and Vernon, R.H. (1988) *Lithos*, **21**, pp.237–245.

Floyd, P.A. and Winchester, J.A. (1975) *Earth and Planetary Science Letters*, **27**, pp. 211–218.

Fonteilles, M. and Pascal, M.-L. (1985) *Compte rendu Académie des sciences Paris*, **300** (Serie II), pp. 1003–1006.

Foster, D.A. and Hyndman, D.W. (1990) In: Anderson, J.L. (ed.), *The Nature and Origin of Cordilleran Magmatism. Geological Society of America Memoir No. 174*, pp. 347–358.

Fountain, J.C. *et al.* (1989) *Journal of Volcanology and Greathermal Research*, **39**, pp. 279–296.

Fourcade, S. and Allègre, C.J. (1981) *Contributions to Mineralogy and Petrology*, **76**, pp. 177–195.

Fourcade, S. and Javoy, M. (1991) In: Didier, J. and Barbarin, B. (eds), *Enclaves and Granite Petrology*. Elsevier, Amsterdam, pp. 345–364.

Fowler, M.B. (1988) *Geology*, **16**, pp. 1026–1030.

France-Lanord, C. and Le Fort, P. (1988) *Transactions of the Royal Society of Edinburgh: Earth Sciences*, **79**, pp. 183–198.

French, W.J. (1966) *Proceedings of the Royal Irish Academy*, **64B**, pp. 303–322.

French, W.J. (1978) *Scientific Proceedings of the Royal Dublin Society*, **A6**, pp. 97–107.

French, W.J. and Pitcher, W.S. (1959) *Geological Magazine*, **96**, pp. 69–74.

Frost, T.P. and Mahood, G.A. (1987) *Bulletin of the Geological Society of America*, **99**, pp. 272–291.

Frutos, J. (1981) *Tectonophysics*, **72**, pp. 21–32.

Fyfe, W.S. (1973) *Tectonophysics*, **17**, 273–283.

Fyfe, W.S. (1988) *Transactions of the Royal Society of Edinburgh: Earth Sciences*, **79**, pp. 339–346.

Fyfe, W.S., Price, N.J. and Thompson, A.B. (1978) *Fluids in the Earth's Crust*. Elsevier, Amsterdam, 383 pp.

Gass, I.G. and Smewing, J.D. (1973) *Nature*, London, **242**, pp. 26–29.

Gastil, G. (1975) *Geology*, **3**, pp. 361–363.

Gastil, G. (1979) *Geology*, **7**, pp. 542–544.

Gastil, G. (1983) In: Roddick, J.A. (ed.), *Circum-Pacific Plutonic Terranes. Geological Society of America Memoir No. 159*, pp. 265–275.

Gastil, G. (1989) In: Walawender, M.J., Gastil, G., Clinkenbeard, J.P., Gunn, S.W., McCormick, W.V., Eastman, B.G., Wernikcke, R.S., Chadwick, B., Wardlaw, M.S., Calhoun, J.M. and Smith, D.W. (1989) In: Anderson, J.L. (ed.), *The Origin and Nature of Cordilleran Magmatism. Geological Society of America Memoir No. 174* pp. 1–17.

Gastil, R.G., Diamond, J. and Knaack, C. (1986) *Geological Society of America, Abstracts with Programs*, **18**, p. 109.

Giggenbach, W.F. (1992) *Earth and Planetary Science Letters*, **113**, pp. 495–510.

Gilluly, J. (Chairman) (1948) *Origin of Granite. Geological Society of America Memoir No. 28*. (Report of the Ottawa Conference of 1947.)

Giret, A. (1990) *Geological Journal*, **25**, pp. 239–247.

Glazner, A.F. (1991) *Geology*, **19**, pp. 784–786.

Goldschmidt, V.M. (1911) *Norske Videnskaps-Akademi Oslo, Skrifter I, Mat.-Nat. Kl, Kristiana* **1**, 226 pp.

Goranson, R.W. (1938) *American Journal of Science*, **35A**, pp. 71–91.

Gray, C.M. (1984) *Earth and Planetary Science Letters*, **70**, pp. 47–60.

Gray, C.M. (1990) *Australian Journal of Earth Science*, **37**, pp. 331–349.

Green, T.H. and Watson, E.B. (1982) *Contributions to Mineralogy and Petrology*, **79**, pp. 96–105.

Gromet, L.P. and Silver, L.T. (1987) *Journal of Petrology* **28**, pp. 75–125.

Grossenbacher, K. and Marsh, B.D. (1991) *Transactions of the American Geophysical Union*, **72**, pp. 315–316.

Grout, F.F. (1918) *Journal of Geology*, **26**, pp. 481–499.

Grout, F.F. (1945) *American Journal of Science*, **243A**, pp. 260–284.

Guillet, P., Bouchez, J.L. and Vigneresse, J.L. (1985) *Bulletin de la Société géologique de France*, **8**, pp. 503–513.

Hall, A. (1967) *Contributions to Mineralogy and Petrology*, **16**, pp. 156–171.

Halliday, A.N. (1984) *Nature*, **307**, pp. 229–233.

Halliday, A.N. (1985) *Nature*, **315**, p. 274.

Halliday, A.N., Stephens, W.E. and Harmon, R.S. (1980) *Journal of the Geological Society of London*, **137**, pp. 329–348.

Halliday, A.N., Mahood, G.A., Holden, P., Metz, J.M., Dempster, T.J. and Davidson, J.P. (1989) *Earth and Planetary Science Letters*, **94**, pp. 274–290.

Hamidullah, S. and Bowes, D.R. (1987) *Acta Universitatis Carolinae-Geologica*, **4**, pp. 295–396.

Hamilton, W. and Myers, W.B. (1967) *US Geological Survey Professional Paper No. 554-C*, 30 pp.

Hammarstrom, J.M. and Zen, E-an. (1986) *American Mineralogist*, **71**, pp. 1297–1313.

Hanna, S.S. and Fry, N. (1979) *Journal of Structural Geology*, **1**, pp. 155–162.

Hannah, J.L. and Stein, H.J. (1990) In: Stein, H.J. and Hannah, J.L. (eds), *Ore-bearing Granite Systems. Geological Society of America Special Paper No. 25*, **246**, pp. 1–10.

Hanson, R.B. (1995) *Bulletin of the Geological Society of America*, **107**, pp. 595–611.

Hanson, R.B. and Glazner, A.F. (1995) *Geology*, **23**, pp. 213–216.

Harayama, S. (1992) *Geology*, **20**, pp. 657–660.

Harker, A. (1900) *Journal of Geology*, **8**, pp. 359–399.

Harker, A. (1904) *Tertiary Igneous Rocks of Skye, Memoir of the Geological Survey of Scotland*, 481 pp.

Harker, A. (1909) *The Natural History of Igneous Rocks*. Methuen, London, 384 pp.

Harmon, R.S., Halliday, A.N., Clayburn, J.A.P. and Stephens, W.E. (1984) *Philosophical Transactions of the Royal Society of London*, **A310**, pp. 709–742.

Harrison, T.M. and Clarke, G.K.C. (1979) *Canadian Journal of Earth Science*, **16**, pp. 411–420.

Harrison, T.M. and McDougall, I. (1980) *Geochimica et Cosmochimica Acta*, **44**, pp. 1985–2003.

Harrison, T.M. and McDougall, I. (1981) *Earth and Planetary Science Letters*, **55**, pp. 123–149.

Harrison, T.M., Duncan, I. and McDougall, I. (1985) *Geochimica et Cosmochimica Acta*, **49**, pp. 2461–2468.

Harry, W.T. and Emeleus, H. (1960) *Proceedings of the 21st International Geological Congress, Copenhagen*, Vol. 14, pp. 172–181.

Hibbard, M.J. (1979) *Bulletin of the Geological Society of America*, **90**, pp. 1047–1062.

Hibbard, M.J. (1981) *Contributions to Mineralogy and Petrology*, **76**, pp. 158–170.

Hibbard, M.J. (1987) *Journal of Geology*, **95**, pp. 543–561.

Hibbard, M.J. and Watters, R.J. (1985) *Lithos*, **18**, pp. 1–12.

Hietanen, A. (1975) *Journal of Research US Geological Survey*, **3**, pp. 631–645.

Hildreth, W. (1981) *Journal of Geophysical Research*, **86**, pp. 10153–10192.

Hildreth, W. and Moorbath, S. (1988) *Contributions to Mineralogy and Petrology*, **98**, pp. 455–489.

Hill, R.I. (1988) *Journal of Geophysical Research*, **93**, pp. 10325–10348.

Hill, R.I., Silver, L.T., Chappell, B.W. and Taylor, H.P. (1985) *Nature*, **313**, pp. 643–646.

Hodge, D.S., Abbey, D.A., Harbin, M.A., Patterson, J.L., Ring, M.J. and Sweeney, J.F. (1982) *American Journal of Science*, **282**, pp. 1289–1324.

Hofmann, C. (1983) *Neues Jahrbuch für Mineralogie, Abh.* **146**, pp. 151–169.

Holder, M.T. (1979) In: Atherton, M.P. and Tarney, J. (eds), *Origin of Granite Batholiths: Geochemical Evidence*, Shiva, Nantwich, Cheshire, pp. 116–128.

Holden, P., Halliday, A.N. and Stephens, W.E. (1987) *Nature*, **330**, 53–56.

Holgate, N. (1954) *Journal of Geology*, **62**, pp. 439–480.

Hollister, L.S., Grissom, G.C., Peters, E.K., Stowell, H.H. and Sisson, V.B. (1987) *American Mineralogist*, **72**, pp. 231–239.

Holloway, J.R. (1987) In: Carmichael, I.S.E. and Eugster, H.P. (eds), *Thermodynamic Modelling of Geological Materials, Fluids and Melts. Mineralogical Society of America Reviews in Mineralogy*, **17**, pp. 211–234.

Holmes, A. (1945) *Nature*, **155**, pp. 412–415.

Holmquist, P.J. (1907) *Geologiska Foreningens Stockholm Forhandlingar*, **29**, pp. 313–354.

Holtedahl, O. (1963) *Norske Videnskaps-Akademi Oslo, Skrifter I, Mat.-Nat. Kl*, **12**, pp. 1–24.

Holtz, F. *et al.* (1995) *American Mineralogist*, **80**, pp. 94–108.

Hopson, R.F. and Ramseyer, K. (1990) *Geology*, **18**, pp. 336–339.

Hoschek, G. (1976) *Neues Jahrbuch für Mineralogie, Mh.*, **126**, pp. 79–83.

Huhma, H. (1986) *Geological Survey of Finland, Bulletin No. 337*, pp. 1–48.

Hunter, R.H. and Sparks, R.S.J. (1987) *Contributions to Mineralogy and Petrology*, **95**, pp. 451–461.

Huppert, H.E. and Sparks, R.J. (1988) *Journal of Petrology*, **29**, pp. 599–624.

Huppert, H.E. and Turner, J.S. (1991) *Journal of Petrology*, **32**, pp. 851–854.

Hutchinson, C.S. (1983) *Economic Deposits and their Tectonic Setting*. Macmillan, London, 365 pp.

Hutchison, W.W. (1982) *Geological Survey of Canada, Memoir No. 394*, 116 pp.

Hutton, D.H.W. (1982) *Journal of the Geological Society of London*, **139**, pp. 615–631.

Hutton, D.H.W. (1988a) *Transactions of the Royal Society of Edinburgh: Earth Sciences*, **79**, pp. 245–255.

Hutton, D.H.W. (1988b) *Bulletin of the Geological Society of America*, **100**, pp. 1392–1399.

Hutton, D.H.W. (1991) *Transactions of the Royal Society of Edinburgh: Earth Sciences*, **83**, pp. 383–386.

Hutton, D.H.W. (1992) *Transactions of the Royal Society of Edinburgh: Earth Sciences*, **83**, pp. 377–382.

Hutton, D.H.W. and Ingram, G.M. (1992) *Transactions of the Royal Society of Edinburgh: Earth Sciences*, **83**, pp. 383–386.

Hutton, D.H.W., Dempster, T.J., Brown, P.E. and Becker, S.D. (1990) *Nature*, **343**, pp. 452–455.

Huang, W.L. and Wyllie, P.J. (1974) *American Journal of Science*, **274**, pp. 378–395.

Hyndman, D.W. (1981) *Geology*, **9**, pp. 244–249.

Iddings, J.P. (1909) *Bulletin of the Philosophical Society*, Washington, **11**, pp. 65–113.

Iddings, J.P. (1909) *Igneous Rocks*. John Wiley and Sons, New York.
Ishihara, S. (1977) *Mining Geology Japan*, **27**, pp.293–300.
Iyer, H.M. (1984) *Philosophical Transactions of the Royal Society of London*, **A310**, pp.473–510.

Jacobson, R.R.E., Macleod, W.N. and Black, R. (1958) *Ring-Complexes in the Younger Granite Province of Northern Nigeria. Geological Society of London Memoir No. 1*, 72 pp.
Jahns, R.H. (1953) *American Mineralogist*, **38**, pp.563–598.
Jahns, R.H. and Burnham, C.W. (1969) *Economic Geology*, **64**, pp.843–864.
Jennings, D.J. and Sutherland, F.L. (1969) *Technical Reports of the Department of Mines, Tasmania*, **13**, pp.45–82.
Joesten, R. and Fisher, G. (1988) *Bulletin of the Geological Society of America*, **100**, pp.714–732.
Johannes, W. (1983) In: Atherton M.P. and Gribble, C.D. (eds), *Migmatites, Melting and Metamorphism*. Shiva, Nantwich, Cheshire, pp.27–36; 234–248.
Johannes, W. and Holtz, F. (1992) *Transactions of the Royal Society of Edinburgh: Earth Sciences*, **83**, pp.417–422.
Johannes, W. and Holtz, F. (1996) *Petrogenesis and Experimental Petrology of Granitic Rocks*. Springer-Verlag, Berlin, 335 pp.
John, B.E. (1988) *Geology*, **16**, pp.613–617.
Johnston, A.D. and Wyllie, P.J. (1988) *Contributions to Mineralogy and Petrology*, **98**, pp.352–362.
Jones, P.R. (1981) In: Kulm, L.D., Dymond, J., Dasch, E.J. and Hussong, D.M. (eds), *Nazca Plate: Crustal Formation and Andean Convergence. Geological Society of America Memoir No. 154*, pp.423–443.
Joplin, G.A. (1959) *Geological Magazine*, **96**, pp.361–373.
Jurewicz, S.R. and Watson, E.B. (1984a) *Contributions to Mineralogy and Petrology*, **85**, pp.125–129.
Jurewicz, S.R. and Watson, E.R. (1984b) *Geochimica et Cosmochimica Acta*, **49**, pp.1109–1121.

Kalakay, T.J. and Snoke, A.W. (1995) In: *Origin of Granitic and Related Rocks, US Geological Survey Circular*, **1129**, p.77.
Kanaris-Sotiriou, R. and Gibb, F.G.F. (1989) *Journal of the Geological Society of London*, **146**, pp.607–610.
Kanisawa, S. (1983) In: *Circum-Pacific Plutonic Terranes. Geological Society of America Memoir No. 159*, pp.129–134.
Kerrick, D.M. (1991) (ed.), *Contact Metamorphism. Mineralogical Society of America Reviews in Mineralogy*, **26**, 847 pp.
King, B.C. (1965) In: Pitcher, W.S. and Flinn, G.W. (eds), *Controls of Metamorphism*. Oliver and Boyd, Edinburgh, pp.219–234.
King, R.F. (1966) *Geological Journal*, **5**, pp.43–66.
Kistler, R.W. (1974) *Annual Review of Earth and Planetary Sciences*, **2**, pp.404–418.
Kistler, R.W. (1990) In: Anderson, J.L. (ed.), *The Nature and Origin of Cordilleran Magmatism. Geological Society of America Memoir No. 174*, pp.271–281.
Kistler, R.W. and Fleck, R.J. (1994) *US Geological Survey Open-file Report*, 94-267, 50 pp. Referenced in Paterson, S.R. and Vernon, R.H. (1995).
Kistler, R.W. and Peterman, Z.E. (1973) *Bulletin of the Geological Society of America*, **84**, pp.3489–3512.
Kovalenko, N.I. (1979) *Nauka*, Moscow, 152 pp. (in Russian).

Koyaguchi, T. (1987) *Earth and Planetary Science Letters*, **84**, pp.339–344.

Kumar, S.C. (1988) *Journal of Southeast Asian Earth Science*, **2**, pp.109–121.

Kuroda, Y., Suzuoki, T. and Matsuo, S. (1983) In: Roddick, J.A. (ed.), *Circum-Pacific Plutonic Terranes. Geological Society of America Memoir No. 159*, pp.123–128.

Lacroix, A. (1898) *Bulletin du Service de la carte géologique de France*, **64**, 10, pp.241–306.

Lacy, E.D. (1960) *21st International Geological Congress, Norden*, Pt 14, pp.7–15.

Lameyre, J. (ed.) (1983) *Le Plutonisme Oceanique intraplaque: Exemple de Isles Kerguelen. Comité national Français des Recherches antarctiques*, **54**, 290 pp.

Lameyre, J. and Bonin, B. (1991) In: Didier, J. and Barbarin, B. (eds), *Enclaves and Granite Petrology*. Elsevier, Amsterdam, pp.3–17.

Lameyre, J. and Bowden, P. (1982) *Journal of Volcanology and Geothermal Research*, **14**, pp.169–186.

Lameyre, J., Black, R., Bowden, P. and Giret, A. (1984) In: Xu Kegin and Tu Guangchi (eds), *Geology of Granites and their Metallogenic Relations. Proceedings of the International Symposium, Nanjing University*. Science Press, Beijing, pp.241–253.

Lange, R.L. and Carmichael, I.S.E. (1990) In: Nicholls, J, and Russell, J.K. (eds), *Modern Methods of Igneous Petrology. Mineralogical Society of America, Reviews in Mineralogy*, **24**, pp.25–64.

Langmuir, C.H. (1989) *Nature*, **340**, pp.199–205.

Langmuir, C.H., Vocke, R.D., Hanson, G.N. and Hart, S.R. (1977) *Earth and Planetary Science Letters*, **37**, pp.380–392.

Laouar, R., Boyce, A.J., Fallick, A.E. and Leake, B.E. (1990) *Geological Journal*, **25**, pp.359–370.

Lapadu-Harques, P. (1945) *Bulletin de la Société géologique de France*, **5**(15), pp.255–310.

La Roche, H. de. (1986) *Bulletin de la Société géologique de France*, **8**(11), pp.337–353.

La Roche, H. de, Leterrier, J., Grandclaude, P. and Marchal, M. (1980) *Chemical Geology*, **29**, pp.183–210.

Larsen, L.M. and Sorensen, H. (1987) In: Fitton, J.G. and Upton, B.G.J. (eds), *Alkaline Igneous Rocks. Geological Society of London Special Publication No. 30*, pp.473–488.

Larson, R.L. (1991) *Geology*, **19**, pp.547–550.

Law, R.D., Morgan, S.S., Casey, M., Sylvester, A.G. and Nyman, M. (1992) *Transactions of the Royal Society of Edinburgh: Earth Sciences*, **83**, pp.361–376.

Leake, B.E. (1990) *Journal of the Geological Society of London*, **147**, pp.579–589.

Leake, B.E. and Ahmed Said, Y. (1994) *Mineralogy and Petrology*, **51**, pp.243–250.

Leake, B.E. and Cobbing, J. (1993) *Scottish Journal of Geology*, **29**, pp.177–182.

Le Bel, L.M. (1985) In: Pitcher, W.S., Atherton, M.P., Cobbing, E.J. and Beckinsale, R.D. (eds), *Magmatism at a Plate Edge: The Peruvian Andes*. Blackie, Glasgow, pp.250–260.

Le Breton, N. and Thompson, A.B. (1988) *Contributions to Mineralogy and Petrology*, **99**, pp.226–237.

Le Fort, P. (1981) *Journal of Geophysical Research*, **86**, pp.10545–10568.

Le Fort, P., Coney, M., Deniel, C., France-Lanord, C., Sheppard, S.M.F., Upreti, B.N. and Vidal, P. (1987) *Tectonophysics*, **134**, p.39057.

Leger, J.-M. (1985) *Journal of African Earth Science*, **3**, pp.89–96.

Lehmann, B. (1990) *Metallogeny of Tin. Lecture Notes in Earth Sciences No. 32.* Springer-Verlag, Berlin, 212 pp.

Lesher, C.E. (1994) *Journal of Geophysical Research*, **99**, pp.9585–9604.

Leventhal, J.A., Reid, M.R., Montana, A. and Holden, P. (1995) *Geology*, **23**, pp.399–402.

Levi, B. and Aguirre, L. (1981) *Journal of the Geological Society of London*, **138**, pp.75–81.

Lindgren, W. (1900) *American Journal of Science*, **4**, pp.269–282.

Lindsley, D.H. and Anderson, D.J. (1983) *Journal of Geophysical Research*, **88**, A887–906.

Lipman, P.W. (1988) *Transactions of the Royal Society of Edinburgh: Earth Sciences*, **79**, pp.265–288.

Liu Yingjun, Zhang Jingrong, Sun Chengyuan, Ma Dongsheng, Qiao Enquang and Chen Jun (1984) *Geology of Granites and their Metallogenic Relations. Proceedings of the International Symposium, Nanjing University.* Science Press, Beijing, pp.753–770.

Livenson, A.A. (1974) *Introduction to Exploration Geochemistry.* Applied Publishing, Calgary, p.318.

Loewinson-Lessing, F.Y. (1936) *A Historical Survey of Petrology.* (Translation by Tomkeieff, S.I.) Oliver and Boyd, Edinburgh, 112 pp.

Lofgren, G.E. (1980) In: Hargraves, R.B. (ed.), *Physics of Magmatic Processes.* Princeton University Press, Princeton, NJ, pp.487–551.

Lofgren, G.E. and Donaldson, C.H. (1975) *Contributions to Mineralogy and Petrology*, **49**, pp.309–319.

Loiselle, M.C. and Wones, D.R. (1979) *Geological Society of America, Abstr. Prog.*, **11** (7), 468.

London, D. (1990) In: Stein, H.J. and Hannah, J.L. (eds), *Ore-bearing Granite Systems. Geological Society of America Special Paper No. 25*, pp.35–50.

Loomis, T.P. (1982) *Contributions to Mineralogy and Petrology*, **81**, pp.219–229.

Loomis, T.P. and Welber, P.W. (1982) *Contributions to Mineralogy and Petrology*, **81**, pp.230–239.

Lovera, O., Richter, F. and Harrison, M. See: Appenzeller, T. (1991) *Science*, **257**, pp.1588–1590.

Lundqvist, L. (1995) *Terra Nova*, **7**, *Abstract Supplement No. 1*, XI-2, p.143.

Luth, W.C. (1976) In: Bailey, D.K. and Macdonald, R. (eds), *The Evolution of the Crystalline Rocks.* Academic Press, New York, pp.333–417.

Lynn, H.B., Hale, L.D. and Thompson, G.A. (1981) *Journal of Geophysical Research*, **86**, pp.10633–10638.

Maaløe, S. (1985) *Principles of Igneous Petrology.* Springer-Verlag, Berlin, 374 pp.

Maaløe, S. and Johnson, A.D. (1986) *Contributions to Mineralogy and Petrology*, **93**, pp.449–458.

Maaløe, S. and Wyllie, P.J. (1975) *Contributions to Mineralogy and Petrology*, **52**, pp.175–191.

Mahon, K.I., Harrison, T.M. and Drew, D.A. (1988) *Journal of Geophysical Research*, **93**, pp.1175–1188.

Mahood, G.A. and Cornejo, P.C. (1992) *Transactions of the Royal Society of Edinburgh: Earth Sciences,* **83,** pp.63–69.

Mandlebrot, B.B. (1982) *The Fractal Geometry of Nature.* W.H. Freeman, San Francisco, 460 pp.

Maniar, P.D. and Piccoli, P.M. (1989) *Geological Society of America Bulletin,* **101,** pp.635–643.

Manning, D.A.C. (1981) *Contributions to Mineralogy and Petrology,* **76,** pp.206–215.

Manning, D.A.C. and Pichavant, M. (1983) In: Atherton, M.P. and Gribble, C.D. (eds), *Migmatites, Melting and Metamorphism.* Shiva, Nantwich, Cheshire, pp.94–109.

Marakushev, A.A. and Shapovalov, Yu.B. (1994) *Petrology,* **2,** 1–18. Translated from *Petrologiya,* **2**(1), 1994, pp.4–23.

Marmo, V. (1971) *Granite Petrology and the Granite Problem.* Elsevier, Amsterdam, 244 pp.

Marre, J. (1986) *The Structural Analysis of Granitic Rocks* (English translation). North Oxford Academic, Kogan Page, London, 123 pp.

Marre, J. and Pons, J. (1973) Structurologie. *Encyclopaedia Universalis,* **15,** pp.455–460.

Marsh, B.D. (1981) *Contributions to Mineralogy and Petrology,* **78,** pp.85–98.

Marsh, B.D. (1982) *American Journal of Science,* **282,** pp.808–855.

Marsh, B.D. (1989) *Journal of Petrology,* **30,** pp.479–530.

Marsh, B.D. (1991) *Journal of Petrology,* **32,** pp.855–860.

Marsh, B.D. (1996) *Mineralogical Magazine,* **60,** pp.5–40.

Marshall, S.D. and DePaolo, D.J. (1989) *Geochimica et Cosmochimica Acta,* **53,** pp.917–953.

Martin, D. and Nokes, R. (1988) *Nature,* **332,** pp.534–536.

Martin, H. (1986) *Geology,* **14,** pp.753–756.

Martin, H. (1987) In: *International Symposium on Granites and Associated Mineralizations, Extended Abstracts, Salvador, Bahia, Brazil,* pp.23–24.

Martin, M.R. (1953) *Quarterly Journal of the Geological Society of London,* **108,** pp.311–342.

Martin, R.F. and Bonin, B. (1976) *Canadian Mineralogist,* **14,** pp.228–237.

Mason, D.R. and McDonald, J.A. (1978) *Economic Geology,* **73,** pp.857–877.

Mason, G.H. (1985) In: Pitcher, W.S., Atherton, M.P., Cobbing, E.J. and Beckinsale, R.D. (eds), *Magmatism at a Plate Edge: The Peruvian Andes.* Blackie, Glasgow, pp.156–166.

Matsuhisa, Y., Goldsmith, J.R. and Clayton, R.N. (1979) *Geochimica et Cosmochimica Acta,* **43,** pp.1131–1140.

McBirney, A.R. (1979) In: Yoder, H.S. (ed.), *The Evolution of the Igneous Rocks,* Princeton University Press, Princeton, NJ, pp.307–338.

McBirney, A.R. (1980) *Journal of Volcanological and Geothermal Research,* **7,** pp.357–371.

McBirney, A.R. and Murase, T. (1984) *Annual Review of Earth and Planetary Sciences,* **12,** pp.337–357.

McBirney, A.R. and Nakamura, Y. (1974) *Carnegie Institution of Washington Year Book,* **73,** 348–352.

McCaffrey, K.J.W. (1992) *Journal of the Geological Society of London,* **149,** pp.221–235.

McCarthy, T.S. and Hasty, R.A. (1976) *Geochimica et Cosmochimica Acta,* **40,** pp.1351–1358.

McCourt, W.J. (1981) *Journal of the Geological Society of London*, **138**, pp. 407–420.
McCulloch, M.T. and Chappell, B.W. (1982) *Earth and Planetary Science Letters*, **58**, pp. 51–64.
MacDonald, R., McGarvie, D.W., Pinkerton, H., Smith, R.L. and Palacz, Z.A. (1990) *Journal of Petrology*, **31**, 429–459.
McDougall, I. (1964) *Journal of the Geological Society of Australia*, **11**, 107–143.
McGregor, V.R. (1973) *Philosophical Transactions of the Royal Society of London*, **A273**, pp. 343–358.
McKenzie, D.P. (1978) *Earth and Planetary Science Letters*, **40**, pp. 25–32.
McKenzie, D.P. (1984) *Journal of Petrology*, **25**, pp. 713–765.
McKenzie, D.P. (1985) *Earth and Planetary Science Letters*, **74**, pp. 81–91.
McKenzie, D.P. and Bickle, M.J. (1988) *Journal of Petrology*, **29**, pp. 625–679.
McLellan, A.G. (1980) *The Classical Thermodynamics of Deformable Materials*. Cambridge University Press.
McLellan, E. (1984) *Geological Magazine*, **121**, pp. 339–345.
McNutt, R.H., Crocket, J.H., Clark, A.H., Caelles, J.C., Farrar, E., Haynes, S.J. and Zentilli, M. (1975) *Earth and Planetary Science Letters*, **27**, pp. 305–313.
Mehnert, K.R. (1968) *Migmatites and the Origin of Granitic Rocks*. Elsevier, Amsterdam, 393 pp.
Mehnert, K.R. (1987) *Fortschritte der Mineralogie*, **65**(2), pp. 285–306.
Mehnert, K.R. and Büsch, W. (1985) *Neues Jahrbuch für Mineralogie*, Abh., **151**, pp. 229–259.
Mehnert, K.R., Büsch, W. and Schneider, G. (1973) *Neues Jahrbuch für Mineralogie*, Abh., **4**, pp. 165–183.
Meighan, I.G., Gibson, D. and Hood, D.N. (1984) *Mineralogical Magazine*, **48**, pp. 351–363.
Meyer, C. (1985) *Science*, **227**, pp. 1421–1428.
Miller, C.F. and Barton, M.D. (1989) In: Kay, S.M. and Rapela, D.W. (eds), *Plutonism from Alaska to Antarctica. Geological Society of America Special Paper No. 241*, pp. 213–232.
Miller, C.F. and Bradfish, L.J. (1980) *Geology*, **8**, pp. 412–416.
Miller, C.F., Stoddard, E.F., Bradfish, L.J. and Dollase, W.A. (1981) *Canadian Mineralogist*, **19**, pp. 25–34.
Miller, C.F., Watson, E.B. and Harrison, T.M. (1988) *Transactions of the Royal Society of Edinburgh: Earth Sciences*, **79**, pp. 135–156.
Mitchell, A.H.G. and Garson, M.S. (1981) *Mineral Deposits and Global Tectonic Settings*. Academic Press, 405 pp.
Möller, P. (1989) Referenced in Lehmann, B. (1990).
Molyneux, S.J. (1995) Abstracts, Tectonic Studies Group, Geological Society of London, Cardiff Meeting, Convenors, Gayer, R.A. and Lisle, R.J., University of Wales, Cardiff.
Moorbath, S., Taylor, P.N. and Goodwin, R. (1981) *Geochimica et Cosmochimica Acta*, **45**, pp. 1051–1060.
Moorbath, S., Thompson, R.N. and Oxburgh, E.R. (1984) *Philosophical Transactions of the Royal Society of London*, **A310**, pp. 437–480.
Moore, J.G. (1959) *Journal of Geology*, **67**, pp. 198–210.
Moore, J.G. and Lockwood, J.P (1973) *Bulletin of the Geological Society of America*, **84**, pp. 1–20.
Moreau, C. (1982) Quoted in Bowden, P. *et al.* (1987).
Morgan, G.B. and London, D. (1987) *American Mineralogist*, **72**, pp. 1097–1121.

Morris, J.D. *et al.* (1990) *Nature*, **844**, pp.51–56.
Mukasa, S.B. (1986) *Bulletin of the Geological Society of America*, **97**, pp.241–254.
Mukasa, S.B. and Tilton, G.R. (1985) In: Pitcher, W.S., Atherton, M.P., Cobbing, E.J. and Beckinsale, R.D. (eds), *Magmatism at a Plate Edge: The Peruvian Andes*. Blackie, Glasgow, pp.235–240.
Mullan, H.S. and Bussell, M.A. (1977) *Geological Magazine*, **114**, pp.265–280.
Muller, O.H. and Pollard, D.D. (1977) *Pageoph.*, **115**, pp.69–86.
Murry, J.D. (1979) In: Abbott, P.L. and Todd, V.R. (eds), *Geological Society of America Guidebook, 1979*, pp.163–176.

Nabelek, C.R. and Lindsley, D.H. (1985) *Geological Society of America, Abstr. Prog. No. 68384*, p.673.
Nachit, H. *et al.* (1985) *Comptes rendu de l'Académie des Science*, Paris, **301**, pp.813–818.
Naney, M.T. (1983) *American Journal of Science*, **283**, pp.993–1033.
Natland, J. (1991) In: Floyd, P.A. (ed.), *Oceanic Basalts*. Blackie, Glasgow, pp.63–93.
Neiva, A.M.R. (1986) *Chemical Geology*, **63**, pp.299–317.
Nekvasil, H. (1988) *American Mineralogist*, **73**, pp.966–982.
Nekvasil, H. (1991) *American Mineralogist*, **76**, pp.1279–1290.
Nekvasil, H. (1992) *Transactions of the Royal Society of Edinburgh: Earth Sciences*, **83**, pp.399–407.
Nicholls, J. and Russell, J.K. (1990) Pearce element ratios – an overview, example and bibliography. In: Russell, J.K. and Stanley, C.R. (eds), *Theory and Application of Pearce Element Ratios to Geochemical Data Analysis. Geological Association of Canada, Short Course Notes*, **8**, pp.11–21.
Nicholls, J., Russell, J.K. and Stout, M.Z. (1986) In: Scarfe, C.M. (ed.), *Short Course in Silicate Melts, No. 12. Mineralogical Association of Canada*, pp.210–235.
Nickel, E., Kock, H. and Nangässer, W. (1967) *Schweizerische Mineralogische und Petrographische Mitteilungen*, **47**, pp.399–498.
Noble, S.R. and Searle, M.P. (1995) *Geology*, **23**, pp.1135–1138.
Nockolds, S.R. (1933) *Journal of Geology*, **41**, pp.561–589.
Nockolds, S.R. (1934) *Geological Magazine*, **71**, pp.31–39.
Nockolds, S.R. (1946) *Geological Magazine*, **83**, pp.206–216.
Nockolds, S.R. and Allen, R. (1956) *Geochimica et Cosmochimica Acta* **9**, pp.34–77.
Nockolds, S.R. and Mitchell, R.L. (1948) *Transactions of the Royal Society of Edinburgh*, **61**, pp.533–575.
Norton, D.L. (1982) In: Titley, S.R. (ed.), *Advances in Geology of the Porphyry Copper Deposits: Southwestern North America*. University of Arizona Press, Tucson, pp.59–71.
Nozawa, T. (1983) In: Roddick, J.A. (ed.), *Circum-Pacific Plutonic Terranes. Geological Society of America Memoir No. 159*, pp.105–120.
Nozawa, T. and Tainasho, Y. (1990) In: Shimizu, M. and Gastil, G. (eds), *Recent Advances in Concepts Concerning Zoned Plutons in Japan and Southern and Baja California. Nature and Culture*, No. 2. University Museum, University of Tokyo, pp.101–114.

Odé, H. (1957) *Bulletin of the Geological Society of America*, **68**, pp.567–575.
Oftedahl, C. (1953) *Norske Videnskaps-Akademie Oslo, Mat.-Nat. Kl.3*.

Oglethorpe, R. (1987) *A Mineralogical and Chemical Study of the Interaction between Granite Magma and Pelitic Country Rock, Thorr Pluton, Co. Donegal, Eire.* PhD Thesis, University of Liverpool.

O'Hara, M.J. (1977) *Nature*, **266**, pp.503–507.

O'Hara, M.J. and Mathews, R.E. (1981) *Journal of the Geological Society of London*, **138**, 237–277.

O'Hara, M.J. (1993) In: Prichard, H.M. *et al.* (eds), *Magmatic Processes and Plate Tectonics*, Geological Society of London Special Publication No. 76, pp.39–60.

Oliver, H.W. (1977) *Bulletin of the Geological Society of America*, **88**, pp.445–461.

Olsen, S.N. (1985) In: Ashworth, J.R. (ed.), *Migmatites*. Blackie, Glasgow, pp.145–178.

O'Neil, J.R. and Chappell, B.W. (1977) *Journal of the Geological Society of London*, **133**, pp.559–571.

O'Nions, R.K. and Grönvold, K. (1973) *Earth and Planetary Science Letters*, **19**, pp.397–409.

O'Nions, R.K. and Pankhurst, R.J. (1978) *Earth and Planetary Science Letters*, **38**, pp.211–236.

Operto, S. and Charvis, P. (1995) *Geology*, **23**, pp.137–140.

Orsini, J.B. (1979) *Comptes rendu de l'Académie des Sciences*, Paris, **289**, pp.981–984.

Pabst, A. (1928) *Observations on Inclusions in the Granitoid Rocks of the Sierra Nevada. University of California Publications, Department of Geological Sciences No. 17*, pp.325–386.

Palivcova, M. (1981) *Geologicky Zbornik, Geologica Carpathica*, **32**(5), pp.559–589.

Palivcova, M. (1982) *Transformists' Petrology*, Theophrastus Publications, Athens, pp.149–175.

Palmason, G. (1986) In: Vogt, P.R. and Tucholke, B.E. (eds), *The Geology of North America: The Western North Atlantic Region*. Geological Society of America, pp.87–97.

Pankhurst, R.J. (1979) In: Atherton, M.P. and Tarney, J. (eds), *Origin of Granite Batholiths: Geochemical Evidence*. Shiva, Nantwich, Cheshire, pp.18–33.

Pankhurst, R.J., Hole, M.J. and Brook, M. (1988) *Transactions of the Royal Society of Edinburgh: Earth Sciences*, **79**, pp.123–134.

Parmentier, M. and Schedl, A. (1981) *Journal of Geology*, **89**, pp.1–22.

Parsons, I. and Butterfield, A.W. (1981) *Journal of the Geological Society of London*, **138**, pp.289–306.

Patchett, J. and Kouvo, O. (1986) *Contributions to Mineralogy and Petrology*, **92**, pp.1–12.

Paterson, B.A. *et al.* (1992) *Transactions of the Royal Society of Edinburgh: Earth Sciences*, **83**, pp.459–471.

Paterson, S.R. and Fowler, T.K. (1993) *Journal of Structural Geology*, **15**, pp.191–206.

Paterson, S.R. and Tobisch, O.T. (1988) *Geology*, **16**, pp.1108–1111.

Paterson, S.R. and Tobisch, O.T. (1992) *Journal of Structural Geology*, **14**, pp.297–300.

Paterson, S.R. and Vernon, R.H. (1995) *Bulletin of the Geological Society of America*, **107**, pp.1356–1380.

Paterson, S.R., Vernon, R.H. and Tobisch, O.T. (1989) *Journal of Structural Geology*, **11**, pp.349–363.

Paterson, S.R., Brudos, T., Fowler, K., Carlson, C. and Bishop, K. (1991) *Geology*, **19**, pp. 324–327.

Paterson, S.R., Vernon, R.H. and Fowler, T.K. (1991) In: Kerrick, D.M. (ed.), *Contact Metamorphism. Mineralogical Society of America Reviews in Mineralogy*, **26**, pp. 673–722.

Patiño-Douce, A.E. and Beard, J.S. (1995) Abstracts III Hutton Symposium, *Origin of Granites and Related Rocks, US Geological Survey Circular*, **1129**, pp. 112–113.

Patiño-Douce, A.E. and Johnstone, A.D. (1991) *Contributions to Mineralogy and Petrology*, **107**, pp. 202–218.

Pearce, J.A. (1987) *Journal of Volcanological and Geothermal Research*, **32**, pp. 51–65.

Pearce, J.A. and Norry, M.J. (1979) *Contributions to Mineralogy and Petrology*, **69**, pp. 33–47.

Pearce, J.A., Harris, N.B.W. and Tindle, A.G. (1984) *Journal of Petrology*, **45**, pp. 956–983.

Pearce, T.H. (1969) *Contributions to Mineralogy and Petrology*, **19**, pp. 142–157.

Percival, J.A. (1991) *Journal of Petrology*, **32**, pp. 1261–1297.

Perfit, M.R. and Lawrence, J.R. (1979) *Earth and Planetary Science Letters*, **45**, pp. 16–22.

Perfit, M.R., Brueckner, H., Lawrence, J.R. and Kay, R.W. (1980) *Contributions to Mineralogy and Petrology*, **73**, pp. 69–87.

Perrin, R. and Roubault, M. (1949) *Journal of Geology*, **57**, pp. 357–379.

Petford, N. (1993) In: Stone, D.B. and Runcorn, S.K. (eds), *Flow and Creep in the Solar System*. Kluwer Academic, The Netherlands, pp. 281–286.

Petford, N. (1995) *Journal of Geophysical Research*, **100**, pp. 15735–15745.

Petford, N. and Atherton, M.P. (1992) *Tectonophysics*, **205**, pp. 171–185.

Petford, N., Byron, D., Atherton, M.P. and Hunter, R.H. (1993) *European Journal of Mineralogy*, **5**, pp. 593–598.

Petford, N., Lister, J.R. and Kerr, R.C. (1994) *Lithos*, **32**, pp. 161–168.

Petford, N., Paterson, B., McCaffrey and Pugliese, S. (1996) *European Journal of Mineralogy*, **8**, 405–412.

Peucat, J.J. (1986) *Journal of the Geological Society*, London, **143**, pp. 875–886.

Pfiffner, O.A. and Ramsay, J.G. (1982) *Journal of Geophysical Research*, **87**, pp. 311–321.

Phillips, W.J., Fuge, R. and Phillips, N. (1981) *Journal of the Geological Society of London*, **138**, pp. 351–366.

Pidgeon, P.T. and Aftalion, M. (1978) *Geological Journal Special Issue No. 10*, pp. 183–284.

Pidgeon, P.T. and Compston, W. (1992) *Transactions of the Royal Society of Edinburgh: Earth Sciences*, **83**, pp. 473–483.

Pin, C. (1991) In: Didier, J. and Barbarin, B. (eds), *Enclaves and Granite Petrology*. Elsevier, Amsterdam, pp. 333–343.

Pin, C. and Sills, J.D. (1986) In: Dawson, J.B. (ed.) *Nature of the Lower Crust. Geological Society of America Special Publication No. 25*, pp. 231–249.

Pin, C., Binon, M., Belin, J.M., Barbarin, B. and Clemens, J.D. (1990) *Journal of Geophysical Research*, **95**, pp. 17821–17828.

Pirsson, L.V. (1905) *US Geological Survey Bulletin No. 237*, pp. 1–208.

Pitcher, W.S. (1952) *Proceedings of the Geologists' Association*, London, **64**, pp. 153–183.

Pitcher, W.S. (1953) *Quarterly Journal of the Geological Society of London*, **108**, pp.413–446.

Pitcher, W.S. (1970) In: Newall, G. and Rast, N. (eds), *Mechanism of Igneous Intrusion. Geological Journal Special Publication No. 2*, pp.123–140.

Pitcher, W.S. (1975) *Journal of the Geological Society, London*, **131**, pp.584–591.

Pitcher, W.S. (1978) (Presidential Address) *Journal of the Geological Society of London*, **135**, pp.157–182.

Pitcher, W.S. (1979) (Presidential Address) *Journal of the Geological Society of London*, **136**, pp.627–662.

Pitcher, W.S. (1982) In: Hsu, K. (ed.), *Mountain Building Processes*. Academic Press, London, pp.19–40.

Pitcher, W.S. (1987) *Geologie Rundschau*, **76**, pp.51–79.

Pitcher, W.S. (1988) *Rendiconti della Società Italiana di Mineralogia e Petrologia*, **43**(2), pp.275–280.

Pitcher, W.S. (1991) In: Didier, J. and Barbarin, B. (eds), *Enclaves and Granite Petrology*. Elsevier, Amsterdam, pp.383–392.

Pitcher, W.S. (1992) *Transactions of the Royal Society of Edinburgh*, **83**, Abstracts, p.497.

Pitcher, W.S. and Berger, A.R. (1972) *The Geology of Donegal: A Study of Granite Emplacement and Unroofing*. Wiley-Interscience, New York, 435 pp.

Pitcher, W.S. and Bussell, M.A. (1977) *Journal of the Geological Society of London*, **133**, pp.249–256.

Pitcher, W.S. and Bussell, M.A. (1985) In: Pitcher, W.S., Atherton, M.D., Cobbing, E.J. and Beckinsale, R.D. (eds), *Magmatism at a Plate Edge: The Peruvian Andes*. Blackie, Glasgow, pp.102–107.

Pitcher, W.S. and Cobbing, E.J. (1985) In: Pitcher, W.S. *et al.* (eds), *Magmatism at a Plate Edge*. Blackie, Glasgow, pp.285–289.

Pitcher, W.S. and Read, H.H. (1952) *Geological Magazine*, **89**, pp.328–336.

Pitcher, W.S. and Read, H.H. with others (1959) *Quarterly Journal of the Geological Society of London*, **114**, pp.259–305.

Pitcher, W.S. and Read, H.H. (1963) *Journal of Geology*, **71**, pp.261–296.

Pitcher, W.S., Atherton, M.P., Cobbing, E.J. and Beckinsale, R.D. (1985) *Magmatism at a Plate Edge: The Peruvian Andes*. Blackie-Halsted Press, Glasgow, 328 pp.

Pitfield, P.E.J., Teoh, L.H. and Cobbing, E.J. (1990) *Geological Journal*, **25**, pp.419–430.

Piwinskii, A.J. (1973) *Neues Jahrbuch für Mineralogie*, Mh., **5**, pp.193–215.

Piwinskii, A.J. and Wyllie, P.J. (1968) *Journal of Geology*, **76**, pp.205–234.

Plant, J.A. *et al.* (1983) *Transactions of the Institution of Mining and Metallurgy*, **B92**, pp.33–42.

Platten, I.M. (1982) *Geological Magazine*, **119**, pp.413–419.

Platten, I.M. (1983) *Geological Magazine*, **120**, pp.37–49.

Platten, I.M. (1984) *Geological Journal*, **19**, pp.209–226.

Power, G.M. (1993) *Journal of the Geological Society of London*, **150**, pp.465–468.

Presnall, D.C. (1969) *American Journal of Science*, **267**, pp.1178–1194.

Presnall, D.C. and Bateman, P.C. (1973) *Bulletin of the Geological Society of America*, **84**, pp.3180–3202.

Pupin, J.P. (1980) *Contributions to Mineralogy and Petrology*, **73**, pp.207–220.

Pupin, J.P. (1985) *Schweizeriche Mineralogische und Petrographische Mitteilungen*, **65**, pp.29–56.

Pupin, J.P. (1988) *Rendiconti della Società Italiana di Mineralogia e Petrologia*, **43**, pp.237–262.

Purdy, J.W. and Jäger, E. (1976) *Memoire degli Instituti di geologia e mineralogia dell'* *Universita di Padova No. 30*, pp. 1–31.

Quick, J.E., Sinigoi, S., Negrini, L., Demarchi, G. and Mayer, A. (1992) *Geology*, **20**, pp. 613–616; **23**, pp. 739–742.

Rae, D.A. and Chambers, A.D. (1988) *Transactions of the Royal Society of Edinburgh*, **79**, pp. 1–12.

Raguin, E. (1946) *Géologie du Granite*. Masson, Paris, 275 pp.

Ramberg, H. (1970) In: Newall, G. and Rast, N. (eds), *Mechanism of Igneous Intrusion. Geological Journal Special Issue No. 2*, pp. 261–286.

Ramberg, H. (1981) *Gravity, Deformation and the Earth's Crust*, 2nd edn. Academic Press, London, 452 pp.

Ramberg, I.B. and Larsen, B.T. (1978) *Norges geologiske undersokelse*, **337**, pp. 55–73.

Ramberg, I.B. and Spjeldnaes, N. (1978) In: Ramberg, I.B. and Neumann, E.-R. (eds), *Tectonics and Geophysics of Continental Rifts*. D. Reidel Publishing Co., Dordrecht, pp. 167–194.

Rämö, O.T. and Haapala, I. (1991) In: Gower, C.F., Rivers, T. and Ryan, B. (eds), *Mid-Proterozoic Laurentia-Baltica. Geological Association of Canada Special Paper No. 38*, pp. 401–415.

Ramsay, C.R., Stoeser, D.B. and Drysdall, A.R. (1986) *Journal of African Earth Sciences*, **4**, pp. 13–20.

Ramsay, J.G. (1975) *Research Institute on African Geology, University of Leeds, Annual Report*, **19**, p. 81.

Ramsay, J.G. (1989) *Journal of Structural Geology*, **11**, pp. 191–209.

Rapp, R.P., Watson, E.B. and Miller, C.F. (1991) *Precambrian Research*, **51**, pp. 1–25.

Read, H.H. (1957) *The Granite Controversy*. Thomas Murby and Co., London, 430 pp.

Reagan, M.K., Herrstrom, E.A. and Murrell, M.R. (1991) *Geological Society of America, Abstracts*, **22**, No. 19780.

Reavy, R.J. (1989) *Journal of the Geological Society of London*, **136**, pp. 649–657.

Regan, P.F. (1985) In: Pitcher, W.S., Atherton, M.P., Cobbing, E.J. and Beckinsale, R.D. (eds), *Magmatism at a Plate Edge: The Peruvian Andes*. Blackie, Glasgow, pp. 72–89.

Reynolds, D.L. (1947) *Geological Magazine*, **84**, pp. 209–223.

Reynolds, D.L. (1954) *American Journal of Science*, **252**, pp. 577–614.

Ribe, N.M. (1987) *Journal of Volcanology and Geothermal Research*, **33**, pp. 241–253.

Rickard, M.P. and Ward, P. (1981) *Journal of the Geological Society of Australia*, **28**, pp. 19–32.

Roberts, J.L. (1970) In: Newall, G. and Rast, N. (eds), *Mechanisms of Igneous Intrusion. Geological Journal Special Issue No. 2*, pp. 287–338.

Robertson, A. and Xenophontos, C. (1993) In: Prichard, H.M. *et al.* (eds), *Magmatic Processes and Plate Tectonics, Geological Society of London Special Publication No. 76*, pp. 85–119.

Robertson, J.K. and Wyllie, P.S. (1971) *American Journal of Science*, **271**, pp. 252–277.

Robinson, P.T. and Malpas, J. (1990) In: Malpas, J. *et al.* (eds), *Ophiolites: Oceanic Crustal Analogues. Cyprus Geological Survey Department*, pp. 13–36.

Rock, N.M.S. (1991) *Lamprophyres*. Blackie, Glasgow, 285 pp.

Roddick, J.A. (ed.) (1983) *Circum-Pacific Plutonic Terranes. Geological Society of America Memoir No. 159.*

Roddick, J.C. and Armstrong, J.E. (1959) *Journal of Geology*, **67**, pp. 603–613.

Roedder, E. (1979) In: Yoder, H.S. (ed.), *The Evolution of the Igneous Rocks: 50th Anniversary Perpectives.* Princeton University Press, Princeton, 588 pp.

Rogers, G. (1985) *Geochemical Traverse across the North Chilean Andes.* PhD Thesis, Open University, Milton Keynes.

Rogers, G. and Dunning, G.R. (1991) *Journal of the Geological Society of London*, **148**, pp. 17–27.

Rogers, G. and Hawkesworth, C.J. (1989) *Earth and Planetary Science Letters*, **91**, pp. 271–283.

Rogers, J.W. and Greenberg, J.K. (1981) *Bulletin of the Geological Society of America*, **92**, pp. 6–9.

Roobol, M.J. (1971) *Geological Magazine*, **108**, pp. 525–531.

Rosenberg, C.L., Berger, A. and Schmid, S.M. (1995) *Geology*, **23**, pp. 443–446.

Rossi, P. and Chevremont, P. (1987) *Géochronique*, **21**, pp. 14–18.

Roturra, A., Caggianelli, A., Campana, R. and Del Moro, A. (1993) *European Journal of Mineralogy*, **5**, pp. 737–754.

Routhier, P. (1967) *Essai critique sur les Méthodes de la Géologie. De l'Objet à la Genèse.* Masson, Paris, 204 pp.

Roycroft, P.D. (1991) *Geology*, **19**, pp. 437–440.

Rubin, A.M. (1995) *Journal of Geophysical Research*, **100**, pp. 5911–5929.

Rushmer, T. (1991) *Contributions to Mineralogy and Petrology*, **107**, pp. 41–59.

Rutter, E.H. and Neumann, D.H.K. (1995) *Journal of Geophysical Research*, **100**, 15697–15715.

Rutter, M.J. and Wyllie, P.J. (1988) *Nature*, **331**, pp. 159–160.

Sabatier, H. (1980) *Bulletin de Mineralogie*, **103**, pp. 507–522.

Sahama, Th.G. (1945) *Bulletin de la Commission géologique de Finlande*, **136**, pp. 15–67.

Saleeby, J.B., Kistler, R.W., Longiaru, S., Moore, J.G. and Nokleberg, W.J. (1990) In: Anderson, J.L. (ed.), *The Nature and Origin of Cordilleran Magmatism. Geological Society of America Memoir No. 174*, pp. 251–268.

Samms, M.S. and Thomas-Betts, A. (1988) *Journal of the Geological Society of London*, **145**, pp. 809–817.

Sanderson, D.J. and Meneilly, A.W. (1981) *Journal of Structural Geology*, **3**, pp. 109–116.

Sawka, W.N. (1988) *Transactions of the Royal Society of Edinburgh: Earth Sciences*, **79**, pp. 157–168.

Sawka, W.N., Chappell, B.W. and Kistler, R.W. (1990) *Journal of Petrology*, **31**, pp. 519–553.

Sawyer, E.W. (1987) *Journal of Petrology*, **28**, pp. 445–473.

Sawyer, E.W. and Barnes, S.J. (1988) *Journal of Metamorphic Geology*, **6**, pp. 437–450.

Scaillet, B., Pichavant, M. and Holtz, F. (1995) *Terra Nova*, **7**, Abstract Supplement No. I, X1–2, p. 136.

Schärer, U., Copeland, P., Harrison, T.M. and Searle, M.P. (1990) *Journal of Geology*, **98**, pp. 233–251.

Scarfe, C. (1986) *Mineralogical Association of Canada, Short Course Handbook 12*, pp. 36–56.

Schermerhorn, L.J.G. (1956) *Tschermaks Minerologische und Petrographische Mitteilungen*, **6**, pp. 74–114.

Schermerhorn, L.J.G. (1987) *Revista Brasileira de Geociências*, **17**, pp.617–618,
Schmidt, C.J., Smedes, H.W. and O'Neill, J.M. (1990) *Geological Journal*, **25**, pp.305–318.
Sederholm, J.J. (1891) *Tschermaks Mineralogische und Petrographische Mitteilungen*, **12**, pp.1–31.
Sederholm, J.J. (1967) *Selected Works: Granites and Migmatites*. Oliver and Boyd, Edinburgh. (Includes appreciation by P. Eskola).
Seedman, J.K. and Donaldson, C.H. (1996) *Mineralogical Magazine*, **60**, pp.115–130.
Shackleton, R.M., Ries, A.C. and Coward, M.P. (1982) *Journal of the Geological Society of London*, **139**, pp.533–541.
Shand, S.J. (1947) *Eruptive Rocks*. T. Murby, London (3rd edn, 1st edn 1927), 488 pp.
Shaw, H.R. (1965) *American Journal of Science*, **263**, pp.120–152.
Shaw, H.R. (1978) *Geochimica et Cosmochimica Acta*, **42**, pp.933–943.
Shaw, H.R. (1980) In: Hargreaves, R.B. (ed.), *Physics of Magmatic Processes*. Princeton University Press, Princeton, NJ, pp.201–264.
Shelly, D. (1964) *American Mineralogist*, **49**, pp.41–52.
Shelly, D. (1970) *Mineralogical Magazine*, **37**, pp.674–681.
Sheppard, S.M.F. (1977) *Journal of the Geological Society of London*, **133**, pp.573–591.
Shimizu, M. and Gastil, G. (eds) (1990) *Recent Advances in Concepts Concerning Zoned Plutons in Japan and Southern and Baja California. Nature and Culture*, No. 2. University Museum, University of Tokyo, 244 pp.
Shinohara, H., Kazahaya, K. and Lowenstern, J.B. (1995) *Geology*, **23**, pp.1087–1090.
Sigurdsson, H. and Sparks, R.L.S. (1981) *Journal of Petrology*, **22**, pp.41–84.
Sillitoe, R.H. (1973) *Economic Geology*, **68**, pp.799–815.
Sillitoe, R.H. (1981) *Bulletin of the Geological Society of America*, **14**, pp.49–70.
Silver, L.T. and Chappell, B.W. (1988) *Transactions of the Royal Society of Edinburgh: Earth Sciences*, **79**, pp.105–121.
Silver, L. and Stolper, E. (1985) *Journal of Geology*, **93**, pp.161–178.
Silver, L.T., Taylor, H.P. and Chappell, B. (1979) In: Abbott, P.L. and Todd, V.R. (eds), *Mesozoic Crystalline Rocks. Geological Society of America 1979 Guidebook*, pp.83–110.
Simpson, C. and Wintsch, R.T. (1989) *Journal of Metamorphic Geology*, **7**, pp.261–275.
Simpson, P.R., Plant, J.A., Watson, J.V., Green, P.M. and Fowler, M.B. (1982) In: *Uranium Exploration Methods*. Nuclear Energy Agency, Paris, pp.157–168.
Skjerlie, K. and Johnston, A.D. (1993) *Journal of Petrology*, **34**, pp.785–815.
Sleep, H.N. (1974) *Bulletin of the Geological Society of America*, **85**, pp.1225–1232.
Smith, J.V. (1974) *Feldspar Minerals 2. Chemical and Textural Properties*. Springer-Verlag, Berlin, pp.282–286.
Smith, R.L. (1960) *Bulletin of the Geological Society of America*, **71**, pp.795–842.
Sobolev, R. (1991a) In: *Proceedings of the International Conference on Rare Earth Minerals and Minerals for Electronic Uses. 1991*. Prince of Songkla University, Hat Yai, Thailand, pp.1–12.
Sobolev, R.N. (1991b) In: Pagal, M. and Leroy, J.L. (eds), *Source, Transport and Deposition of Metals*. Proceedings of the 25th SGA Anniversary Meeting, Balkema, Rotterdam, pp.815–816.
Sobolev, R.N. (1992) *Byuileten' Moskovskogo Obshchestva Ispytatelei Prirody*, **67**(5), pp.109–119.

Soler, P. and Bonhomme, M.G. (1990) In: Kay, S.M. and Rapela, C.W. (eds), *Plutonism from Antarctica to Alaska. Geological Society of America Special Paper No. 241*, pp.173–192.

Soler, P. and Rotach-Toulhoat, N. (1987) In: Rapela, C.W. (ed.), *Decimo Congreso Geológico Argentino, Tucuman, Argentina*, Vol. 6, pp.51–55.

Southwick, D.L. (1991) *Bulletin of the Geological Society of America*, **103**, pp.1385–1394.

Sparks, R.S.J. and Marshall, L. (1986) *Journal of Volcanological and Geochemical Research*, **29**, pp.99–124.

Sparks, S.R., Huppert, H.E. and Turner, J.S. (1984) *Philosophical Transactions of the Royal Society of London*, **A310**, pp.511–534.

Speer, J.A., Naeem, A. and Almohandis, A.A. (1987) *Chemical Geology*, **75**, pp.153–181.

Spencer, K.J. and Lindsley, D.H. (1981) *American Mineralogist*, **66**, pp.1189–1201.

Spera, F.J. (1980) In: Hargraves, R.B. (ed.), *Physics of Magmatic Processes*. Princeton University Press, Princeton, NJ, pp.265–323.

Spera, F.J. (1982) *Science*, **207**, pp.299–301.

Spiegelman, M. (1993) *Philosophical Transactions of the Royal Society of London*, **A342**, pp.23–41.

Stein, H.J. and Hannah, J.L. (1990) (eds), *Ore-bearing Granite Systems: Petrogenesis and Mineralizing Processes. Geological Society of America Special Paper No. 246*, pp.1–364.

Stephansson, O. (1975) *Precambrian Research*, **2**, pp.189–214.

Stephens, W.E. (1992) *Transactions of the Royal Society of Edinburgh: Earth Sciences*, **83**, pp.191–199.

Stephens, W.E. and Halliday, A.N. (1979) In: Atherton, M.P. and Tarney, J. (eds), *Origin of Granite Batholiths: Geochemical Evidence*. Shiva, Nantwich, Cheshire, pp.9–17.

Stephens, W.E. and Halliday, A.N. (1980) *Bulletin of the Geological Society of America*, **91**, pp.165–170.

Stephens, W.E. and Halliday, A.N. (1984) *Transactions of the Royal Society of Edinburgh: Earth Sciences*, **75**, pp.259–273.

Stephenson, P.J. (1990) *Geological Journal*, **25**, pp.325–336.

Stimac, J.A., Clark, A.H., Chen, Y. and Garcia, S. (1995) *Mineralogical Magazine*, **59**, pp.273–296.

Stone, M. (1975) *Proceedings of the Geologists' Association*, London, **86**, pp.155–170.

Stone, M. and Austin, W.G.C. (1961) *Journal of Geology*, **69**, pp.464–472.

Streckeisen, A.L. (1976) *Earth Science Reviews*, **12**, pp.1–33.

Streckeisen, A.L. and Le Maitre, R.W. (1979) *Neues Jahrbuch für Mineralogie*, Abh. **136**, pp.169–206.

Sultan, M., Butiza, R. and Stuchio, N.C. (1986) *Contributions to Mineralogy and Petrology*, **93**, pp.513–523.

Swanson, S.E. (1977) *American Mineralogist*, **62**, pp.966–978.

Tait, R.E. and Harley, S.L. (1988) *Transactions of the Royal Society of Edinburgh: Earth Sciences*, **79**, pp.209–222.

Tait, S.R. and Jaupart, C. (1996) *Mineralogical Magazine*, **60**, pp.99–114.

Tarney, J. and Saunders, A.D. (1979) In: Atherton, M.P. and Tarney, J. (eds), *Origin of Granite Batholiths: Geochemical Evidence*. Shiva, Nantwich, Cheshire, pp.90–105.

Tarney, J., Dalziel, I.W.D. and De Wit, M.J. (1976) In: Windley, B.F. (ed.), *The Early History of the Earth*, Wiley, London, pp.131–146.

Tarney, J., Rex, A.J. and Bartholomew, D.S. (1987) In: *International Symposium on Granites and Associated Mineralizations, Salvador, Bahia, Brazil. Extended Abstracts*, pp.33–34.

Taubeneck, W.H. (1967) *Geological Society of America Special Paper No. 91*, 56 pp.

Taubeneck, W.H. and Poldervaart, A. (1960) *Bulletin of the Geological Society of America*, **71**, pp.1295–1322.

Tauson, L.V. (1977) *Geochemical Types and Potential Ore-bearing Capacity of Granitoids*. Nauka, Moscow, 279 pp. (in Russian).

Tauson, L.V. and Kozlov, Y.D. (1973) *Transactions of the Institution of Mining and Metallurgy*, London, **4**, pp.37–44.

Taylor, H.P. (1977) *Journal of the Geological Society of London*, **133**, pp.509–558.

Taylor, H.P. (1978) *Earth and Planetary Science Letters*, **38**, pp.177–210.

Taylor, H.P. (1988) *Transactions of the Royal Society of Edinburgh: Earth Sciences*, **79**, pp.317–338.

Taylor, H.P. and Forester, R.N. (1971) *Journal of Petrology*, **12**, pp.465–497.

Taylor, H.P. and Forester, R.W. (1979) *Journal of Petrology*, **20**, pp.355–419.

Taylor, H.P. and Silver, L.T. (1978) *US Geological Survey Open File Report 78–701*, pp.423–426.

Taylor, P.N., Jones, N.W. and Moorbath, S. (1984) *Philosophical Transactions of the Royal Society of London*, **A310**, pp.605–625.

Taylor, W.P. (1976) *Journal of Petrology*, **17**, pp.194–218.

Taylor, W.P. (1981) In: Cobbing, E.J. *et al. The Geology of the Western Cordillera of Northern Peru. Institute of Geological Sciences, UK. Overseas Memoir 5*.

Taylor, W.P. (1985) In: Pitcher, W.S., Atherton, M.P., Cobbing, E.J. and Beckinsale, R.D. (eds), *Magmatism at a Plate Edge: The Peruvian Andes*. Blackie, Glasgow, pp.228–234.

Thomas, H.H. and Campbell-Smith, W. (1931) *Quarterly Journal of the Geological Society of London*, **88**, pp.274–296.

Thompson, R.N. and Morrison, M.A. (1988) *Chemical Geology*, **68**, pp.1–15.

Tikoff, B. and Teyssier, C. (1994) *Journal of Structural Geology*, **16**, pp.477–491.

Tischendorf, G. (compiler) (1989) *Silicic magmatism and metallogenesis of the Erzgebirge. Veroffeptlichungen des Zentralinstituts für Physik der Erde*, Potsdam, **107**, pp.1–316.

Tischendorf, G. and Förster, H.-J. (1990) *Geological Journal*, **25**, pp.443–454.

Tischendorf, G. and Pälchen, W. (1985) *Zeitschrift für Geologische Wissenschaften*, Berlin, **13**, pp.615–627.

Tracy, R.J. and McLellan, E.L. (1985) In: Thompson, A.B. and Rubie, D.C. (eds), *Metamorphic Reactions: Kinetics, Textures, and Deformation*. Springer-Verlag, New York, pp.118–137.

Treloar, P.J. (1995) *Geological Journal*, **30**, pp.333–348.

Tribe, I.R. and D'Lemos, R.S. (1996) *Journal of the Geological Society of London*, **153**, pp.127–138.

Troll, G. and Weiss, S. (1985) *Fortschitte der Mineralogie*, **63**(1), pp.1–240.

Tullis, J. and Yund, R.A. (1977) *Journal of Geophysical Research*, **82**, pp.5705–5718.

Turner, D.C. (1963) *Quarterly Journal of the Geological Society of London*, **119**, pp.345–366.

Tuttle, O.F. and Bowen, N.L. (1958) *Origin of Granite in the Light of Experimental*

Studies in the System NaAlSi₃O₈–KAlSi₃O₈–SiO₂–H₂O. Geological Society of America Memoir No. 74, 153 pp.

Vance, J.A. (1961) *Bulletin of the Geological Society of America*, **72**, pp. 1723–1727.
Vance, J.A. (1962) *American Journal of Science*, **260**, pp. 746–760.
Vance, J.A. (1965) *Journal of Geology*, **73**, pp. 636–651.
Van den Bogaard, P. and Schirnick, C. (1995) *Geology*, **23**, pp. 759–762.
Van der Molen, I. and Paterson, M.S. (1979) *Contributions to Mineralogy and Petrology*, **70**, pp. 299–318.
van Moort, J.C. (1966) *Annals Faculté Science Université Clérmont-Ferrand*, **31**, 272 pp.
Vernon, R.H. (1983) *Journal and Proceedings of the Royal Society of New South Wales*, **116**, pp. 77–103.
Vernon, R.H. (1984) *Nature*, **309**, pp. 438–439.
Vernon, R.H. (1985) *Geology*, **13**, pp. 843 845.
Vernon, R.H. (1986) *Earth Science Reviews*, **23**, pp. 1–63.
Vernon, R.H. (1991) *Journal of Structural Geology*, **10**, pp. 979–985.
Vernon, R.H. and Collins, W.J. (1988) *Geology*, **16**, pp. 1126–1129.
Vernon, R.H., Etheridge, M.A. and Wall, V.J. (1988) *Lithos*, **22**, pp. 1–11.
Vidal, C.E. (1985) In: Pitcher, W.S., Atherton, M.P., Cobbing, E.J. and Beckinsale, R.D. (eds), *Magmatism at a Plate Edge: The Peruvian Andes*. Blackie, Glasgow, pp. 243–249.
Vidal, P. (1987) *Revista Brasileira de Geociências*, **17**, pp. 468–472.
Vielzeuf, D. and Holloway, J.R. (1988) *Contributions to Mineralogy and Petrology*, **98**, pp. 257–276.
Vielzeuf, D. and Montel, J.M. (1994) *Mineralogical Magazine*, **58A**, pp. 940–941.
Vigneresse, J.-L. (1984) *Bulletin Société géologique et minéralogique de Bretagne*, **15**(1), pp. 1–15.
Vigneresse, J.-L. (1990) *Geological Journal*, **25**, pp. 249–260.
Vigneresse, J.-L. and Brun, J.-P. (1983) *Bulletin de la Société géologique de France*, **7** (25), pp. 357–366.
Vogt, J. (1908) *Norske Videnskaps Akademi Oslo, Skrifter I, Math.-Nat. Kl., Kristiania*, **10**, pp. 1–104.
Vogt, P.R. (1974) *Earth and Planetary Science Letters*, **21**, p. 235.
von Platen, H. (1965) In: Pitcher, W.S. and Flinn, G.W. (eds), *Controls of Metamorphism*. Oliver and Boyd, Edinburgh, pp. 203–218.
Vorma, A. (1976) *Bulletin of the Geological Survey of Finland*, **285**, 98 pp.
Voshage, H., Hoffmann, A.W. *et al.* (1990) *Nature*, **347**, pp. 731–736.

Wager, L.R. (1960) *Journal of Petrology*, **1**, pp. 368–398.
Wager, L.R. and Bailey, E.B. (1953) *Nature*, **172**, pp. 68–72.
Wager, L.R. and Brown, G.M. (1968) *Layered Igneous Rocks*. Oliver and Boyd, Edinburgh, 588 pp.
Wager, L.R. and Deer, W.A. (1939) Geological investigations in East Greenland. III The petrology of the Skaergaard intrusion, Kangerdlugssuang, East Greenland, *Meddelelser om Grønland*, **105**(4), pp. 1–352.
Wager, L.R., Vincent, E.A., Brown, G.M. and Bell, J.D. (1965) *Philosophical Transactions of the Royal Society of London*, **257**, 273–307.
Walawender, M.J. and Smith, T.E. (1980) *Journal of Geology*, **88**, pp. 233–242.

Walawender, M.J. *et al* (1990) In: Anderson, J.L. (ed.), *Geological Society of America Memoir No. 174*, pp.1–18.

Walker, B.M., Vogel, T.A. and Ehrlich, R. (1972) *Earth and Planetary Science Letters*, **15**, pp.133–139.

Walker, G.P.L. (1963) *Quarterly Journal of the Geological Society of London*, **119**, pp.29–63.

Walker, G.P.L. (1975) *Journal of the Geological Society of London*, **131**, pp.121–142.

Wall, V.J., Clemens, J.D. and Clarke, D.B. (1987) *Journal of Geology*, **95**, pp.731–749.

Walton, M. (1965) *American Journal of Science*, **253**, pp.1–18.

Wark, D.A. and Stimac, J.A. (1991) In: Haapala, I. and Ramo, O.T. (eds), *Symposium on Rapakivi Granites and Related Rocks, Abstract Volume. Geological Survey of Finland Guide No. 34*, p.63.

Watkins, E.D., Gunn, B.N., Nougier, J. and Baksi, A.K. (1974) *Bulletin of the Geological Society of America*, **85**, pp.201–212.

Watson, J.V. (1984) *Journal of the Geological Society of London*, **141**, pp.193–214.

Watterson, J. (1965) *Meddelelser om Grønland*, **172**, 147 pp.

Wegmann, C.E. (1930) *Bulletin de la Commission géologique de Finlande*, **92**, pp.58–76.

Wegmann, C.E. (1935) *Geologisches Rundschau*, **26**, pp.307–350.

Wegmann, C.E. (1950) *18th International Congress*, Pt 3, pp.45–52.

Weinberg, R.F. and Podladchikov, Y. (1994) *Journal of Geophysical Research*, **99**, pp.9543–9559.

Weiss, S. and Troll, G. (1989) *Journal of Petrology*, **30**, pp.1069–1115.

Wells, A.K. and Bishop, A.C. (1955) *Quarterly Journal of the Geological Society of London*, **111**, pp.143–166.

Wells, A.K. and Wooldridge, S.W. (1931) *Proceedings of the Geologists' Association, London*, **42**, pp.178–216.

Whalen, J.B., Currie, K.L. and Chappell, B.W. (1987) *Contributions to Mineralogy and Petrology*, **95**, pp.407–419.

Wheeler, J. (1991) *Terra Nova*, **3**, pp.123–136.

Wheeler, J., Treloar, P.J. and Potts, G.J. (1995) *Geological Journal*, **30**, pp.349–371.

White, A.J.R. and Chappell, B.W. (1983) In: Roddick, J.A. (ed.) *Circum-Pacific Plutonic Terranes. Geological Society of America Memoir No. 159*, pp.21–34.

White, A.J.R. and Chappell, B.W. (1988) *Transactions of the Royal Society of Edinburgh: Earth Sciences*, **79**, pp.169–181.

White, R. and McKenzie, D. (1989) *Journal of Geophysical Research*, **94**, pp.7685–7729.

White, R.S. (1992) In: Alabaster, B.C. and Pankhurst, R.J. (eds), *Magmatism and the Causes of Continental Breakup. Geological Society of London Special Publication No. 68*, pp.1–16.

Whitney, J.A. (1975) *Journal of Geology*, **83**, pp.1–31.

Whitney, J.A. (1988) *Geological Society of America Bulletin*, **100**, pp.1886–1897.

Whitten, E.H.T. (1996) *Journal of the Geological Society of London*, **153**, pp.121–125.

Whitten, E.H.T., Bornhorst, T.J., Li, G., Hicks, D.L. and Beckwith, J.P. (1986) *American Journal of Science*, **287**, pp.332–352.

Wickham, S.M. (1987) *Journal of the Geological Society of London*, **144**, pp.281–298.

Wickman, F.E., Levi, B. and Åberg, G. (1981) *Appendix 2 to Dnr GO177–108*. Naturvetenskapliga Forskningsradet, Stockholm, 6 pp.

Wilkinson, J.J. (1991) *Journal of the Geological Society of London*, **148**, pp. 731–736.
Williams, I.S. (1992) *Transactions of the Royal Society of Edinburgh: Earth Sciences*, **83**, pp. 447–458.
Williams, I.S., Chen, Y., Chappell, B.W. and Compston, W. (1988) *Abstracts of the Geological Society of Australia*, **21**, p. 424.
Wilshire, H.G. (1969) *Mineral Layering in the Twin Lakes Granodiorite, Colorado. Geological Society of America Memoir No. 115*, pp. 235–261.
Wilson, M.R. (1980) *Forhandlingar Geologiska Foreningens Stockholm*, **102**, pp. 167–176.
Wilson, P.A. (1975) *Potassium-Argon Age Studies in Peru with Special Reference to the Emplacement of the Coastal Batholith*. PhD Thesis, University of Liverpool.
Windley, B.F. (1993) *Journal of the Geological Society of London*, **150**, pp. 39–50.
Winkler, H.G.F. (1961) *Geologische Rundschau*, **61**, pp. 347–364.
Winkler, H.G.F. (1979) *Petrogenesis of Metamorphic Rocks*. Springer-Verlag, New York, 237 pp.
Winkler, H.G.F. and Breitbart, R. (1978) *Neues Jahrbuch für Mineralogie*, Mh., **10**, pp. 463–480.
Wolf, M.B. and Wyllie, P.J. (1991) *Contributions to Mineralogy and Petrology*, **44**, pp. 151–179.
Wones, D.R. (1980) In: Wones, D.R. (ed.), *The Caledonides in the USA. Memoir No. 2 of the Virginia Polytechnic Institute and State University*.
Wones, D.R. and Eugster, H.P. (1965) *American Mineralogist*, **50**, pp. 1228–1272.
Wones, D.R. and Gilbert, M.C. (1982) In: Veblen, D.R. and Ribbe, P.H. (eds), *Amphiboles: Petrology and Phase Relations. Reviews in Mineralogy 9B*, Mineralogical Society of America, pp. 355–389.
Worster, M.G. (1991) *Journal of Fluid Mechanics*, **224**, pp. 335–359.
Wyborn, D. and Chappell, B.W. (1986) *Geological Magazine*, **123**, pp. 619–628.
Wyborn, D., Chappell, B.W. and Johnston, R.M. (1981) *Journal of Geophysical Research*, **86**, pp. 10335–10348.
Wyborn, L.A.I. (1988) *Precambrian Research*, **40/41**, pp. 37–60.
Wyllie, P.J. (1977) *Tectonophysics*, **43**, pp. 41–71.
Wyllie, P.J. (1983) In: Atherton, M.P. and Gribble, C.D. (eds), *Migmatites, Melting and Metamorphism*. Shiva, Nantwich, Cheshire, pp. 27–36.

Xu Kegin, Sun Nai, Wang Dezi, Hu Shouxi, Liu Yingjun and Ji Shouyuan (1984) In: Xu Kegin and Tu Guangchi (eds), *Geology of Granites and their Metallogenetic Relations. Proceedings of the International Symposium, Nanjing University*. Science Press, Beijing, pp. 1–3.

Yang Chaoqun (1984) In: Xu Kegin and Tu Guangchi (eds), *Geology of Granites and their Metallogenetic Relations. Proceedings of the International Symposium, Nanjing University*. Science Press, Beijing, pp. 253–276.
Yardley, B.W.D. (1978) *Bulletin of the Geological Society of America*, **89**, pp. 941–951.
Yarr, T.R. (1991) *A Petrological Study of the Appinite Suite Associated with the Ardara Pluton, Co. Donegal, Ireland*. PhD Thesis, University of St. Andrews, Scotland.
Yoder, H.S. (1973) *American Mineralogist*, **58**, 153–171.
Yoder, H.S. and Eugster, H.P. (1955) *Geochimica et Cosmochimica Acta*, **8**, pp. 225–280.
Yoder, H.S. and Tilley, C.E. (1962) *Journal of Petrology*, **3**, pp. 342–532.

Zen, E-an. (1988) *Transactions of the Royal Society of Edinburgh: Earth Sciences*, **79**, pp.223–235.

Zen, E-an. (1992) *Transactions of the Royal Society of Edinburgh: Earth Sciences*, **83**, pp.107–117.

Zen, E-an and Hammarstrom, J.M. (1984) *Geology*, **12**, pp.515–518.

Zhou, J.-X. (1987) *Journal of the Geological Society of London*, **144**, pp.699–706.

Zonensajn, L.P., Kuz'min, M.I. and Moralev, V.M. (1976) *Global Tectonics, Magmatism and Metallogenesis*. Nedra, Moscow, 231 pp. (in Russian).

Zorpi, M.J., Coulon, C., Orsini, J.B. and Cocirta, C. (1989) *Tectonophysics*, **157**, pp.315–329.

Index

Compiled by Hilary Davies